CALL-BY-PUSH-VALUE

Semantic Structures in Computation

Volume 2

Title of the Series:
Semantic Structures in Computation

Series Editor in Chief
Guo-Qiang Zhang
Case Western Reserve University
Department of Electrical Engineering and Computer Science
Olin 610, 10900 Euclid Avenue, Cleveland, OH 44106, USA
E-mail: gqz@eecs.cwru.edu

Scope of the Series

With the idea of partial information and approximation as the starting point, this bookseries focuses on the interplay among computer science, logic, and mathematics through algebraic, order-theoretic, topological, and categorical means, with the goal of promoting cross-fertilization of ideas and advancing interdisciplinary research.

This bookseries provides a distinctive publication forum for collected works and monographs on topics such as domain theory, programming semantics, types, concurrency, lambda calculi, topology and logic in computer science, and especially applications in non-traditional and emerging areas in which the development of formal semantics deepens our understanding of a computational phenomenon.

CALL-BY-PUSH-VALUE

A Functional/Imperative Synthesis

by

Paul Blain Levy
School of Computer Science,
University of Birmingham

SPRINGER SCIENCE+BUSINESS MEDIA, B.V.

A C.I.P. Catalogue record for this book is available from the Library of Congress.

ISBN 978-94-010-3752-5 ISBN 978-94-007-0954-6 (eBook)
DOI 10.1007/978-94-007-0954-6

Printed on acid-free paper

Abstract

Call-by-push-value (CBPV) is a new programming language paradigm, based on the slogan "a value is, a computation does", developing earlier work of Moggi and Filinski. We claim that CBPV provides the semantic primitives from which the call-by-value and call-by-name paradigms are built. The primary goal of the monograph is to present the evidence for this claim, which is found in a remarkably wide range of semantics: from operational semantics, in big-step form and in machine form, to denotational models using domains, possible worlds, continuations and games.

In the first part of the monograph, we come to CBPV and its equational theory by looking critically at the call-by-value and call-by-name paradigms in the presence of general computational effects. We give a Felleisen/Friedman-style CK-machine semantics, which explains how CBPV can be understood in terms of push/pop instructions and leads us to consider stack terms.

In the second part, we give simple CBPV models for printing, divergence, global store, errors, erratic choice and control effects, as well as for various combinations of these effects. We develop the store model into a possible world model for cell generation, and (following Steele) we develop the control model into a "jumping implementation" using a continuation language called Jump-With-Argument (JWA).

We present the Abramsky-Honda-McCusker pointer game model for general storage, in the setting of both JWA and CBPV. We see informally how, because these languages make control flow explicit, we can explain each of the concepts of pointer games in concrete computational terms. In particular, the pointer indicate the time of receipt of the point being jumped to, and the question/answer distinction corresponds to the two kinds of jump in CBPV: forcing and returning.

In the third part of the monograph, we present the categorical semantics for CBPV extended with stack terms: it is an adjunction, resolving the strong monad considered by Moggi. We see how each concrete CBPV model we have treated is indeed an adjunction. Finally, we explain how each connective in both CBPV and JWA is modelled by a representing object for an appropriate functor.

Contents

List of Figures xvii
Preface xxi
Acknowledgments xxv
Introduction xxvii

I.1 Computational Effects xxviii
I.2 Reconciling CBN and CBV xxix
I.2.1 The Problem Of Two Paradigms xxix
I.2.2 A Single Language xxix
I.2.3 "Hasn't This Been Done By ...?" xxxi
I.3 The Case For Call-By-Push-Value xxxii
I.4 Conventions xxxii
I.4.1 Notation and Terminology xxxii
I.4.2 Weakening xxxiii
I.5 A CBPV Primer xxxiii
I.5.1 Basic Features xxxiii
I.5.2 Example Program xxxiv
I.5.3 CBPV Makes Control Flow Explicit xxxvi
I.5.4 Value Types and Computation Types xxxvi
I.6 Structure of Thesis xxxvii
I.6.1 Goals xxxvii
I.6.2 Chapter Outline xxxviii
I.6.3 Chapter Dependence xl
I.6.4 Proofs xl

Part I Language

1. CALL-BY-VALUE AND CALL-BY-NAME 3

1.1 Introduction 3

1.2 The Main Point Of The Chapter 3

1.3 A Simply Typed λ-Calculus 4

1.3.1 The Language 4

1.3.2 Product Types 4

1.3.3 Equations 6

1.3.4 Reversible Derivations 7

1.4 Adding Effects 8

1.5 The Principles Of Call-By-Value and Call-By-Name 8

1.6 Call-By-Value 10

1.6.1 Operational Semantics 10

1.6.2 Denotational Semantics for `print` 11

1.6.3 Scott Semantics 13

1.6.4 The Monad Approach 14

1.6.5 Observational Equivalence 17

1.6.6 Coarse-Grain CBV vs. Fine-Grain CBV 17

1.7 Call-By-Name 18

1.7.1 Operational Semantics 18

1.7.2 Observational Equivalence 18

1.7.3 CBN vs. Lazy 20

1.7.4 Denotational Semantics for `print` 21

1.7.5 Scott Semantics 22

1.7.6 Algebras and Plain Maps 23

1.8 Comparing CBV and CBN 25

2. CALL-BY-PUSH-VALUE: A SUBSUMING PARADIGM 27

2.1 Introduction 27

2.1.1 Aims Of Chapter 27

2.1.2 CBV And CBN Lead To CBPV 28

2.2 Syntax 28

2.3 Operational Semantics Without Effects 30

2.3.1 Big-Step Semantics 30

2.3.2 CK-Machine 32

2.3.3 CK-Machine For Non-Closed Computations 33

2.3.4 Typing the CK-Machine 33

2.3.5 Operations On Stacks 36
2.3.6 Agreement Of Big-Step and CK-Machine Semantics 37
2.4 Operational Semantics for `print` 37
2.4.1 Big-Step Semantics for `print` 37
2.4.2 CK-Machine For `print` 38
2.5 Observational Equivalence 39
2.6 Denotational Semantics 39
2.6.1 Values and Computations 39
2.6.2 Denotational Semantics Of Stacks 41
2.6.3 Monads and Algebras 41
2.7 Subsuming CBV and CBN 42
2.7.1 From CBV to CBPV 42
2.7.2 From CBN to CBPV 43
2.8 CBPV As A Metalanguage 45
2.9 Useful Syntactic Sugar 45
2.9.1 Pattern-Matching 45
2.9.2 Commands 46
2.10 Computations Matter Most 47
3. COMPLEX VALUES AND EQUATIONAL THEORY 49
3.1 Introduction 49
3.2 Complex Values 50
3.3 Equations 51
3.4 CK-Machine Illuminates $\to$ Equations 52
3.5 Adding Complex Values is Computation-Unaffecting 54
3.6 The Problem With the CBPV Equational Theory 56
3.7 Complex Stacks 57
3.8 Equational Theory For CBPV With Stacks 58
3.9 Reversible Derivations 60
3.10 Example Isomorphisms Of Types 61
3.11 Trivialization 61
4. RECURSION AND INFINITELY DEEP CBPV 65
4.1 Introduction 65
4.2 Divergence and Recursion 66
4.2.1 Divergent and Recursive Terms 66

4.2.2 Type Recursion 68

4.3 Denotational Semantics Of Type Recursion 69

4.3.1 Bilimit-Compact Categories 69

4.3.2 Sub-Bilimit-Compact Categories 72

4.3.3 Minimal Invariants 73

4.3.4 Interpreting Recursive Types 75

4.4 Infinitely Deep Terms 77

4.4.1 Syntax 77

4.4.2 Partial Terms and Scott Semantics 78

4.5 Infinitely Deep Types 79

4.5.1 Syntax 79

4.5.2 Partial Types and Scott Semantics 79

4.6 The Inductive/Coinductive Formulation of Infinitely Deep Syntax 80

4.7 Relationship Between Recursion and Infinitely Deep Syntax 82

4.8 Predomain Semantics For CBPV 83

4.8.1 Domains and Predomains 83

4.8.2 Interpreting Type Recursion In Predomains and Domains 84

Part II Concrete Semantics

5. SIMPLE MODELS OF CBPV 89

5.1 Introduction 89

5.2 Semantics of Values 90

5.3 Global Store 91

5.3.1 The Language 91

5.3.2 Denotational Semantics 92

5.3.3 Combining Global Store With Other Effects 94

5.4 Control Effects 95

5.4.1 `letstk` and `changestk` 95

5.4.2 Typing Control Effects 96

5.4.3 Regarding `nil` As A Free Identifier 98

5.4.4 Observational Equivalence 100

5.4.5 Denotational Semantics 101

5.4.6 Combining Control Effects With Other Effects 105

5.5 Erratic Choice 107

5.5.1 The Language 107
5.5.2 Denotational Semantics 107
5.5.3 Finite Choice 109
5.5.4 New Model For May Testing 109
5.6 Errors 111
5.6.1 The Language 111
5.6.2 Denotational Semantics 113
5.7 Summary 114

6. POSSIBLE WORLD MODEL FOR CELL GENERATION 117
6.1 Cell Generation 117
6.2 The Language 118
6.3 Operational Semantics 119
6.3.1 Worlds 119
6.3.2 Stores 120
6.3.3 Operational Rules 121
6.3.4 Observational Equivalence 122
6.4 Cell Types As Atomic Types 122
6.5 Excluding Thunk Storage 123
6.6 Introduction 124
6.6.1 What Is A Possible World Semantics? 124
6.6.2 Contribution Of Chapter 125
6.6.3 Stores 125
6.6.4 Possible Worlds Via CBPV Decomposition 126
6.7 Denotational Semantics 127
6.7.1 Semantics of Values 127
6.7.2 Semantics of Computations 128
6.7.3 A Computation Type Denotes A Contravariant Functor 128
6.7.4 Semantics of Stacks 129
6.7.5 Summary 130
6.8 Combining Cell Generation With Other Effects 131
6.8.1 Introduction 131
6.8.2 Cell Generation + Printing 132
6.8.3 Cell Generation + Divergence 133
6.9 Related Models and Parametricity 135
6.10 Modelling Thunk Storage And Infinitely Deep Types 137

7. JUMP-WITH-ARGUMENT 141
7.1 Introduction 141
7.2 The Language 142
7.3 Jumping 143
7.3.1 Intuitive Reading of Jump-With-Argument 143
7.3.2 Graphical Syntax For JWA 144
7.3.3 Execution 145
7.4 The StkPS Transform 148
7.4.1 StkPS Transform As Jumping Implementation 148
7.4.2 Related Transforms 149
7.5 C-Machine 151
7.5.1 C-Machine for Effect-Free JWA 151
7.5.2 Adapting the C-Machine for `print` 153
7.5.3 Relating Operational Semantics 154
7.5.4 Observational Equivalence 155
7.6 Equations 155
7.6.1 Equational Theory 155
7.6.2 Example Isomorphisms 156
7.7 JWA As A Type Theory For Classical Logic 157
7.8 The StkPS Transform Is An Equivalence 160
7.8.1 The Main Result 160
7.8.2 Complex Values 161
7.8.3 Equational Theory For CBPV+Control 162
7.8.4 Representing CBPV+Stacks In CBPV+Control 163
7.8.5 Stacks and Complex Values Are Computation-Unaffecting 164
7.8.6 The StkPS Transform Between Equational Theories 166
8. POINTER GAMES 169
8.1 Introduction 169
8.1.1 Pointer Games And Their Problems 169
8.1.2 Contribution Of This Chapter 170
8.1.3 More Is Simpler 171
8.1.4 Cells As Objects 171
8.1.5 Related Work On Abstract Machines 173
8.1.6 Structure Of Chapter 174
8.2 Arenas In The Literature 174

8.2.1 Player/Opponent Labelling 174
8.2.2 Tokens Of An Arena 174
8.2.3 Arenas Versus Arena Families 175
8.2.4 Forests Versus Graphs 176
8.3 Pointer Game Semantics For Infinitary JWA + Storage 176
8.3.1 Types 176
8.3.2 Terms 180
8.3.3 η-Expansion And Copycat Behaviour 188
8.4 Semantics Of Terms 189
8.4.1 Categorical Structure 190
8.4.2 Cell Generation 192
8.4.3 Claims 193
8.5 Pointer Game For Infinitary CBPV + Control + Storage 194
8.5.1 Pointer Game Via StkPS 194
8.5.2 Introducing Questions and Answers 195
8.5.3 Answer-Move Pointing To Answer-Move 197
8.6 Removing Control 198

Part III Categorical Semantics

9. SEMANTICS IN ELEMENT STYLE 207
9.1 Countable vs. Finite 207
9.2 Introduction 207
9.3 Categorical Preliminaries 208
9.3.1 Modules 208
9.3.2 Homset Functors 209
9.3.3 Some Notation 210
9.3.4 Locally Indexed Categories 210
9.3.5 OpGrothendieck Construction And Homset Functors 213
9.3.6 Locally Indexed Modules 215
9.3.7 Locally Indexed Natural Transformations 216
9.4 Modelling Effect-Free Languages 216
9.4.1 Cartesian Categories and Substitution Lemma 216
9.4.2 Products, Exponentials and Distributive Coproducts 217
9.5 Categorical Structures For Judgements Of JWA And CBPV 220

9.5.1 Introduction 220
9.5.2 Judgement Model of JWA 220
9.5.3 Stacks In CBPV 221
9.5.4 Computations In CBPV 222
9.5.5 Enrichment 223
9.6 Connectives 225
9.6.1 Right Adjunctives—Interpreting U 225
9.6.2 Jumpwiths—Interpreting $\neg$ 226
9.6.3 Left Adjunctives—Interpreting F 226
9.6.4 Products—Interpreting $\prod$ 227
9.6.5 Exponentials—Interpreting $\rightarrow$ 229
9.6.6 Distributive Coproducts—Interpreting $\sum$ 231
9.6.7 Tying It All Together 234
9.6.8 Adjunctions 235
9.7 Examples 236
9.7.1 Trivial Models Of CBPV 236
9.7.2 Eilenberg-Moore Models Of CBPV 236
9.7.3 Between JWA and CBPV 237
9.7.4 Erratic Choice 239
9.7.5 Global Store 240
9.7.6 Possible Worlds: Part 1 241
9.8 Families 243
9.8.1 Composite Connectives 243
9.8.2 Families Construction For JWA 246
9.8.3 Families Construction For CBPV 247
10. ALL MODELS ARE CATEGORICAL MODELS 249
10.1 Introduction 249
10.2 All Models Of $\times$-Calculus Are Cartesian Categories 249
10.2.1 The Theorem, Vaguely Stated 249
10.2.2 Fixing The Object Structure 250
10.2.3 What Is A Model Of $\times$-Calculus? 251
10.2.4 The Theorem, Precisely Stated 253
10.3 All Models Of CBPV Are CBPV Adjunction Models 254
10.4 All Models Of JWA Are JWA Module Models 259
10.5 Enrichment 260
11. REPRESENTING OBJECTS 261
11.1 Introduction 261

11.2 Categorical Structures 262
11.2.1 Joint vs. Separate Naturality 262
11.2.2 Context Extension Functors 263
11.3 Representable Functors 264
11.3.1 Ordinary Category 264
11.3.2 Locally Indexed Category 266
11.3.3 Cartesian Category 267
11.3.4 Category With Right Modules 269
11.3.5 Cartesian Category With Left Modules 271
11.4 Parametrized Representability 273
11.5 Possible Worlds: Part 2 277
11.6 Adjunctions and Monads 278
11.6.1 Definitions of Adjunction 278
11.6.2 Terminality of Eilenberg-Moore 283
11.6.3 Kleisli and Co-Kleisli Adjunctions 284
11.6.4 CBV Is Kleisli, CBN Is Co-Kleisli 287

Part IV Conclusions

12. CONCLUSIONS, COMPARISONS AND FURTHER WORK 293
12.1 Summary of Achievements and Drawbacks 293
12.2 Simplifying Algebra Semantics 294
12.3 Advantages Of CBPV Over Monad And Linear Logic Decompositions 294
12.4 Beyond Simple Types 296
12.4.1 Dependent Types 296
12.4.2 Polymorphism 297
12.5 Further Work 297

Appendices 298
Technical Treatment of CBV and CBN 299
A.1 The Jumbo λ-Calculus 300
A.1.1 Introduction 300
A.1.2 Tuple Types 300
A.1.3 Function Types 301
A.2 Languages and Translations 302
A.3 Call-By-Value 304
A.3.1 Coarse-Grain Call-By-Value 304

A.3.2 Fine-Grain Call-By-Value 305
A.3.3 From CG-CBV To FG-CBV 308
A.4 Call-By-Name 311
A.5 The Lazy Paradigm 312
A.6 Subsuming FG-CBV and CBN 314
A.6.1 From FG-CBV to CBPV 314
A.6.2 From CBPV Back To FG-CBV 314
A.6.3 From CBN to CBPV 319
A.6.4 From CBPV Back To CBN 320
Models In The Style Of Power-Robinson 327
B.1 Introduction 328
B.2 Actions of Monoidal Categories 328
B.3 Freyd Categories 329
B.4 Judgement Model 330
B.5 Enrichment 331
B.6 Connectives 331
B.7 Modelling CBPV 332
B.8 The Full Reflection 333
B.9 Theories 334
B.10 Conservativity 336

References 337

Index 345

List of Figures

I.1	Chapter and Section Dependence	xli
1.1	Terms of λ `bool+`	5
1.2	Big-Step Semantics for CBV with `print`	10
1.3	Big-Step Semantics for CBN with `print`	19
2.1	Terms of Basic Language	29
2.2	Big-Step Semantics for CBPV	31
2.3	CK-Machine For CBPV, With Types	34
2.4	CK-Machine For Non-Closed Computations	35
2.5	Typing Stacks	35
2.6	Translation of CBV types, values and returners	43
2.7	Translation of CBN types and terms	44
3.1	Varieties of CBPV	49
3.2	Complex Values	50
3.3	CBPV equations, using conventions of Sect. I.4.2	53
3.4	Definitions used in proof of Prop. 25	56
3.5	Complex Values and Stacks	57
3.6	Equational laws for CBPV + stacks, using operations of Sect. 2.3.5	59
3.7	Reversible derivations for CBPV with stacks	60
3.8	The $\times\sum\prod\rightarrow$-Calculus	62
5.1	Typing Rules for Control Operators	97
5.2	Varieties of CBPV with control	97
5.3	CK-Machine With Control	99
5.4	Typing Stacks in CBPV + Control	100

5.5	Translating CK-machine for CBPV into CK-machine for CBPV+control	100
5.6	Semantics of types—two equivalent presentations	102
5.7	Continuations and Stacks	104
5.8	Big-Step Rules for Errors in CBPV	112
5.9	Summary of simple CBPV models	115
5.10	Induced semantics for CBV and CBN function types	116
7.1	Terms of Jump-With-Argument, and its embedding into CBPV+Ans	144
7.2	Examples of Graphical Syntax for JWA	146
7.3	Example Program With Repeated Jumps—Please Try This	149
7.4	The StkPS transform from CBPV+control to JWA, with printing	150
7.5	C-Machine for Jump-With-Argument	152
7.6	Complex Values For JWA	156
7.7	JWA equations, using conventions of Sect. I.4.2	156
7.8	Reversible Derivations For JWA	157
7.9	The StkPS transform between various languages	160
7.10	Complex Values For CBPV+Control	162
7.11	Equational Theory For CBPV+Control	163
7.12	Representing $\vdash^{\mathsf{k}}$	164
7.13	Definitions used in proof of Prop. 83	165
7.14	StkPS Transform On Complex Values	166
7.15	Syntactic Isomorphisms α and β used in proof of Prop. 80	167
8.1	Example Program and Context	181
8.2	Command That η-Expands to Fig. 8.1	188
10.1	Rules For A Signature s	256
10.2	Dismantling and Concatenation (Including Complex Stacks And Operation Symbols)	256
10.3	Adjunction Model From A Direct Model (s, θ)	259
A.1	Syntax and Equations of Jumbo λ-calculus	302
A.2	Effectful Languages	303
A.3	Big-Step Semantics for CG-CBV—No Effects	305
A.4	Terms and Big-Step Semantics for FG-CBV	306
A.5	CBV equations, using conventions of Sect. I.4.2	308

A.6	Translation from CG-CBV to FG-CBV	309
A.7	Big-Step Semantics for CBN—No Effects	311
A.8	CBN equations, using conventions of Sect. I.4.2	312
A.9	Translation from lazy to FG-CBV: types, returners, values	313
A.10	Translation of FG-CBV types, values and returners	315
A.11	The Reverse Translation $-^{\mathsf{v}^{-1}}$	317
A.12	Translation of CBN types and terms	319
A.13	The Reverse Translation $-^{\mathsf{n}^{-1}}$	323

Preface

This research monograph explores the surprising consequences of combining, within a single programming language, imperative and functional features. It was originally written as a PhD thesis [Levy, 2001] in the Department of Computer Science, Queen Mary, University of London, submitted (after corrections) in March 2001. Since then, much of the theory—especially categorical semantics—has been simplified. As detailed below, I have revised the text to incorporate these changes.

Although it is not a textbook, the monograph should be accessible to a broad range of readers. At one extreme, a category theorist can read Chap. 9 and Chap. 11 and omit all the discussions of syntax. At the other extreme, a reader interested in operational ideas is recommended the following trail.

1 Read the CBPV primer (Sect. I.5), and understand the example program informally.

2 Look at the CK-machine (Fig. 2.3) and run the example program on it—as well as on the big-step semantics (Fig. 2.2). Don't worry about the arithmetic—just calculate sums when you please.

3 Learn the 3 operations on stacks (Sect. 2.3.5)

4 Read the description of Jump-With-Argument in Sect. 7.2–7.3, omitting the motivating discussion, and follow the example execution. Compare this with the C-machine (Fig. 7.5). Try the exercise in Fig. 7.3.

5 To grasp the StkPS transform from CBPV to JWA (Fig. 7.4), apply it to the example CBPV program of Sect. I.5.2, and execute the resulting JWA program, both using the C-machine (compare this to the CK-machine) and the jumping machine. See how the latter

precisely describes the informal understanding in Sect. I.5.2, where both `force` and `return` instructions cause a jump.

6 Finally, notice that JWA is a fragment of CBPV (extended with a free type identifier Ans)—as shown in Fig. 7.1, and the C-machine follows accordingly from the CK-machine.

For readers between these two extremes, it is the interplay between operational and denotational aspects of CBPV that provides the greatest interest. This is most apparent in the chapters on possible worlds and pointer games.

Historical Background

- [Landin, 1966] explained how to combine functional and imperative features in a call-by-value setting, influencing the Scheme and ML programming languages.

- [Moggi, 1991] presented a denotational framework for this combination using monads.

- [Filinski, 1996] modified this to incorporate call-by-name, using a "generalized `let`" construct, which evolved into the sequencing construct of CBPV.

Changes Since The PhD Thesis

In the version submitted as a PhD thesis, the only terms considered were values and computations, and there was a mismatch between the equational theory and categorical semantics using adjunctions. To achieve a match, various alternative but complicated categorical structures were proposed in Part III.

The problem was solved in [Levy, 2003], where it was shown that introducing stack terms gives a perfect match with adjunction semantics. This development made Part III of the thesis obsolete. So a much shorter and simpler Part III has now been substituted. (A short treatment of one of the alternative categorical structures survives as Appendix B.) Stack terms are introduced in Chap. 2, and treated throughout the book, along with values and computations.

Another change is the introduction of *sub-bilimit-compact* categories (such as **Cpo**, and functor categories to **Cpo**) in Sect. 4.3.2. Using this concept, the semantics of thunk storage in Sect. 6.10 has been greatly simplified.

The treatment of pointer games in Chap. 8 has been revised. The previous version attempted to first explain the CBPV model to readers who had not read about Jump-With-Argument, and only afterwards explain the link. But JWA is so fundamental to pointer games that the first account of pointer games divorced from JWA was unconvincing. Consequently, JWA is now central to our account, and it is the CBPV model (with its question/answer distinction) that is treated afterwards. JWA has also been integrated into Part III.

A newly discovered model for may-testing has been added at the last minute—it appears in Sect. 5.5.4.

We have made some changes in our notation to bring it into line with common usage.

old	new
produce	return (like in Java)
producer	returner
outside	stack
consumer	continuation (like in ML)
continuation	jump-point
OPS transform	StkPS transform
`os`	`stk`
`letcos`	`letstk`
`changecos`	`changestk`
enriched-compact	bilimit-compact
exponent	exponential
value/producer structure	Freyd category
SE domain	domain
SEAM predomain	predomain
revised simply typed λ-calculus	jumbo λ-calculus
Revised-λ	jumbo λ-calculus
isomorphism style	naturality style
representation	representing object
$\langle \ldots, M_i, \ldots \rangle$	$\lambda\{\ldots, i.M_i, \ldots\}$
`let x be` V `in` M	`let` V `be x.` M
`print` c; M	`print` c. M (like in CCS)
M `to x in` M	M `to x.` N
$\Gamma \vdash^{\mathsf{k}}_{\underline{C}} K : \underline{B}$	$\Gamma \vert \underline{D} \vdash^{\mathsf{k}} \underline{C}$
`neverused`	`ignored`
SEAM	**Predom**
$\mathbf{SE}_{\mathsf{strict}}$	**DomStr**
$\mathbf{SEAM}_{\mathsf{partmin}}$	**Partmin**

We use "return" rather than "produce", because it is common usage: "this program returns 3". We write `pm`, short for "pattern-match", rather than `case`, because sometimes there is no case analysis, as there is only one pattern.

The definition of *CBPV adjunction model* has changed. In the PhD thesis, there was no requirement for the function

$$\begin{array}{rcl} \mathcal{D}_{X\times\sum_{i\in I}A_i}(\underline{Y},\underline{Z}) & \longrightarrow & \prod_{i\in I}\mathcal{D}_{X\times A_i}(\underline{Y},\underline{Z}) \text{ for all } X,\underline{Y},\underline{Z} \\ f & \longmapsto & \lambda i.((X\times \mathsf{in}_i)^* f) \end{array}$$

to be an isomorphism. Although the theorems about adjunction models were all correct, this omission was clearly a mistake. In the new version, this requirement is used in Sect. 10.3 to prove the match between adjunction models and the CBPV equational theory with stack terms. Incorporating this requirement in a non-*ad hoc* way has led to a new presentation of products and exponentials as well as distributive coproducts.

Finally, it is worth mentioning that two ways in which we develop CBPV

- adding complex values and an equational theory (Chap. 3)
- allowing infinitely deep syntax (Chap. 4)

must, at the present time, be regarded as incompatible. For our justification for adding complex values is that they can be removed from any computation (Prop. 25). The proof of this uses induction over terms, so it relies on terms being finitely deep. It is hoped that future work will remedy this.

Acknowledgments

I am most grateful to my PhD supervisor, Peter O'Hearn, for his constant assistance and advice, and to numerous colleagues and friends at Queen Mary, Boston University, Brandeis, Paris 7, Birmingham and elsewhere. Discussions with Adam Eppendahl, Julian Gilbey, Martin Hyland, Jim Laird, Olivier Laurent, Michael Marz, Guy McCusker, Paul-André Melliès, Eugenio Moggi, John Power, Uday Reddy, Paul Taylor (whose macros are used for many of the diagrams here), Hayo Thielecke, my examiners Chris Hankin and Gordon Plotkin, and many others have shaped the material in all kinds of ways.

Finally, I am grateful to my parents, Ruth and David Levy, for their love and support.

Introduction

I.1 Computational Effects

Much attention has been devoted to functional languages with divergence (nontermination), such as PCF [Plotkin, 1977], FPC [Plotkin, 1985], Haskell, Scheme and ML, and to their models. After all, it is well-known that a language with full recursion must have divergent programs. Yet some of the most fundamental semantic issues involved in these languages are far more general than is often realized, and (as we will argue in Sect. 1.2) can be correctly understood only in the light of this generality. They arise whenever we combine imperative and functional features within a single language.

It may strike the reader as strange to regard divergence as an imperative feature, for what could be more "purely functional" than PCF? But it is a remarkable fact that the same core theory (the subject of Chap. 1) applies when we add to the simply typed λ-calculus any of the following [Moggi, 1991]:

- divergence
- reading and assigning to a storage cell
- input and output
- erratic choice
- generating a new name or cell
- halting with an error message
- control effects (we shall explain these in Chap. 5)

Thus, whether or not `diverge` really is a command, it certainly behaves like one. We call all these features *computational effects* or just *effects*. In our exposition, we shall use `print` commands as our leading example of an effect, rather than the more familiar `diverge`, because doing so makes important distinctions clearer (as we shall argue in Sect. 1.2).

The first issue that arises for effectful languages is that—by contrast with the simply typed λ-calculus—order of evaluation matters. Whether a program converges or diverges, whether a program prints `hello` or prints `goodbye`, will depend on the evaluation order that the language uses.

A priori, there are many evaluation orders that could be considered. But two of them have been found to be significant, in the sense that they possess a wide range of elegant semantics[1]: *call-by-name* (CBN) and *call-by-value* (CBV).

I.2 Reconciling CBN and CBV

I.2.1 The Problem Of Two Paradigms

Researchers have developed many semantics for CBN and CBV. But each time a new form of semantics is introduced, each time a technical result is proved, each time an analysis of semantic issues is presented, we have to perform the work twice: once for CBN, once for CBV.

Here are some examples of this situation.

- In many introductory textbooks on the semantics of programming languages (e.g. [Gunter, 1995]), we are first shown a CBN language and its operational semantics. We are given a denotational semantics using domains and this is proved adequate. Next, we are shown a CBV language and its operational semantics. We are given a denotational semantics and this too is proved adequate.
- In [Hyland and Ong, 2000; Nickau, 1996] a game semantics is presented for a CBN language, and various technical properties are proved. Then, in [Abramsky and McCusker, 1998; Honda and Yoshida, 1997], a game semantics is presented for a CBV language, and similar technical properties are proved.
- In [Streicher and Reus, 1998], a machine semantics and continuation semantics are presented and their agreement proved: first for a CBV language, then for a CBN language.
- In [Oles, 1982], a functor category semantics for a CBN language is presented. Then, in [Moggi, 90; Stark, 1994] a functor category semantics for a CBV language is presented.

This duplication of work is tiresome. Furthermore, it makes the languages involved seem inherently arbitrary. We would prefer to study a single, canonical language.

I.2.2 A Single Language

How can the situation described in Sect. I.2.1 be remedied? A first suggestion is as follows.

[1]A third method of evaluation, *call-by-need*, is useful for implementation purposes. But it lacks a clean denotational semantics—at least for effects other than divergence and erratic choice whose special properties are exploited in [Hennessy, 1980] to provide a call-by-need model. So we shall not consider call-by-need.

> Suppose we have a language $\mathfrak{L}$ in which both CBN and CBV programs can be written. Then we need only give semantics for $\mathfrak{L}$, and this automatically provides semantics for CBN and CBV.

There is a basic problem with this approach. As an extreme example, consider that CBN and CBV languages can both be translated into π-calculus [Sangiorgi, 1999] so π-calculus is "a language in which both CBN and CBV programs can be written". But this does not relieve us of the responsibility to provide (say) Scott semantics for CBN and CBV, because we do not have such a semantics for π-calculus. For this reason, it is essential for $\mathfrak{L}$ to have a Scott semantics, a game semantics, a continuation semantics, an operational semantics, etc.

But even where $\mathfrak{L}$ does have all these semantics, there can be a more subtle problem. As an example, consider that CBN can be translated into CBV [Hatcliff and Danvy, 1997], and so CBV itself is "a language in which both CBN and CBV programs can be written". Now in this case $\mathfrak{L}$ certainly has a Scott semantics, a game semantics, a continuation semantics etc., and so, as proposed, we obtain semantics for CBN. But the Scott semantics for CBN thus obtained is not the traditional CBN Scott semantics. (For example, whereas in the traditional semantics λ`x.diverge` and `diverge` have the same denotation, in the new semantics they have different denotations.) So the translation of CBN into CBV does not relieve us of the task of constructing the traditional semantics for CBN.

We see that, in order to remedy the situation described in Sect. I.2.1, we have to be sure that the Scott semantics for $\mathfrak{L}$ induces the *usual* Scott semantics for CBN and CBV, rather than a more complicated one; and likewise for game semantics, continuation semantics, operational semantics etc. Another way of saying this is that the translations from CBV and from CBN into $\mathfrak{L}$ must *preserve* Scott semantics, game semantics etc.

Of course, it will be impossible to show that the translations preserve *every* semantics we might wish to study. But the semantics we have mentioned are extremely diverse, and if these are all preserved then this makes a strong case that we do not need to continue investigating CBN and CBV as independent entities.

We say then that the language $\mathfrak{L}$ *subsumes* CBN and CBV, and that the two translations into it are *subsumptive*. Notice that subsumptiveness is not a formal, technical property of a translation (such as full abstraction). Rather, it says, informally, that there is no reason to regard the source language as anything more than an arbitrary fragment of the target language.

I.2.3 "Hasn't This Been Done By ... ?"

There are many claims in the literature that a given language contains both CBN and CBV:

- Moggi's monadic metalanguage [Benton and Wadler, 1996; Moggi, 1991] and Filinski's variant [Filinski, 1996]
- CBV itself [Hatcliff and Danvy, 1997]—in the presence of general effects, Moggi called this the *computational* λ*-calculus*, or λ_{c}*-calculus* [Moggi, 1988]
- π-calculus [Sangiorgi, 1999]
- continuation languages [Plotkin, 1976], as well as their polarized counterpart LLP [Laurent, 1999]
- languages based on Girard's linear logic [Benton and Wadler, 1996; Girard, 1987]
- SFL [Marz et al., 1999] and SFPL [Marz, 2000].

Indeed these translations, representations and encodings have provided much insight and, in many cases, new models. But we argue that they do not achieve what we want.

We explained in Sect. I.2.2 that CBV itself does not contain CBN in our sense because the equation $\lambda \mathtt{x.diverge} = \mathtt{diverge}$ is invalidated by the translation from CBN into CBV. The same criticism applies to the translation from CBN to Moggi's metalanguage (although not Filinski's variant). Thus Moggi's language does not subsume CBN in our sense; it is not a solution to our problem.

Filinski's variant does not have this drawback, but both Moggi's language and Filinski's variant have the drawback that they lack operational semantics, essentially because every term is a value. (Our value/computation distinction in CBPV will remedy this.)

We explained in Sect. I.2.2 that π-calculus does not have Scott semantics (at least, not the kind we want), so the translations into it are not subsumptive. Continuation calculi (and LLP) do have Scott semantics, but the CPS transforms into them do not preserve direct semantics, so the CPS transforms (from source languages without control effects) are not subsumptive. The linear λ-calculus of [Benton and Wadler, 1996] does have Scott semantics, but, as remarked there, the language is based on the assumption that effects are "commutative", so effects such as printing are excluded.

The translation from CBV to the language SFL does not preserve the η-law for sum types (explained in Sect. 1.8). By contrast, the successor language SFPL does indeed subsume CBV and CBN. It is in a sense equivalent to the computation part of CBPV, although less simple.

In Chap. 12.3, we make further criticisms of Moggi's decomposition

$$A \rightarrow_{\mathbf{CBV}} B = A \rightarrow TB$$

and of the linear decomposition

$$A \rightarrow_{\mathbf{CBN}} B = !A \multimap B$$

I.3 The Case For Call-By-Push-Value

In this book we introduce a new language paradigm, *call-by-push-value* (CBPV). It is based on Filinski's variant of Moggi's "monadic metalanguage" [Filinski, 1996; Moggi, 1991], but we do not assume familiarity with these calculi (discussed in Sect. I.2.3 and Sect. 12.3).

We shall see that CBPV satisfies all that was required in Sect. I.2.2. Consequently, what we previously regarded as the primitives of CBN and CBV can now be seen as just idioms built from the genuine primitives of CBPV.

For this reason, we argue that CBPV deserves the attention *even of those who are interested only in CBN* (or only in CBV) and not in the above problem of reconciling CBN and CBV. Such people benefit from using CBPV because the CBN semantics they are studying will exhibit the decomposition of CBN primitives into CBPV primitives, so CBPV is closer to the semantics. We will see numerous examples of this in Part II.

We mention also that, although full abstraction is neither a necessary nor a sufficient condition for subsumptiveness, we prove (Cor. 149 and Cor. 159) that the translations into CBPV are indeed fully abstract. Thus, semanticists initially concerned with fully abstract models for CBV and CBN can obtain them from fully abstract models for CBPV.

I.4 Conventions

I.4.1 Notation and Terminology

We write $V`M$ for "M applied to V". This operand-first notation has some advantages over the traditional notation MV. In particular, it allows the "push" reading of Sect. I.5.1.

For any natural number n, we write $\$n$ for $\{0, \ldots, n-1\}$, the canonical set of size n.

In a category, we write $f;g$ for the composite of f and g in diagrammatic order.

We avoid the ambiguous word "variable", using instead the following distinct terms:

- An *identifier* is a syntactic symbol whose binding does not change, as in λ-calculus and predicate logic. A *context* is a list of distinct identifiers, with associated types. A list giving the binding of each identifier in a given context is called an *environment*.

- A *cell* is a memory location whose contents can change through time. A list giving the contents of each cell in memory is called a *store*.

This crucial distinction between environment and store, attributed by [Tennent and Ghica, 2000] to Park and his contemporaries [Park, 1968], is maintained throughout our treatment.

I.4.2 Weakening

All the systems we treat have the *weakening* property: if $\Gamma \vdash M : B$ then $\Gamma, \mathtt{x} : A \vdash M : B$. We will often write the latter term not as M but as ${}^{\mathtt{x}}M$. This explicit indication of weakening has two advantages.

- M and ${}^{\mathtt{x}}M$ have different denotations: an environment for the latter must provide a binding for $\mathtt{x}$, even though this binding is not used.

- Writing ${}^{\mathtt{x}}M$ implicitly includes the assumption that $\mathtt{x} \notin \Gamma$, because otherwise $\Gamma, \mathtt{x} : A$ would not be a context.

As an example, the familiar η-law for functions is written

$$\frac{\Gamma \vdash M : A \to B}{\Gamma \vdash M = \lambda \mathtt{x}.(\mathtt{x}\,\text{`}\,{}^{\mathtt{x}}M) : A \to B} \tag{I.1}$$

We do not need to state explicitly that $\mathtt{x} \notin \Gamma$, as is traditionally done.

In fact, to reduce clutter when writing equations, we will generally omit the context, turnstile and type, and omit too all the assumptions required to make the equations well typed. So, instead of writing (I.1), we will just write

$$M = \lambda \mathtt{x}.(\mathtt{x}\,\text{`}\,{}^{\mathtt{x}}M)$$

I.5 A CBPV Primer

I.5.1 Basic Features

The aim of this section is to enable the reader to understand a simple example program. For this purpose, it suffices to explain the following basic features.

- CBPV has a simple imperative reading in terms of a stack.
  - $\lambda \mathtt{x}$ is understood as the command "pop $\mathtt{x}$".
  - V\` is understood as the command "push V".

- CBPV terms are classified into computations and values. A computation *does* something, whereas a value is something like a boolean or number which can be passed around.

  Slogan A value is, a computation does.

- Only a value can be pushed or popped. In other words, only a value can be an operand. However, we can "freeze" a computation M into a value; this value is called the *thunk* of M [Ingerman, 1961]. Later, when desired, this thunk can be *forced* (i.e. executed).

- Certain computations return values, or at least that is their goal; they are called *returners*. If M is a returner, and N another computation, then we can *sequence* them into the computation M `to x.` N. This means: first obey M until finally it returns a value V, then obey N with `x` bound to V.

I.5.2 Example Program

The easiest way to grasp these basic ideas is to see an example program explained very informally. The following program uses the computational effect of printing messages to the screen.

```
print "hello0".
let 3 be x.
let thunk (
        print "hello1".
        λz.
        print "we just popped "z.
        return x + z
     ) be y.
print "hello2".
( print "hello3".
  7'
  print "we just pushed 7".
  force y
) to w.
print "w is bound to "w.
return w + 5
```

We give a blow-by-blow account of execution—the program executes the commands in order. First it prints `hello0`, then binds `x` to 3, then binds `y` to a thunk. If the word `thunk` were omitted, the program would not typecheck, because an identifier can be bound only to a value—a computation is too active to sit in an environment (list of bindings for identifiers).

Next, the program prints `hello2` and the computation enclosed in parentheses (from `print "hello3"` to `force y`), which as we shall see is a returner, commences execution. This returner first prints `hello3`, pushes 7, and prints `we just pushed 7`. Then it forces the thunk `y`. So it prints `hello1` and pops `z` from the stack; i.e. it removes the top entry (which is 7) from the stack and binds `z` to it. It reports `we just popped 7` and returns `x` + `z` which is 10.

So the returner enclosed in parentheses (from `print "hello3"` to `force y`) has had the overall effect of printing several messages and returning 10. Thus `w` becomes bound to 10 and the program prints `w is bound to 10`. Finally the program returns `w` + 5 which is 15.

In summary the program outputs as follows

```
hello0
hello2
hello3
we just pushed 7
hello1
we just popped 7
w is bound to 10
```

and finally returns the value 15.

In more familiar terms, `y` is a procedure, 7 is the parameter that is passed to it and it returns 10. But, compared to a typical high-level language, CBPV is unusually liberal in that

- it allows the command `print "we just pushed 7"` to intervene between pushing the parameter 7 and calling the procedure;

- it allows the command `print "hello 3"` to intervene between the start of the procedure and popping the parameter.

In a practical sense, this liberality is of little benefit, for the program would have the same observable behaviour if the lines

```
7‘
print "we just pushed 7".
```

were exchanged, and likewise if the lines

```
print "hello 3".
λz.
```

were exchanged. But it is this flexibility that allows CBPV to give a fine-grain analysis of types.

I.5.3 CBPV Makes Control Flow Explicit

It is worth noticing that there are two lines in this program which cause execution to move to another part of the program, rather than to the next line. These are `force y`, where execution jumps to the body of the thunk that `y` is bound to, and `return x + z`, where the value 10 is returned to just after the line `force y`. This illustrates a general phenomenon.

> Only `force` and `return` cause execution to move to another part of the program.

This kind of information about control flow is much less explicit in CBV and CBN. For this reason, as we shall see in Chap. 7 and Chap. 8, it is beneficial to use CBPV when studying semantics that describes interaction between different parts of a program, such as continuation semantics or game semantics.

Although `force` and `return` both cause jumps, there is an important difference between the two. A `force` instruction will specify where control moves to: in the example program, to `y`. By contrast, a `return` instruction will cause a jump to a point at the top of a stack, so it is not specified explicitly. This distinction is apparent in both continuation semantics and game semantics.

In Chap. 7, we look at a language called *Jump-With-Argument* where, unlike CBPV, the destination of *every* jump is explicit. The translation from CBPV to JWA (called the stack-passing-style transform) makes explicit the destination of a `return` instruction.

I.5.4 Value Types and Computation Types

We said in Sect. I.5.1 that CBPV has 2 disjoint classes of terms: values and computations. It is therefore unsurprising that it has 2 disjoint classes of types: value types and computation types.

- A value has a value type.
- A computation has a computation type.

For example, `nat` and `bool` are value types.

The two classes of types are given by

$$\begin{array}{lrl} \text{value types} & A ::= & U\underline{B} \mid \sum_{i\in I} A_i \mid 1 \mid A \times A \\ \text{computation types} & \underline{B} ::= & FA \mid \prod_{i\in I} \underline{B}_i \mid A \to \underline{B} \end{array}$$

where each set I of tags is finite. (We will also be concerned with *infinitely wide* CBPV in which I may be countably infinite—see Sect. 4.1 for more discussion.) We underline computation types for clarity.

We explain these types as follows—notice how this explanation maintains the principle "a value is, a computation does".

- A value of type $U\underline{B}$ is a *thunk* of a computation of type $\underline{B}$.
- A value of type $\sum_{i\in I}A_i$ is a pair (i, V), where $i \in I$ and V is a value of type A_i.
- A value of type 1 is an empty tuple $()$.
- A value of type $A \times A'$ is a pair (V, V'), where V is a value of type A and V' is a value of type A'.
- A computation of type FA *returns* a value of type A.
- A computation of type $\prod_{i\in I}\underline{B}_i$ *pops* a tag $i \in I$ from the stack, and then behaves as a computation of type $\underline{B}_i$.
- A computation of type $A \to \underline{B}$ *pops* a value of type A from the stack, and then behaves as a computation of type $\underline{B}$.

We have apparently excluded types such as `bool` and `nat` from this, but `bool` can be recovered as $1 + 1$. If we add type recursion as in Sect. 4.2.2, we can recover `nat` as $\mu\mathtt{X}.(1 + \mathtt{X})$. In infinitely wide CBPV, we can recover `nat` as $\sum_{i\in\mathbb{N}}1$.

Looking at the example computation of Sect. I.5.2, we can see that its type is F`nat` because it returns a natural number. The identifier `y` has type U(`nat` $\to F$`nat`). This means that the value to which `y` is bound is a thunk (U) of a computation that pops a natural number (`nat` $\to$) and then returns (F) a natural number (`nat`).

I.6 Structure of Thesis

I.6.1 Goals

This book has two goals.

1. The main goal is to argue the claim made in Sect. I.3—to persuade the reader that CBPV is significant by exhibiting a wide range of elegant semantics, from which CBN and CBV semantics can be recovered. This goal is the subject of Part II; the preliminary work is done in Chap. 2.

2. The second goal is to understand the categorical semantics of the CBPV equational theory, introduced in Chap. 3. This goal is the subject of Part III. We do not claim that this provides any further motivation for CBPV.

I.6.2 Chapter Outline

We begin by describing an intellectual journey that leads to CBPV. We do this briefly—a full technical treatment is given in the Appendix. The journey starts in Chap. 1, where we explain the language design choices that characterize the CBV and CBN paradigms, and we examine the consequences of these choices. We give both operational (big-step) semantics and denotational semantics in the presence of printing, our leading example of a computational effect. (We look at divergence too, because of its familiarity.) We learn, most importantly, that CBV types and CBN types denote different kinds of things—sets and $\mathcal{A}$-sets (which we will define in Sect. 1.7.4) respectively. By considering which equations are valid as observational equivalences, we see that function types behave well in CBN but not in CBV, whereas sum types behave well in CBV but not in CBN.

This critical exploration of CBV and CBN leads us, in Chap. 2, to CBPV. Because of the denotational difference between CBV types and CBN types, it is clear that the subsuming language must have two kinds of type. We give big-step and denotational semantics (for printing and for divergence) and we explain how CBV and CBN are subsumed in CBPV. This completes the journey. In addition, we give another form of operational semantics, the *CK-machine* [Felleisen and Friedman, 1986], which makes precise the push/pop reading that we illustrated in Sect. I.5.2. This machine gives us another class of terms—the *stacks*.

Whereas function types behave badly in CBV and sum types behave badly in CBN, in CBPV all types behave well. We see this in Chap. 3 by giving an equational theory for CBPV. We actually give 2 theories: without stacks and with stacks (the latter conservatively extends the former). To present this theory, we have to extend CBPV with *complex values*. However, as we explain in Sect. 3.5, this is not a problem, because complex values can always be removed from computations and from closed values.

Chap. 4, the last chapter of Part I, looks at recursion and infinitary CBPV. There is little original here (except the notion of "sub-bilimit-compact category"); it merely sets up material that is needed later in Chap. 6, where we need to understand the semantics of type recursion in order to be able to interpret thunk storage, and in Chap. 8, where we need infinitely deep syntax in order to obtain definability results for types.

Having seen the operational ideas and the equational theory we are ready to give the concrete semantics of the former (Part II) and the categorical semantics of the latter (Part III).

In Chap. 5 we look at denotational semantics for a range of effects: global store, control, errors, erratic choice, printing, divergence and various combinations of these. We learn that a useful heuristic for creating CBPV semantics is to guess the form of the soundness theorem. Again and again, throughout these examples, we see traditional semantics for CBV primitives decomposing naturally into CBPV semantics. As for the corresponding CBN models, some are new while others were known but previously appeared mysterious—the CBPV decomposition makes the structure of these semantics clear.

Both the store model and the control model from Chap. 5 are developed further in the subsequent chapters.

- The store model is developed in Chap. 6 into a possible world model for cell generation, which captures some interesting intuitions about CBPV: in particular, the idea that a thunk is something that can be forced at any future time. This model is surprising in itself, as previous possible world models (with the exception of [Ghica, 1997]) allow only the storage of ground values, whereas this model treats storage of all values. But it also provides a good example of the benefits of CBPV, because in the corresponding model for CBV [Levy, 2002] the semantics of functions is unwieldy—CBPV decomposes it into manageable pieces.

- Based on the control model, we introduce in Chap. 7 another language called Jump-With-Argument (JWA), equivalent to—but much simpler than—CBPV with control effects. Unlike CBPV, Jump-With-Argument is not really new, as it is essentially Steele's CPS intermediate language [Steele, 1978]. This language enables us to see that the StkPS transform (the CBPV analogue of the CPS transform) provides a jumping implementation for CBPV, just as Steele explained for CBV. We see how the explicit control flow described in Sect. I.5.3 makes CBPV a good starting point for such an analysis.

Our final piece of evidence for the advantages of CBPV and JWA is Hyland-Ong-style game semantics, which we discuss in Chap. 8. We see once again that CBPV is much closer to the semantics than the usual CBN languages (PCF and Idealized Algol) because of its explicit control flow described in Sect. I.5.3. Key notions of game semantics—the question/answer distinction, pointers between moves, the bracketing condition—become clearer from a CBPV and JWA viewpoint. For example, we see that "asking a question" corresponds to forcing, while "answering" corresponds to returning.

The goal of Part III is to present and analyze the structures required to model

- effect-free language (simply typed λ-calculus with sum, product and exponential types)
- CBPV with stacks
- JWA.

Of course the effect-free language is treated in the literature, so our treatment is essentially a recapitulation, presented in such a way as to smooth the transition to the effectful languages.

Many of these categorical structures, such as products, can be given an elementary definition using a universal property ("element style"), and a more abstract definition using naturality. In Chap. 9 we present all the required structures in an element style, and explain how we can use this to interpret CBPV with stacks in a *CBPV adjunction model*, and JWA in a *JWA module model*. We see how all the concrete denotational semantics from Parts I–II are instances of these structures.

Each of the following 2 chapters is a sequel to Chap. 9. Chap. 10 develops the relationship between the categorical structures and the syntax, showing that *every* model of the equational theory is a CBPV adjunction model.

In Chap. 11, we show that each connective of CBPV is a *representing object* for a suitable functor, which gives us a naturality style definition. Furthermore, we show that our so-called "CBPV adjunction models" are indeed adjunctions in the usual sense.

Finally, in Chap. 12, we look at some problems and possible directions for further work.

I.6.3 Chapter Dependence

The dependence between chapters and sections is shown in Fig. I.1. The target of each arrow depends on its source. The diagram does not cover examples, only the main text. Thus, we will sometimes describe an example that depends on a chapter not indicated here.

The diagram indicates that Sect. 7.6—the JWA equational theory—can be read without any preparation. That is true, if one refers to the term syntax in Fig. 7.1. However, this leaves out the motivation for JWA.

I.6.4 Proofs

We have usually omitted proofs which are straightforward inductions, and we have omitted proofs of results for CBV and CBN where we have given corresponding results for CBPV.

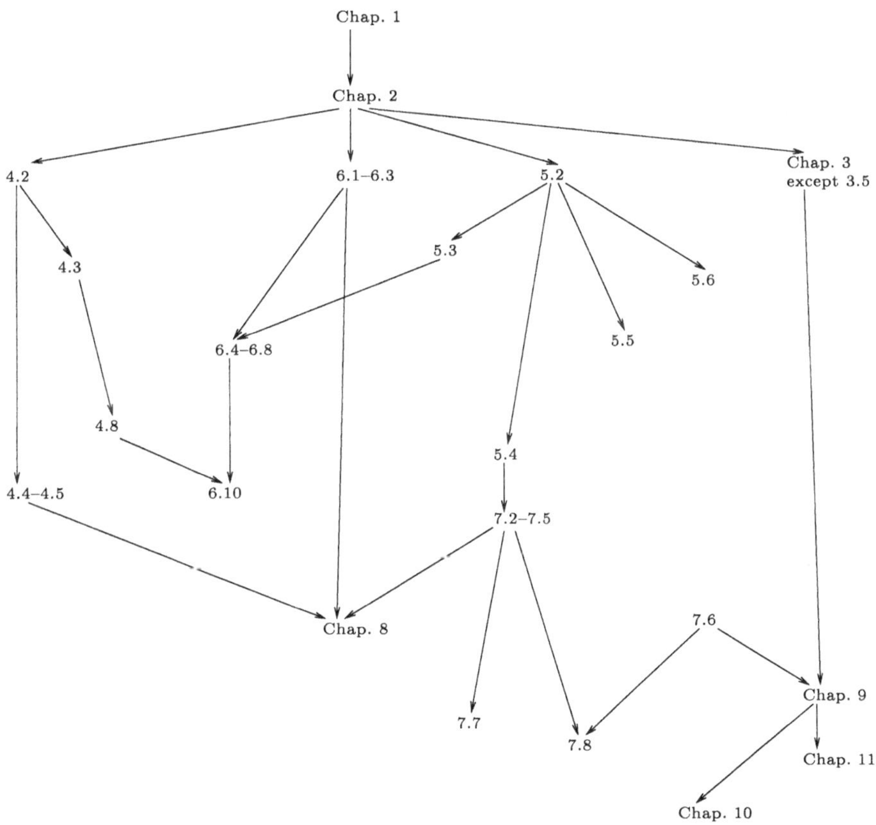

Figure I.1. Chapter and Section Dependence

I

LANGUAGE

Chapter 1

CALL-BY-VALUE AND CALL-BY-NAME

1.1 Introduction

There is a certain tension in the presentation of CBPV: how much attention shall we devote to CBN and CBV? On the one hand, we are claiming that CBPV "subsumes" CBN and CBV, and to see how that is achieved we have to discuss CBN and CBV, at least to some extent. On the other hand, a thorough study of CBN and CBV would be a waste of effort, since a primary purpose of CBPV is to relieve us of that task. Once we have CBPV (which we want to introduce as early as possible), CBV and CBN are seen to be just particular fragments of it.

The CBN/CBV material is therefore organized as follows. In this chapter, we look informally at the key concepts and properties of these paradigms, assuming no prior knowledge of them. This provides background for CBPV which we introduce in Chap. 2. But for the interested reader, we provide in Appendix A a thorough, technical treatment of CBN and CBV and their relationship to CBPV. This chapter can thus be seen as a synopsis of Appendix A.

1.2 The Main Point Of The Chapter

The main point of the chapter is this:

CBV types and CBN types denote different kinds of things.

This is true both for printing semantics and for Scott semantics:

- In our printing semantics, a CBV type denotes a set whereas a CBN type denotes an $\mathcal{A}$-set (which we will define in Sect. 1.7.4).
- In Scott semantics, a CBV type denotes a cpo whereas a CBN type denotes a pointed cpo.

The importance of this main point is that it makes it clear why CBPV, the subsuming paradigm, will need to have two disjoint classes of type.

Unfortunately, the Scott semantics obscures this main point, because a pointed cpo is a special kind of cpo. By contrast, the class of sets and the class of $\mathcal{A}$-sets are (as will be apparent once we have defined $\mathcal{A}$-sets) disjoint. Thus printing illustrates our main point much better than divergence does. This is why we have chosen printing as our leading example of a computational effect.

1.3 A Simply Typed λ-Calculus

1.3.1 The Language

Before we look at any computational effects, we study an effect-free language. We look at the simply typed λ-calculus whose only ground type is a boolean type, extended with binary sum types; we call this $\lambda\,\mathtt{bool}+$. Its types are

$$A ::= \ \mathtt{bool} \mid A + A \mid A \to A$$

and its terms are given in Fig. 1.1. We write $\mathtt{pm}$ for "pattern-match"; and recall that we write ' for operand-first application. While $\mathtt{let}$ is not strictly necessary ($\mathtt{let}\ M\ \mathtt{be}\ \mathtt{x}.\ N$ can be desugared as $M`\lambda\mathtt{x}N$), it is convenient to include it as a primitive.

$\lambda\,\mathtt{bool}+$ has a straightforward semantics where types denote sets and terms denote functions. A closed term of type $\mathtt{bool}$ denotes either true or false; there is an easy decision procedure to find which. We call $\mathtt{bool}$ the ground type.

1.3.2 Product Types

Suppose we want to add product types $A \times A'$ to $\lambda\,\mathtt{bool}+$. It is clear what the introduction rule should be:

$$\frac{\Gamma \vdash M : A \quad \Gamma \vdash M' : A'}{\Gamma \vdash (M, M') : A \times A'}$$

But we have a choice as to the form of the elimination rule. Either we can use *projections*:

$$\frac{\Gamma \vdash M : A \times A'}{\Gamma \vdash \pi M : A} \qquad\qquad \frac{\Gamma \vdash M : A \times A'}{\Gamma \vdash \pi' M : A'}$$

Or we can use *pattern-matching*:

$$\frac{\Gamma \vdash M : A \times A' \quad \Gamma, \mathtt{x} : A, \mathtt{y} : A' \vdash N : B}{\Gamma \vdash \mathtt{pm}\ M\ \mathtt{as}\ (\mathtt{x}, \mathtt{y}).N : B}$$

$$\frac{}{\Gamma, \mathtt{x} : A, \Gamma' \vdash \mathtt{x} : A} \qquad \frac{\Gamma \vdash M : A \quad \Gamma, \mathtt{x} : A \vdash N : B}{\Gamma \vdash \mathtt{let}\ M\ \mathtt{be}\ \mathtt{x}.\ N : B}$$

$$\frac{}{\Gamma \vdash \mathtt{true} : \mathtt{bool}} \qquad \frac{}{\Gamma \vdash \mathtt{false} : \mathtt{bool}}$$

$$\frac{\Gamma \vdash M : \mathtt{bool} \quad \Gamma \vdash N : B \quad \Gamma \vdash N' : B}{\Gamma \vdash \mathtt{if}\ M\ \mathtt{then}\ N\ \mathtt{else}\ N' : B}$$

$$\frac{\Gamma \vdash M : A}{\Gamma \vdash \mathtt{inl}\ M : A + A'} \qquad \frac{\Gamma \vdash M : A'}{\Gamma \vdash \mathtt{inr}\ M : A + A'}$$

$$\frac{\Gamma \vdash M : A + A' \quad \Gamma, \mathtt{x} : A \vdash N : B \quad \Gamma, \mathtt{x} : A' \vdash N' : B}{\Gamma \vdash \mathtt{pm}\ M\ \mathtt{as}\ \{\mathtt{inl}\ \mathtt{x}.N, \mathtt{inr}\ \mathtt{x}.N'\} : B}$$

$$\frac{\Gamma, \mathtt{x} : A \vdash M : B}{\Gamma \vdash \lambda \mathtt{x}.M : A \to B} \qquad \frac{\Gamma \vdash M : A \quad \Gamma \vdash N : A \to B}{\Gamma \vdash M\text{`}N : B}$$

Figure 1.1. Terms of λ bool+

At first sight, the choice between projections and pattern-matching seems unimportant, because these two forms of elimination rule are equivalent:

$$\begin{array}{rcl} \pi M & = & \mathtt{pm}\ M\ \mathtt{as}\ (\mathtt{x}, \mathtt{y}).\ \mathtt{x} \\ \pi' M & = & \mathtt{pm}\ M\ \mathtt{as}\ (\mathtt{x}, \mathtt{y}).\ \mathtt{y} \\ \mathtt{pm}\ M\ \mathtt{as}\ (\mathtt{x}, \mathtt{y}).N & = & \mathtt{let}\ \pi M\ \mathtt{be}\ \mathtt{x}, \pi' M\ \mathtt{be}\ \mathtt{y}.\ N \end{array}$$

In the presence of effects (CBN and CBV), however, these equations are not necessarily valid and the choice does matter. This is discussed in Appendix A.

Notice

- the resemblance between a pattern-match product and a sum type—each of these types has an elimination rule using pattern-matching
- the resemblance between a projection product and a function type from $\{0, 1\}$. For we can think of a tuple (M, M') of projection product type as a function taking 0 to M and 1 to M'. To emphasize this

resemblance, we use a novel notation: we write

$$\begin{array}{rcl} (M, M') & \text{as} & \lambda\{0.M, 1.M'\} \\ \pi M & \text{as} & 0`M \\ \pi' M & \text{as} & 1`M \end{array}$$

Because of these resemblances, we can understand the key issues in CBV and CBN without having to include products. That is why, in this chapter, we will not consider product types further. They are dealt with fully in Appendix A.

There are more type constructors we could include while remaining simply typed, and in Appendix A we will include them so that our treatment of CBV and CBN there is as thorough as possible. The type system used in this chapter, therefore, provides only a fragment of the full type system that a (simply typed) CBN or CBV language can allow. However, our aim here is just to explain the key ideas, and the types in λ bool+ are quite sufficient for that purpose.

1.3.3 Equations

Each type constructor ($\rightarrow$, bool, $+$) has two associated equations called the β-law and the η-law.

We look first at the β-laws, which are straightforward.

- The β-law for $A \rightarrow B$:

$$M`\lambda \mathtt{x}.N = N[M/\mathtt{x}]$$

- The β-laws for bool:

$$\begin{array}{r} \mathtt{if\ true\ then}\ N\ \mathtt{else}\ N' = N \\ \mathtt{if\ false\ then}\ N\ \mathtt{else}\ N' = N' \end{array}$$

- The β-laws for $A + A'$:

$$\begin{array}{r} \mathtt{pm\ inl}\ M\ \mathtt{as}\ \{\mathtt{inl\ x}.N, \mathtt{inr\ x}.N'\} = N[M/\mathtt{x}] \\ \mathtt{pm\ inr}\ M\ \mathtt{as}\ \{\mathtt{inl\ x}.N, \mathtt{inr\ x}.N'\} = N'[M/\mathtt{x}] \end{array}$$

- There is also a β-law for let :

$$\mathtt{let}\ M\ \mathtt{be\ x}.\ N = N[M/\mathtt{x}]$$

We next turn to the η-laws. They are more subtle than the β-laws, but they are important in understanding the CBV/CBN issues. So we look at them carefully and understand why they are valid in the set semantics.

- The η-law for $A \to B$: any term M of type $A \to B$ can be expanded
$$M = \lambda \mathtt{x}(\mathtt{x}\,`\,{}^{\mathtt{x}}M)$$
- The η-law[1] for `bool`: if M has a free identifier $\mathtt{z} : \mathtt{bool}$ then for any term N of type `bool`
$$M[N/\mathtt{z}] = \mathtt{if}\ N\ \mathbf{then}\ M[\mathtt{true}/\mathtt{z}]\ \mathtt{else}\ M[\mathtt{false}/\mathtt{z}] \qquad (1.3)$$
Intuitively, this holds because, in a given environment (list of bindings for identifiers), N denotes either true or false.
- The η-law for $A + A'$: if M has a free identifier $\mathtt{z} : A + A'$ then for any term N of type $A + A'$
$$M[N/\mathtt{z}] = \mathtt{pm}\ N\ \mathtt{as}\ \{\mathtt{inl}\ \mathtt{x}.\ {}^{\mathtt{x}}M[\mathtt{inl}\ \mathtt{x}/\mathtt{z}], \mathtt{inr}\ \mathtt{x}.\ {}^{\mathtt{x}}M'[\mathtt{inr}\ \mathtt{x}/\mathtt{z}]\}$$

Notice the similarity between sum types and `bool`.

1.3.4 Reversible Derivations

As a consequence of these equations, we have *reversible derivations*, e.g. for $\to$
$$\frac{\Gamma, A \vdash B}{\Gamma \vdash A \to B}$$
This means that from a term of the form $\Gamma, \mathtt{x} : A \vdash M : B$ (assuming $\mathtt{x} \notin \Gamma$) we can construct a term of the form $\Gamma \vdash N : A \to B$ and vice versa and that these operations are inverse up to provable equality. The reversible derivation is given by

- the operation $\theta : M \mapsto \lambda\mathtt{x}.M$, which turns a term $\Gamma, \mathtt{x} : A \vdash M : B$ into a term $\Gamma \vdash N : A \to B$
- the context $\theta^{-1} : N \mapsto \mathtt{x}`N$, which turns a term $\Gamma \vdash N : A \to B$ into a term $\Gamma, \mathtt{x} : A \vdash M : B$

[1]Some authors e.g. [Girard et al., 1988] give the name "η-law" to the weaker equation
$$M = \mathtt{if}\ M\ \mathtt{then}\ \mathtt{true}\ \mathtt{else}\ \mathtt{false} \qquad (1.1)$$
together with the *commuting conversion* law
$$M[\mathtt{if}\ N\ \mathtt{then}\ P\ \mathtt{else}\ P'/\mathtt{z}] = \mathtt{if}\ N\ \mathtt{then}\ M[P/\mathtt{z}]\ \mathtt{else}\ M[P'/\mathtt{z}] \qquad (1.2)$$
or a variant of this. (1.1)–(1.2) together are equivalent to our η-law (1.3).

and it is clear that these operations are inverse up to provable equality, using the β- and η- laws for $\rightarrow$.

The reversible derivations for `bool` and $+$ are

$$\frac{\Gamma \vdash B \quad \Gamma \vdash B}{\Gamma, \texttt{bool} \vdash B} \qquad \frac{\Gamma, A \vdash B \quad \Gamma, A' \vdash B}{\Gamma, A + A' \vdash B}$$

Readers familiar with categorical semantics will see that these reversible derivations provide important information about the categorical structure of an equational theory.

1.4 Adding Effects

We now wish to add a computational effect to λ `bool+`, and we will use output as our example. We suppose that to any term we can prefix a command such as `print "hello"`. For example, the term `print "hello". true` when evaluated, prints `hello` and then returns `true`.

We write $\mathcal{A}$ for the set of characters that can be printed, $\mathcal{A}^*$ for the set of finite strings of characters, $*$ for concatenation of strings and `""` for the empty string. Formally, we add to the term syntax the following rule, for every character $c \in \mathcal{A}$:

$$\frac{\Gamma \vdash M : B}{\Gamma \vdash \texttt{print}\ c.\ M : B}$$

There are other, equivalent, syntactic possibilities, e.g. in CBV we could let commands have type 1, as in ML. But we prefer commands to be prefixes. This is for the sake of consistency between CBN, CBV, CBPV and the Jump-With-Argument language discussed in Chap. 7.

1.5 The Principles Of Call-By-Value and Call-By-Name

The first consequence of adding effects is that, by contrast with pure λ `bool+`, the order of evaluation matters. Given a closed term of type `bool`, the output depends on the evaluation order. For some effects the answer too will depend on the evaluation order, but for output that is not the case. The two evaluation orders we will look at are CBV and CBN. We describe the principles of these two paradigms.

Firstly, in both CBV and CBN, we do not evaluate under λ. Thus a λ-abstraction is *terminal*, in the sense that it requires no further evaluation.

Having decided not to evaluate under λ, we have numerous choices still to make.

1 To evaluate the term $\texttt{let}\ M\ \texttt{be x.}\ N$, do we

   (a) evaluate M to T and then evaluate $N[T/\mathtt{x}]$, or
   (b) leave M alone, and just evaluate $N[M/\mathtt{x}]$?

2 To evaluate a term M of sum type, we must first evaluate it to $\texttt{inl}\ N$ or $\texttt{inr}\ N$. Do we then

   (a) continue to evaluate N to T, so that M evaluates to $\texttt{inl}\ T$ or $\texttt{inr}\ T$, or
   (b) stop there, declaring that $\texttt{inl}\ N$ and $\texttt{inr}\ N$ are always terminal?

3 To evaluate an application such as $M`N$, we certainly need to evaluate N, giving say $\lambda\mathtt{x}.P$. Besides this, do we

   (a) evaluate M to T (either before or after evaluating N) and then evaluate $P[T/\mathtt{x}]$, or
   (b) leave M alone, and just evaluate $P[M/\mathtt{x}]$?

In fact, there is one fundamental question whose answer will determine how we answer questions (1)–(3): what may we substitute for an identifier? At one extreme, we allow only a terminal term to replace an identifier—this defines the CBV paradigm. At the other extreme, we allow only a totally unevaluated term to replace an identifier—this defines the CBN paradigm.

To each of questions (1)–(3), CBV requires us to answer (a) and CBN requires us to answer (b). In the case of questions (1) and (3), this is clear because substitution is involved. For question (2), it requires some explanation. Consider a term such as

$$\texttt{pm}\ M\ \texttt{as}\ \{\texttt{inl x.}N, \texttt{inr y.}N'\}$$

To evaluate this we must first evaluate M. Suppose we evaluate M to $\texttt{inl}\ M'$. In CBN we must not proceed to evaluate M', because we want to substitute it for $\mathtt{x}$ in N, so it must be completely unevaluated. In CBV we must evaluate M' to a terminal term T, so that we can substitute T for $\mathtt{x}$ in N.

In conclusion, we emphasize the following point:

> The essential difference between CBV and CBN is not (as is often thought) the way that application is evaluated; rather it is what an identifier may be bound to.

Notice that within the CBV paradigm, we have the choice of whether, when evaluating an application $M`N$, we evaluate M before N or vice versa. In fact, so long as we are consistent, the semantic theory is essentially unaffected. Arbitrarily, we stipulate that we evaluate the operand M before the operator N.

1.6 Call-By-Value

1.6.1 Operational Semantics

In CBV a closed term which is terminal is called a *closed value*. These are given by

$$V ::= \mathtt{true} \mid \mathtt{false} \mid \mathtt{inl}\ V \mid \mathtt{inr}\ V \mid \lambda \mathtt{x}.M$$

Every closed term M prints a string of characters m and then returns a closed value V. We write $M \Downarrow m, V$. This is defined inductively in Fig. 1.2.

$$\frac{M \Downarrow m, V \quad N[V/\mathtt{x}] \Downarrow m', W}{\mathtt{let}\ M\ \mathtt{be}\ \mathtt{x}.\ N \Downarrow m * m', W}$$

$$\frac{}{\mathtt{true} \Downarrow \mathtt{""}, \mathtt{true}} \qquad \frac{}{\mathtt{false} \Downarrow \mathtt{""}, \mathtt{false}}$$

$$\frac{M \Downarrow m, \mathtt{true} \quad N \Downarrow m', V}{\mathtt{if}\ M\ \mathtt{then}\ N\ \mathtt{else}\ N' \Downarrow m * m', V} \qquad \frac{M \Downarrow m, \mathtt{false} \quad N' \Downarrow m', V}{\mathtt{if}\ M\ \mathtt{then}\ N\ \mathtt{else}\ N' \Downarrow m * m', V}$$

$$\frac{M \Downarrow m, V}{\mathtt{inl}\ M \Downarrow m, \mathtt{inl}\ V} \qquad \frac{M \Downarrow m, V}{\mathtt{inr}\ M \Downarrow m, \mathtt{inr}\ V}$$

$$\frac{M \Downarrow m, \mathtt{inl}\ V \quad N[V/\mathtt{x}] \Downarrow m', W}{\mathtt{pm}\ M\ \mathtt{as}\ \{\mathtt{inl}\ \mathtt{x}.N, \mathtt{inr}\ \mathtt{x}.N'\} \Downarrow m * m', W}$$

$$\frac{M \Downarrow m, \mathtt{inr}\ V \quad N'[V/\mathtt{x}] \Downarrow m', W}{\mathtt{pm}\ M\ \mathtt{as}\ \{\mathtt{inl}\ \mathtt{x}.N, \mathtt{inr}\ \mathtt{x}.N'\} \Downarrow m * m', W}$$

$$\frac{}{\lambda \mathtt{x}.M \Downarrow \mathtt{""}, \lambda \mathtt{x}.M} \qquad \frac{M \Downarrow m, V \quad N \Downarrow m', \lambda \mathtt{x}.N' \quad N'[V/\mathtt{x}] \Downarrow m'', W}{M`N \Downarrow m * m' * m'', W}$$

$$\frac{M \Downarrow m, V}{\mathtt{print}\ c.\ M \Downarrow c * m, V}$$

Figure 1.2. Big-Step Semantics for CBV with `print`

Proposition 1 For every M there is a unique m, V such that $M \Downarrow m, V$. □

The proof is similar to that of Prop. 10.

1.6.2 Denotational Semantics for `print`

Definition 1.1 ▪ The following CBV terms are called *values*:

$$V ::= \ \mathtt{x} \mid \mathtt{true} \mid \mathtt{false} \mid \mathtt{inl}\ V \mid \mathtt{inr}\ V \mid \lambda\mathtt{x}.M$$

(This generalizes the notion of "closed value" we have already used.)

▪ All CBV terms are called *returners*. (This is because their goal is to return a value.)

□

Notice that a term is a value iff every closed substitution instance (substituting only closed values) is a closed value.

We now describe and explain a denotational semantics for the CBV printing language. The key principle is that

> each type A denotes a set $[\![A]\!]$ whose elements are the denotations of closed *values* of type A.

Thus the type `bool` denotes the 2-element set $\{\mathsf{true}, \mathsf{false}\}$ because there are two closed values of type `bool`. Likewise the type $A + A'$ denotes $[\![A]\!] + [\![A']\!]$ because a closed value of type $A + A'$ must be either of the form `inl` V, where V is a closed value of type A, or of the form `inr` V, where V is a closed value of type A'. We shall come to $A \to B$ presently.

Given a closed value V of type A, we write $[\![V]\!]^{\mathsf{val}}$ for the element of $[\![A]\!]$ that it denotes. Given a closed returner M of type A, we recall that M prints a string of characters $m \in \mathcal{A}^*$ and then returns a closed value V of type A. So M will denote an element $[\![M]\!]^{\mathsf{ret}}$ of $\mathcal{A}^* \times [\![A]\!]$. Thus a closed value V will have two denotations $[\![V]\!]^{\mathsf{val}}$ and $[\![V]\!]^{\mathsf{ret}}$ related by

$$[\![V]\!]^{\mathsf{ret}} = (\texttt{""}, [\![V]\!]^{\mathsf{val}})$$

A closed value of type $A \to B$ is of the form $\lambda\mathtt{x}.M$. This, when applied to a closed *value* of type A gives a closed *returner* of type B. So $A \to B$ denotes $[\![A]\!] \to (\mathcal{A}^* \times [\![B]\!])$. It is true that the syntax appears to allow us to apply $\lambda\mathtt{x}.M$ to any returner N of type A, not just to a value. But N will be evaluated before it interacts with $\lambda\mathtt{x}.M$, so $\lambda\mathtt{x}.M$ is really only applied to the value that N returns.

Given a context $\Gamma = \mathtt{x}_0 : A_0, \ldots, \mathtt{x}_{n-1} : A_{n-1}$, an environment (list of bindings for identifiers) associates to each $\mathtt{x}_i$ a closed value of type A_i. So the environment denotes an element of $[\![A_0]\!] \times \cdots \times [\![A_{n-1}]\!]$, and we write $[\![\Gamma]\!]$ for this set.

Given a value $\Gamma \vdash V : B$, we see that V, together with an environment, gives (by substitution) a closed value of type B. So V denotes a function $[\![V]\!]^{\mathsf{val}}$ from $[\![\Gamma]\!]$ to $[\![B]\!]$.

Given a returner $\Gamma \vdash M : B$, we see that M, together with an environment, gives (by substitution) a closed returner of type B. So M denotes a function $[\![M]\!]^{\mathsf{ret}}$ from $[\![\Gamma]\!]$ to $\mathcal{A}^* \times [\![B]\!]$.

Generalizing what we saw for closed values, an arbitrary value V will have two denotations $[\![V]\!]^{\mathsf{val}}$ and $[\![V]\!]^{\mathsf{ret}}$ related by

$$[\![V]\!]^{\mathsf{ret}}\rho = ("", [\![V]\!]^{\mathsf{val}}\rho)$$

for each environment ρ.

In summary, the denotational semantics is organized as follows.

- A type A denotes a set $[\![A]\!]$.
- A context $\mathtt{x}_0 : A_0, \ldots, \mathtt{x}_{n-1} : A_{n-1}$ denotes the set $[\![A_0]\!] \times \cdots \times [\![A_{n-1}]\!]$.
- A value $\Gamma \vdash V : B$ denotes a function $[\![V]\!]^{\mathsf{val}} : [\![\Gamma]\!] \longrightarrow [\![B]\!]$
- A term $\Gamma \vdash M : B$ denotes a function $[\![M]\!]^{\mathsf{ret}} : [\![\Gamma]\!] \longrightarrow \mathcal{A}^* \times [\![B]\!]$

The denotations of types is given by

$$\begin{array}{rcl} [\![\mathsf{bool}]\!] & = & \{\mathsf{true}, \mathsf{false}\} \\ [\![A + A']\!] & = & [\![A]\!] + [\![A']\!] \\ [\![A \to B]\!] & = & [\![A]\!] \to (\mathcal{A}^* \times [\![B]\!]) \end{array}$$

The denotations of values—some example clauses:

$$\begin{array}{rcl} [\![\mathtt{x}]\!]^{\mathsf{val}}(\rho, \mathtt{x} \mapsto x, \rho') & = & x \\ [\![\mathtt{true}]\!]^{\mathsf{val}}\rho & = & \mathsf{true} \\ [\![\mathtt{inl}\ V]\!]^{\mathsf{val}}\rho & = & \mathsf{inl}\ [\![V]\!]^{\mathsf{val}}\rho \\ [\![\lambda\mathtt{x}.M]\!]^{\mathsf{val}}\rho & = & \lambda x.[\![M]\!]^{\mathsf{ret}}(\rho, \mathtt{x} \mapsto x) \end{array}$$

The denotations of returners—some example clauses:

$$\begin{array}{rcl} [\![\mathtt{x}]\!]^{\mathsf{ret}}(\rho, \mathtt{x} \mapsto x, \rho') & = & ("", x) \\ [\![\mathtt{true}]\!]^{\mathsf{ret}}\rho & = & ("", \mathsf{true}) \\ [\![\mathtt{inl}\ M]\!]^{\mathsf{ret}} & = & (m, \mathsf{inl}\ v) \text{ where } [\![M]\!]^{\mathsf{ret}}\rho = (m, v) \\ [\![\mathtt{if}\ M\ \mathtt{then}\ N\ \mathtt{else}\ N']\!]\rho & = & \left\{ \begin{array}{l} (m * m', v) \text{ if } [\![M]\!]^{\mathsf{ret}}\rho = (m, \mathsf{true}) \\ \quad \text{and } [\![N]\!]^{\mathsf{ret}}\rho = (m', v) \\ (m * m', v) \text{ if } [\![M]\!]^{\mathsf{ret}}\rho = (m, \mathsf{false}) \\ \quad \text{and } [\![N']\!]^{\mathsf{ret}}\rho = (m', v) \end{array} \right. \\ [\![\lambda\mathtt{x}.M]\!]^{\mathsf{ret}}\rho & = & ("", \lambda x.[\![M]\!]^{\mathsf{ret}}(\rho, \mathtt{x} \mapsto x)) \\ [\![M`N]\!]^{\mathsf{ret}}\rho & = & (m * m' * m'', w) \text{ where } [\![M]\!]^{\mathsf{ret}}\rho = (m, v) \\ & & \text{and } [\![N]\!]^{\mathsf{ret}}\rho = (m', f) \text{ and } v`f = (m'', w) \end{array}$$

Notice how strongly these clauses resemble the corresponding clauses of Fig. 1.2.

Proposition 2 (soundness) If $M \Downarrow m, V$ then $[\![M]\!]^{\mathsf{ret}} = (m, [\![V]\!]^{\mathsf{val}})$ □

Corollary 3 (by Prop. 1) For any closed ground returner (i.e. returner of ground type) M, we have $M \Downarrow m, \mathtt{return}\ n$ iff $[\![M]\!] = (m, n)$. □

1.6.3 Scott Semantics

The reader may be familiar with Scott semantics for CBV. This has appeared in two forms.

1 In the older form, types denote pointed cpos, returners denote strict functions and values denote strict, bottom-reflecting functions.

2 In the more recent form, due to Plotkin [Plotkin, 1985], types denote (unpointed) cpos, values denote total functions and returners denote partial functions.

Although these two semantics are equivalent, (2) is more natural, and it agrees with the key principle of our printing semantics: the elements of $[\![A]\!]$ are denotations of closed *values* of type A. We outline the Scott semantics in form (2)[2] to make apparent its similarity to the printing semantics.

Definition 1.2 1 A *cpo* $(X, \leqslant)$ is a poset with joins of all directed subsets.

2 A function between cpos is *continuous* when it preserves all directed joins.

3 We write **Cpo** for the category of cpos and continuous functions.

□

The semantics is organized as follows.

- A type A denotes a cpo $[\![A]\!]$. Our intention is that a closed value of type A will denote an element of $[\![A]\!]$.
- A context $\Gamma = A_0, \ldots, A_{n-1}$ denotes the cpo $[\![\Gamma]\!] = [\![A_0]\!] \times \cdots \times [\![A_{n-1}]\!]$. Intuitively, this is the set of environments for Γ, because an identifier can be bound only to a closed value.

[2] We make the slight modification of using total functions to lifted cpos instead of partial functions.

- For a value $\Gamma \vdash V : A$, we see that given an environment we obtain (by substitution) a closed value. So V denotes a continuous function $[\![V]\!]^{\mathsf{val}}$ from $[\![\Gamma]\!]$ to $[\![A]\!]$.
- For a returner $\Gamma \vdash M : A$, we see that, given an environment, we obtain (by substitution) a closed returner, which either diverges or returns a value of type A. So M denotes a continuous function $[\![M]\!]^{\mathsf{ret}}$ from $[\![\Gamma]\!]$ to $[\![A]\!]_{\perp}$.

The semantics of types is given as follows:

- `bool` denotes the flat 2-element cpo $\{\mathsf{true}, \mathsf{false}\}$, because a closed value of type `bool` is either `true` or `false`.
- $A + A'$ denotes the disjoint union of $[\![A]\!]$ and $[\![A']\!]$.
- $A \to B$ denotes the cpo of continuous functions from $[\![A]\!]$ to $[\![B]\!]_{\perp}$, because a closed value of type $A \to B$ is of the form $\lambda\mathtt{x}.M$, and this, when applied to a closed value of type A gives (by substitution) a closed returner of type B.

We omit the semantics of terms.

1.6.4 The Monad Approach

We briefly mention a categorical viewpoint on CBV due to Moggi [Moggi, 1991]. Readers unfamiliar with this viewpoint or with category theory may omit this section.

Products, Strong Monads, Exponentials, Distributive Coproducts

We review some categorical definitions used in Moggi's semantics of CBV.

Definition 1.3 Let $\mathcal{C}$ be a category. A *product* for a family of objects $\{A_i\}_{i \in I}$ is an object V (the *vertex*) together with, for each $i \in I$, a morphism $V \xrightarrow{\pi_i} A_i$ such that, for any family of morphisms $X \xrightarrow{f_i} A_i$, there is a unique morphism $X \xrightarrow{g} V$ such that for each $i \in I$ the diagram

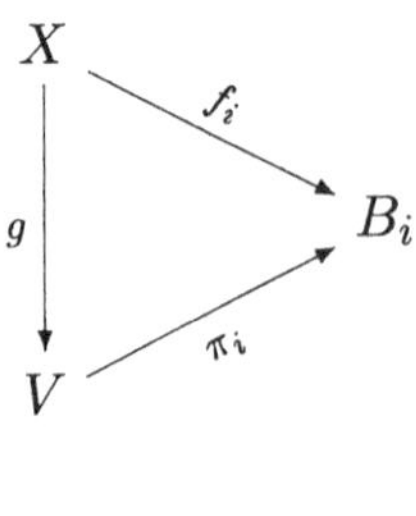

commutes.

A category with all finite products is called *cartesian*. □

Definition 1.4 Let $\mathcal{B}$ be a category. A *monad* on $\mathcal{B}$ consists of

- an endofunctor T on $\mathcal{B}$;
- a natural transformation η from $\mathsf{id}_{\mathcal{B}}$ to T;
- a natural transformation μ from T^2 to T

such that the following diagrams commute:

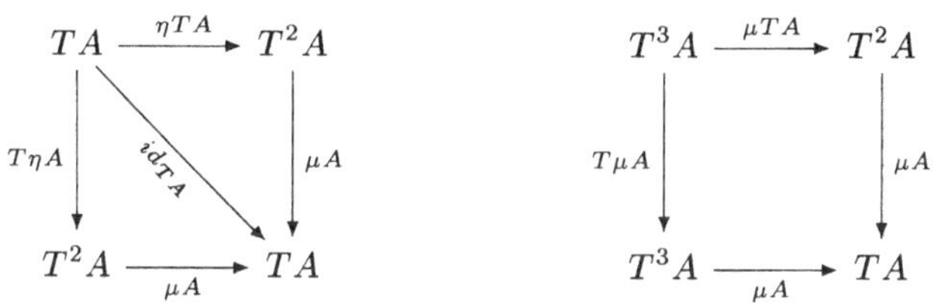

Given such a monad, the *Kleisli category* $\mathcal{C}_T$ has the same objects as $\mathcal{C}$; a $\mathcal{C}_T$-morphism from A to B is a $\mathcal{C}$-morphism from A to TB. The identity on A is η_A while the composite of $A \xrightarrow{f} TB$ with $B \xrightarrow{g} TC$ is $f; Tg; \mu_C$. □

Definition 1.5 Let $\mathcal{C}$ be a cartesian category. A *strength* for a monad (T, η, μ) on $\mathcal{C}$ consists of a natural transformation

$$A \times TB \xrightarrow{t(A,B)} T(A \times B)$$

such that the following diagrams commute:

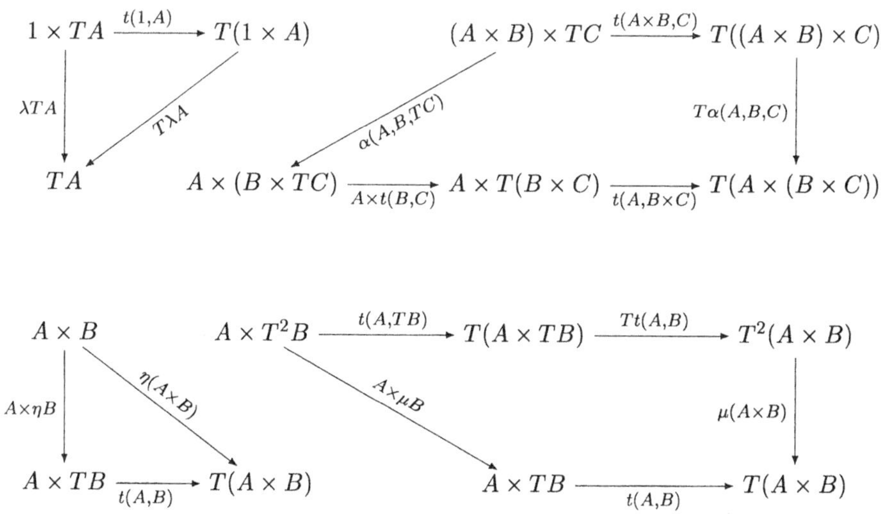

A monad together with a strength is called a *strong monad.* We often refer to a strong monad (T, η, μ, t) just as T. □

Definition 1.6 Let $\mathcal{C}$ be a cartesian category. An *exponential* from an object A to an object B is an object V (the *vertex*) together with a morphism $V \times A \xrightarrow{\mathsf{ev}} B$ such that, for any morphism $X \times A \xrightarrow{f} B$ there is a unique morphism $X \xrightarrow{g} V$ such that

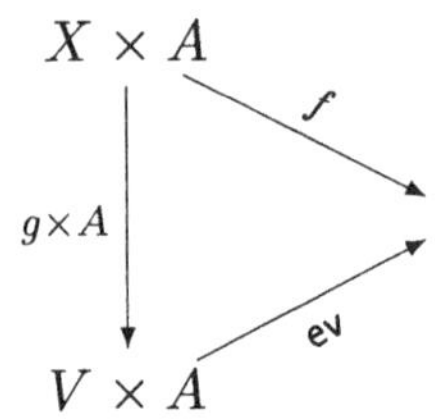

commutes.

A cartesian category with all exponentials is called *exponential.* □

The following is adapted from [Carboni et al., 1993; Cockett, 1993].

Definition 1.7 Let $\mathcal{C}$ be a cartesian category. A *distributive coproduct* for a family of objects $\{A_i\}_{i \in I}$ is an object V (*the vertex*) together with a morphism $A_i \xrightarrow{\mathsf{in}_i} V$ for each $i \in I$, such that, for any family of morphisms $X \times A_i \xrightarrow{f_i} Y$, there is a unique morphism $X \times V \xrightarrow{g} Y$ such that for each $i \in I$ the diagram

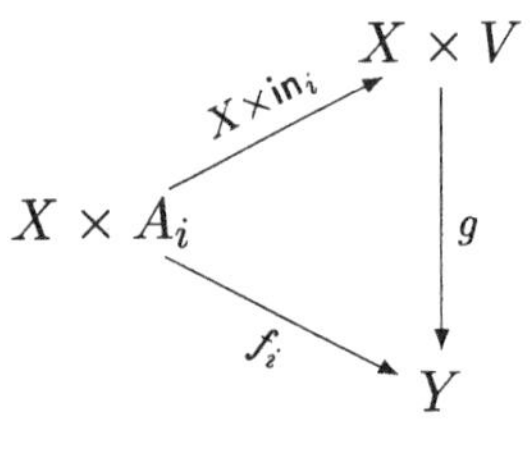

commutes.

A cartesian category with all finite distributive coproducts is called *distributive.* □

Here are some well-known results about distributive coproducts, adapted from [Carboni et al., 1993; Cockett, 1993].

Proposition 4 Let $\mathcal{C}$ be a cartesian category.

1 A distributive coproduct for $\{A_i\}_{i \in I}$ must be a coproduct for $\{A_i\}_{i \in I}$. If $\mathcal{C}$ is cartesian closed, the converse is true also.

2 Each of the morphisms $A_i \xrightarrow{\mathsf{in}_i} V$ in a distributive coproduct is monic.

3 Suppose 0 is a *distributive initial object* i.e. a distributive coproduct for the empty family. Then any morphism to 0 is an isomorphism. So parallel morphisms to 0 must be equal.

□

A cartesian closed category which is also distributive (or, equivalently, which has all finite coproducts) is called *bicartesian closed*.

Monad Semantics for Call-By-Value

Moggi's semantics for call-by-value is as follows. Let $\mathcal{C}$ be a distributive category. Suppose we are given a strong monad (T, η, μ, t) on $\mathcal{C}$ and, for each $A, B \in \mathsf{ob}\,\mathcal{C}$, an exponential from A to TB (this is called a *Kleisli exponential*).

We then obtain a CBV model: a CBV value denotes a morphism of $\mathcal{C}$, and a CBV returner denotes a morphism of the Kleisli category $\mathcal{C}_T$.

Both of our denotational models are instances of this situation. For the printing semantics, $\mathcal{C}$ is **Set** and T is the strong monad $\mathcal{A}^* \times -$ (Moggi's "interactive output" monad). For the cpo semantics, $\mathcal{C}$ is **Cpo** (the category of cpos and continuous functions) and T is the lifting monad.

1.6.5 Observational Equivalence

Definition 1.8 A *context* $\mathcal{C}[\cdot]$ is a term with zero or more occurrences of a hole $[\cdot]$. If $\mathcal{C}[\cdot]$ is of ground type we say it is a *ground context*. □

Definition 1.9 Given two terms $\Gamma \vdash M, M' : B$, we say that

1 $M \simeq_{\mathsf{ground}} M'$ when for all ground contexts $\mathcal{C}[\cdot]$, $\mathcal{C}[M] \Downarrow m, i$ iff $\mathcal{C}[M'] \Downarrow m, i$

2 $M \simeq_{\mathsf{anytype}} M'$ when for all contexts $\mathcal{C}[\cdot]$ of any type, $\mathcal{C}[M] \Downarrow m, V$ for some V iff $\mathcal{C}[M'] \Downarrow m, V$ for some V

□

Proposition 5 The two relations $\simeq_{\mathsf{ground}}$ and $\simeq_{\mathsf{anytype}}$ are the same. □

This is an important feature of CBV: it does not matter whether we allow observation at ground type or every type.

Cor. 3 implies that terms with the same denotation are observationally equivalent.

1.6.6 Coarse-Grain CBV vs. Fine-Grain CBV

The form of CBV we have looked at is called *coarse-grain CBV*. It suffers from two problems.

1 Our decision to evaluate operand before operator was arbitrary.

2 A value V has two denotations: $[\![V]\!]^{\mathsf{val}}$ and $[\![V]\!]^{\mathsf{ret}}$.

We can eliminate these problems by presenting a more refined calculus called *fine-grain CBV* in which (partially based on [Moggi, 1991]) we make a syntactic distinction between values and returners. This calculus is more suitable for formulating an equational theory. For although Moggi in [Moggi, 1988] provided a theory for coarse-grain CBV, which he called "λ_{c}-calculus", this theory was not purely equational but required an additional predicate to assert that a term is a value.

We will not trouble to study fine-grain CBV in this chapter, because in the next chapter we will present CBPV, which is even more fine-grain. A treatment of fine-grain CBV and its equational theory is given in the Appendix.

1.7 Call-By-Name

1.7.1 Operational Semantics

In CBN the following terms are terminal:

$$T ::= \ \mathtt{true} \mid \mathtt{false} \mid \mathtt{inl}\ M \mid \mathtt{inr}\ M \mid \lambda\mathtt{x}.M$$

Every closed term M prints a string of characters and terminates at a terminal term T. We write $M \Downarrow m, T$. This is defined inductively in Fig. 1.3.

Proposition 6 For every M there is a unique m, T such that $M \Downarrow m, T$. □

The proof is similar to that of Prop. 10.

1.7.2 Observational Equivalence

The denotational semantics for CBN is more subtle than that for CBV, so we first look at observational equivalences.

Definition 1.10 A *context* $\mathcal{C}[\cdot]$ is a term with zero or more occurrences of a hole $[\cdot]$. If $\mathcal{C}[\cdot]$ is of ground type we say it is a *ground context*. □

Definition 1.11 Given two terms $\Gamma \vdash M, M' : B$, we say that

1 $M \simeq_{\mathsf{ground}} M'$ when for all ground contexts $\mathcal{C}[\cdot]$, $\mathcal{C}[M] \Downarrow m, i$ iff $\mathcal{C}[M'] \Downarrow m, i$

2 $M \simeq_{\mathsf{anytype}} M'$ when for all contexts $\mathcal{C}[\cdot]$ of any type, $\mathcal{C}[M] \Downarrow m, T$ for some T iff $\mathcal{C}[M'] \Downarrow m, T$ for some T.

$$\frac{N[M/\mathtt{x}] \Downarrow m, T}{\mathtt{let}\ M\ \mathtt{be}\ \mathtt{x}.\ N \Downarrow m, T}$$

$$\frac{}{\mathtt{true} \Downarrow \mathtt{""}, \mathtt{true}} \qquad \frac{}{\mathtt{false} \Downarrow \mathtt{""}, \mathtt{false}}$$

$$\frac{M \Downarrow m, \mathtt{true} \quad N \Downarrow m', T}{\mathtt{if}\ M\ \mathtt{then}\ N\ \mathtt{else}\ N' \Downarrow m * m', T} \qquad \frac{M \Downarrow m, \mathtt{false} \quad N' \Downarrow m', T}{\mathtt{if}\ M\ \mathtt{then}\ N\ \mathtt{else}\ N' \Downarrow m * m', T}$$

$$\frac{}{\mathtt{inl}\ M \Downarrow \mathtt{""}, \mathtt{inl}\ M} \qquad \frac{}{\mathtt{inr}\ M \Downarrow \mathtt{""}, \mathtt{inr}\ M}$$

$$\frac{M \Downarrow m, \mathtt{inl}\ M' \quad N[M'/\mathtt{x}] \Downarrow m', T}{\mathtt{pm}\ M\ \mathtt{as}\ \{\mathtt{inl}\ \mathtt{x}.N, \mathtt{inr}\ \mathtt{x}.N'\} \Downarrow m * m', T}$$

$$\frac{M \Downarrow m, \mathtt{inr}\ M' \quad N'[M'/\mathtt{x}] \Downarrow m', T}{\mathtt{pm}\ M\ \mathtt{as}\ \{\mathtt{inl}\ \mathtt{x}.N, \mathtt{inr}\ \mathtt{x}.N'\} \Downarrow m * m', T}$$

$$\frac{}{\lambda\mathtt{x}.M \Downarrow \mathtt{""}, \lambda\mathtt{x}.M} \qquad \frac{N \Downarrow m, \lambda\mathtt{x}.N' \quad N'[M/\mathtt{x}] \Downarrow m', T}{M`N \Downarrow m * m', T}$$

$$\frac{M \Downarrow m, T}{\mathtt{print}\ c.\ M \Downarrow c * m, T}$$

Figure 1.3. Big-Step Semantics for CBN with `print`

□

By contrast with Prop. 5 we have the following.

Proposition 7 The relation $\simeq_{\mathsf{anytype}}$ is strictly finer than $\simeq_{\mathsf{ground}}$. □

Perhaps the simplest example of this proposition is

$$\mathtt{print\ "hello"}.\ \lambda\mathtt{x}.M \simeq_{\mathsf{ground}} \lambda\mathtt{x}.(\mathtt{print\ "hello"}.\ M) \tag{1.4}$$

Obviously the trivial context $[\cdot]$, which is not ground, distinguishes the two sides. An intuitive explanation (not a rigorous proof) of (1.4) is that, inside a ground CBN term, the only way to cause a subterm of type $A \to B$ to be evaluated is to apply it.

If we use divergence rather than printing, the analogous example is the equivalence

$$\mathtt{diverge} \simeq_{\mathsf{ground}} \lambda\mathtt{x}.\mathtt{diverge}$$

This general CBN phenomenon can be described as "effects commute with λ". Various authors studying CBN languages with only ground types and function types have exploited it by allowing certain features at ground type only, as the corresponding features at function type are then definable: erratic choice [Hennessy and Ashcroft, 1980], control effects [Laird, 1997] and conditional branching [Plotkin, 1977].

For both printing and divergence (indeed for all effects), we have, for similar reasons, the η-law for functions: any term M of type $A \to B$ can be expanded

$$M \simeq_{\mathsf{ground}} \lambda\mathtt{x}.(\mathtt{x}`M) \tag{1.5}$$

where $\mathtt{x}$ is not in the context Γ of M.

1.7.3 CBN vs. Lazy

Definition 1.12 ▪ We use the term "CBN" (equations, models etc.) to refer to the CBN operational semantics together with $\simeq_{\mathsf{ground}}$.

▪ We use the term "lazy" (equations, models etc.) to refer to the CBN operational semantics together with $\simeq_{\mathsf{anytype}}$.

□

Thus in a CBN model the η-law (1.5) must be validated, whereas in a lazy model (1.5) must not be validated.

Our usage of "lazy" follows [Abramsky, 1990; Ong, 1988]. However, the terminology in the literature is not consistent, and the reader should beware the following.

1 "Lazy" is widely used to describe call-by-need.

2 "Call-by-name" is sometimes used to mean (our sense of) "lazy", especially in the continuation literature [Hatcliff and Danvy, 1997; Plotkin, 1976] and the monad literature [Moggi, 1991].

3 In the *untyped* λ-calculus literature, the phrase "call-by-name" is used with a slightly different meaning from ours, and necessarily so because there is no ground type—there is just one type and it is a function type. In order for (1.5) to hold despite observation at this type, a different operational semantics is used: reduction continues to *head normal form* [Wadsworth, 1976].

The lazy paradigm is treated in Appendix A. We see there that it is subsumed in CBV, so its denotational semantics is straightforward.

By contrast, in the CBN paradigm much of the big-step semantics is not observable. For example, the two sides of (1.4) have different operational behaviour, yet that difference cannot be observed. So a denotational semantics, in order to validate (1.4), must conceal part of the big-step semantics. For this reason, the CBN paradigm is more subtle than CBV.

1.7.4 Denotational Semantics for `print`

Definition 1.13 An $\mathcal{A}$-set $(X, *)$ consists of a set X together with a function $*$ from $\mathcal{A} \times X$ to X. We call X the *carrier* and $*$ the *structure*. □

Each CBN type B denotes an $\mathcal{A}$-set $[\![B]\!] = (X, *)$. Our intention is that a closed term M of type B will then denote an element $[\![M]\!]$ of X, and `print` $c.\ M$ will denote $c * [\![M]\!]$. Thus $*$ provides a way of "absorbing" the effect into X. Given an $\mathcal{A}$-set $(X, *)$ we can extend $*$ to a function from $\mathcal{A}^* \times X$ to X in the evident way—we call the extension $*$ too. This extension allows us to interpret `print` $m.\ M$ directly.

Here are some ways of constructing $\mathcal{A}$-sets.

Definition 1.14 1 For any set X, the *free* $\mathcal{A}$-set on X has carrier $\mathcal{A}^* \times X$ and we set $c * (m, x)$ to be $(c * m, x)$.

2 For an $i \in I$-indexed family of $\mathcal{A}$-sets $(X_i, *)$, we define the $\mathcal{A}$-set $\prod_{i \in I}(X_i, *)$ to have carrier $\prod_{i \in I} X_i$ and structure given pointwise: $\hat{\imath}`(c * f) = c * (\hat{\imath}`f)$.

3 For any set X and $\mathcal{A}$-set $(Y, *)$, we define the $\mathcal{A}$-set $X \to (Y, *)$ to have carrier $X \to Y$ and structure given pointwise: $x`(c * f) = c * (x`f)$. □

The semantics of types is as follows:

- `bool` denotes the free $\mathcal{A}$-set on $\{\mathsf{true}, \mathsf{false}\}$. This is because any closed term of type `bool` prints a string and then terminates as `true` or `false`, and this behaviour is observable.
- If $[\![A]\!] = (X, *)$ and $[\![A']\!] = (X', *)$ then $A + A'$ denotes the free $\mathcal{A}$-set on $X + X'$. This is because any closed term of this type prints a string and then terminates as `inl` M or `inr` M, and this behaviour is observable.
- If $[\![A]\!] = (X, *)$ and $[\![B]\!] = (Y, *)$ then $A \to B$ denotes $X \to (Y, *)$. This is because any closed term of this type is equivalent to some $\lambda\mathtt{x}.M$, and prefixing with a `print` command is then given by (1.4).

Given a context $\mathtt{x}_0 : A_0, \ldots, \mathtt{x}_{n-1} : A_{n-1}$, an environment consists of a sequence of closed terms $M_0, \ldots, M_{n-1}$. Writing $(X_i, *)$ for $[\![A_i]\!]$, this environment denotes an element of $X_0 \times \cdots \times X_{n-1}$ We say that the context denotes this set. (There is no need to retain the structure.)

Given a term $\Gamma \vdash M : B$, from any environment we obtain (by substitution) a closed term of type B. So M will denote a function from $[\![\Gamma]\!]$ to the carrier of $[\![B]\!]$.

Semantics of terms—some example clauses:

$$\begin{array}{rcl}
[\![\mathtt{x}]\!](\rho, \mathtt{x} \mapsto x, \rho') & = & x \\
[\![\mathtt{let}\ M\ \mathtt{be\ x.}\ N]\!] & = & [\![M]\!](\rho, \mathtt{x} \mapsto [\![N]\!]\rho) \\
[\![\mathtt{true}]\!]\rho & = & (\text{""}, \mathsf{true}) \\
[\![\mathtt{if}\ M\ \mathtt{then}\ N\ \mathtt{else}\ N']\!] & = & \left\{ \begin{array}{l} m * [\![N]\!]\rho \text{ if } [\![M]\!]\rho = (m, \mathsf{true}) \\ m * [\![N']\!]\rho \text{ if } [\![M]\!]\rho = (m, \mathsf{false}) \end{array} \right. \\
[\![\mathtt{inl}\ M]\!]\rho & = & (\text{""}, \mathsf{inl}\ [\![M]\!]\rho) \\
[\![\mathtt{pm}\ M\ \mathtt{as}\ \{\mathtt{inl\ x.}N, \mathtt{inr\ y.}N'\}]\!] & = & \left\{ \begin{array}{l} m * [\![N]\!](\rho, \mathtt{x} \mapsto a) \\ \quad \text{if } [\![M]\!]\rho = (m, \mathsf{inl}\ a) \\ m * [\![N']\!](\rho, \mathtt{x} \mapsto a') \\ \quad \text{if } [\![M]\!]\rho = (m, \mathsf{inr}\ a') \end{array} \right. \\
[\![\lambda \mathtt{x}.M]\!] & = & \lambda x.[\![M]\!](\rho, \mathtt{x} \mapsto x) \\
[\![M`N]\!] & = & ([\![M]\!]\rho)`([\![N]\!]\rho) \\
[\![\mathtt{print}\ c.\ M]\!]\rho & = & c * [\![M]\!]\rho
\end{array}$$

Proposition 8 (soundness) If $M \Downarrow m, T$ then $[\![M]\!] = m * [\![T]\!]$. □

Corollary 9 (by Prop. 6) For any closed ground term M (term of ground type), $M \Downarrow m, \mathtt{return}\ n$ iff $[\![M]\!] = (m, n)$. Hence terms with the same denotation are observationally equivalent. □

Notice the sequencing of effects in the semantics of `if` and `pm`. This is characteristic of CBN: pattern-matching is what finally causes evaluation to happen.

1.7.5 Scott Semantics

The reader may be familiar with Scott semantics for CBN, where types denote pointed cpos and terms denote continuous functions. We recall this semantics here, to make apparent its similarity to the printing semantics.

Definition 1.15 A cpo is *pointed* iff it has a least element, which we call $\perp$. We use *cppo* as an abbreviation for "pointed cpo". □

Each CBN type A denotes a cppo. Our intention is that a closed term M of type A will denote an element of X, and if it diverges then it will denote $\bot$. (A convergent term also may denote $\bot$.) Thus $\bot$ provides a way of "absorbing" the effect into A.

We recall some familiar ways of constructing cppos:

1 For a cpo X, its *lift* $X_\bot$ consists of X together with an additional element $\bot$, which is below all the elements of X.

2 For an $i \in I$-indexed family of cppos $(X_i, \leqslant, \bot)$ the *product* of this family $\prod_{i \in I}(X_i, \leqslant, \bot)$ is the set $\prod_{i \in I} X_i$, ordered pointwise. Its least element is given pointwise $\hat{\imath}`\bot = \bot$.

3 Given a cpo $(X, \leqslant)$ and a cppo $(Y, \leqslant, \bot)$ the *exponential* $(X, \leqslant) \to (Y, \leqslant, \bot)$ is the set of continuous functions from $(X, \leqslant)$ to $(Y, \leqslant)$, ordered pointwise. Its least element is given pointwise $\mathtt{x}`\bot = \bot$.

The semantics of types is given as follows:

- `bool` denotes the lift of the cpo $\{\mathsf{true}, \mathsf{false}\}$. This is because a closed term of type `bool` either diverges or terminates as `true` or `false`, and this behaviour is observable.
- If A denotes $(X, \leqslant, \bot)$ and A' denotes $(X', \leqslant, \bot)$ then $A + A'$ denotes the lift of the cpo $(X, \leqslant) + (X', \leqslant)$. ("CBN sum denotes lifted sum.") This is because every closed term of this type either diverges or terminates as `inl` M or `inr` M, and this behaviour is observable.
- If A denotes $(X, \leqslant, \bot)$ and B denotes $(Y, \leqslant, \bot)$ then $A \to B$ denotes $(X, \leqslant) \to (Y, \leqslant, \bot)$. This is because every closed term M of this type is equivalent to some $\lambda\mathtt{x}.N$, and if M diverges then it is equivalent to $\lambda\mathtt{x}.\mathtt{diverge}$.

Given a context $\mathtt{x}_0 : A_0, \ldots, \mathtt{x}_{n-1} : A_{n-1}$, an environment consists of a sequence of closed terms $M_0, \ldots, M_{n-1}$. Writing $(X_i, \leqslant, \bot)$ for $[\![A_i]\!]$, this environment denotes an element of the cpo $(X_0, \leqslant) \times \cdots \times (X_{n-1}, \leqslant)$. We say that the context denotes this cpo.

Given a term $\Gamma \vdash M : B$, from any environment we obtain (by substitution) a closed term of type B. So M will denote a continuous function from $[\![\Gamma]\!]$ to $[\![\underline{B}]\!]$. We omit the semantics of terms.

1.7.6 Algebras and Plain Maps

We continue the categorical discussion of Sect. 1.6.4. Again, this section may be omitted by readers unfamiliar with category theory.

The two denotational models of CBN that we have seen are both cartesian closed categories:

- The printing semantics for CBN is the cartesian closed category in which an object is an $\mathcal{A}$-set $(X, *)$ and a morphism from $(X, *)$ to $(Y, *)$ is a function from X to Y.

- The cpo semantics for CBN is the cartesian closed category of cppos and continuous functions.

In fact, every model for CBN must be a cartesian closed category, as it must validate the β- and η-laws for functions.

These two cartesian closed categories are instances of a general construction, based on the notion of algebra for a monad.

Definition 1.16 [Mac Lane, 1971] Let (T, η, μ) be a monad on a category $\mathcal{C}$. A T-*algebra* consists of a pair (X, θ), where X is an object of $\mathcal{C}$, θ is a morphism from TX to X, and

$$\begin{array}{ccccc} X & \xrightarrow{\eta X} & TX & \xleftarrow{\mu X} & T^2X \\ & \searrow^{id} & \downarrow{\scriptstyle\theta} & & \downarrow{\scriptstyle T\theta} \\ & & X & \xleftarrow[\theta]{} & TX \end{array} \tag{1.6}$$

commutes. X is called the *carrier* and θ the *structure map* of the algebra. □

Suppose, as in Sect. 1.6.4, that we have a distributive category $\mathcal{C}$ with a strong monad T. Then we form the categoryof "algebras and plain maps", in which an object is a T-algebra (X, θ) and a morphism from (X, θ) to (Y, ϕ) is *any* $\mathcal{C}$-morphism from X to Y. Assuming sufficient exponentials, this category must be cartesian closed [Simpson, 1992], as we explain presently.

To see that our two models are instances of this construction, notice that

- an algebra for the $\mathcal{A}^* \times -$ monad on **Set** is precisely an $\mathcal{A}$-set;

- an algebra for the lifting monad on **Cpo** is precisely a cppo.

The interpretation of CBN in the category of algebras and plain maps makes use of several ways of constructing algebras.

Definition 1.17 Let (T, η, μ, t) be a strong monad on a cartesian category $\mathcal{C}$.

1 For an object X, the *free* T-algebra on X is the algebra $(TX, \mu X)$.

2 For a family of T-algebras $\{(X_i, \theta_i)\}_{i \in I}$, suppose $\{X_i\}_{i \in I}$ has a given product in $\mathcal{C}$. Then the *product* T-algebra for $\{(X_i, \theta_i)\}_{i \in I}$ is the algebra with carrier $\prod_{i \in I} X_i$ and the obvious structure map.

3 For a $\mathcal{C}$-object X and T-algebra (Y, θ), suppose there is a given exponential in $\mathcal{C}$ from X to Y. Then the *exponential* T-algebra from X to (Y, θ) is the T-algebra with carrier $X \to Y$ and structure map corresponding to

$$X \times T(X \to Y) \xrightarrow{t_{X, X \to Y}} T(X \times (X \to Y)) \xrightarrow{T\mathsf{ev}} TY \xrightarrow{\quad \theta \quad} Y$$

□

We can see how to interpret CBN types using these constructions, and that our constructions of cppos and of $\mathcal{A}$-sets are special cases. For example, the boolean type denotes the free algebra on $1 + 1$. If A denotes (X, θ) and B denotes (Y, ϕ) then $X \to B$ denotes the exponential algebra from X to (Y, ϕ). It is now easy to see that if exponentials to carriers of T-algebras always exist—which implies, by the way, that Kleisli exponentials exist—then the category of algebras and plain maps is cartesian closed.

We emphasize that in the category of algebras and plain maps, an object is an algebra (X, θ), not just a $\mathcal{C}$-object X on which there exists a structure map θ. Although this latter definition would give a cartesian closed category equivalent to ours, it would not give a semantics of CBN. To see this, look at the semantics for pm: it uses the specific structure θ. In summary,

a CBV type denotes an object of $\mathcal{C}$; a CBN type denotes a T-algebra.

We are not claiming that *every* model of CBN arises from a monad in this way, just that the printing model and the Scott model do.

1.8 Comparing CBV and CBN

Because CBN satisfies the β-law and η-law for functions, it is sometimes suggested that CBN is "mathematically better behaved but practically less useful" than CBV. While the claim about the practical inferiority of CBN is valid, the claim about its mathematical superiority is not valid.

For although the CBN *function type* is mathematically superior to the CBV function type, the CBN *sum type* (and boolean type) is inferior to the CBV sum type. We saw that in the printing semantics the CBV sum simply denotes the sum of sets, while the CBN sum denotes a more complex construction on $\mathcal{A}$-sets. Similarly, in the Scott semantics, the sum of cpos is a much simpler construction than the lifted sum of cppos. In particular, the CBV sum is associative whereas the CBN sum is not.

This situation is seen not just in denotational semantics but also in equations. Consider the following equivalence—a special case of the η-law for `bool`. If M has a free identifier `z` : `bool` then

$$M = \texttt{if z then } M[\texttt{true/z}] \texttt{ else } M[\texttt{false/z}]$$

holds in CBV but not in CBN. It holds in CBV because `z` can be bound only to a value, `true` or `false`. It fails in CBN because `z` can be bound to a term such as `print "hello". true`.

Thus, the perception of CBN's superiority is actually due to the fact that function types have been considered more important, and hence have received more attention, than sum types or even ground types.

A consequence of this bias (towards function types and towards CBN) has been the promotion of cartesian closed category as a significant structure, in the semantics of programming languages and even in the semantics of intuitionistic logic. The latter is especially inappropriate, because the type theory to which intuitionistic logic corresponds is effect-free rather than CBN or CBV, so its models must be *bicartesian* closed categories.

Chapter 2

CALL-BY-PUSH-VALUE: A SUBSUMING PARADIGM

2.1 Introduction

2.1.1 Aims Of Chapter

In this chapter we present the CBPV language and its operational and denotational semantics for our example effects of printing and divergence, and we show how it contains both CBV and CBN. Operational semantics is presented in two forms: the familiar big-step form and the CK-machine of [Felleisen and Friedman, 1986].

We will introduce the following vocabulary, all of which will be used in subsequent chapters:

- value, computation, terminal computation
- value type, computation type
- thunk, forcing a thunk
- configuration, stack
- sequenced computation, returner, continuation.

We suggest the following slogans and mnemonics.

- A value is, a computation does.
- U types are thUnk types, F types are returner types.
- For cpos, U means nUthing, F means "liFt".

2.1.2 CBV And CBN Lead To CBPV

Because we have used printing rather than divergence as our leading example of an effect, our exploration of CBV and CBN leads us naturally to the types of CBPV. As we explained in Sect. 1.2, in the printing semantics, a CBV type denotes a set whereas a CBN type denotes an $\mathcal{A}$-set. Thus we will need two disjoint classes of type in the subsuming language: types denoting sets and types denoting $\mathcal{A}$-sets. These are precisely the value types and computation types (respectively) of CBPV. Furthermore, the subsuming language should provide type constructors corresponding to the various ways we have seen of constructing $\mathcal{A}$-sets, and CBPV indeed does this. For example,

- FA denotes the free $\mathcal{A}$-set on $[\![A]\!]$
- $U\underline{B}$ denotes the carrier of $[\![\underline{B}]\!]$.

Inventing the term calculus of the subsuming language is not as easy as the types, but we can obtain helpful guidelines by recalling equational facts about CBV and CBN and the informal reasons for them.

- In CBV, the η-law for sum types (and boolean type) holds because identifiers are bound to values.
- In CBN, the η-law for function types holds because a term of function type can be made to evaluate only by applying it.

We would like our subsuming language to have both of these properties, and indeed CBPV does.

2.2 Syntax

The types of CBPV (as stated in Sect. I.5.4) are given by

$$\begin{array}{rl} A ::= & U\underline{B} \mid \sum_{i\in I} A_i \mid 1 \mid A \times A \\ \underline{B} ::= & FA \mid \prod_{i\in I} \underline{B}_i \mid A \to \underline{B} \end{array}$$

where each set I of tags is finite. (We will also be concerned with *infinitely wide* CBPV in which I may be countably infinite—see Sect. 4.1 for more discussion.) We will often write $\hat{\imath}$ for a particular element of I. We usually omit terms, equations etc. for the type 1, because it is just the nullary analogue of $\times$.

Since identifiers can be bound only to values, they must have value type. So we have the following:

Definition 2.1 A *context* Γ is a finite sequence of identifiers with value types $\mathtt{x}_0 : A_0, \ldots, \mathtt{x}_{n-1} : A_{n-1}$. Sometimes we omit the identifiers and write Γ as a list of value types. □

The calculus has two kinds of judgement

$$\Gamma \vdash^{\mathsf{c}} M : \underline{B} \qquad\qquad \Gamma \vdash^{\mathsf{v}} V : A$$

for computations and values respectively. We emphasize that, in each case, only value types appear on the left of $\vdash$. The terms of basic (i.e. effect-free CBPV) are defined by Fig. 2.1. We will freely add whatever syntax is required for the various computational effects that we look at.

$$\frac{}{\Gamma, \mathtt{x} : A, \Gamma' \vdash^{\mathsf{v}} \mathtt{x} : A} \qquad \frac{\Gamma \vdash^{\mathsf{v}} V : A \quad \Gamma, \mathtt{x} : A \vdash^{\mathsf{c}} M : \underline{B}}{\Gamma \vdash^{\mathsf{c}} \mathtt{let}\ V\ \mathtt{be}\ \mathtt{x}.\ M : \underline{B}}$$

$$\frac{\Gamma \vdash^{\mathsf{v}} V : A}{\Gamma \vdash^{\mathsf{c}} \mathtt{return}\ V : FA} \qquad \frac{\Gamma \vdash^{\mathsf{c}} M : FA \quad \Gamma, \mathtt{x} : A \vdash^{\mathsf{c}} N : \underline{B}}{\Gamma \vdash^{\mathsf{c}} M\ \mathtt{to}\ \mathtt{x}.\ N : \underline{B}}$$

$$\frac{\Gamma \vdash^{\mathsf{c}} M : \underline{B}}{\Gamma \vdash^{\mathsf{v}} \mathtt{thunk}\ M : U\underline{B}} \qquad \frac{\Gamma \vdash^{\mathsf{v}} V : U\underline{B}}{\Gamma \vdash^{\mathsf{c}} \mathtt{force}\ V : \underline{B}}$$

$$\frac{\Gamma \vdash^{\mathsf{v}} V : A_{\hat{\imath}}}{\Gamma \vdash^{\mathsf{v}} (\hat{\imath}, V) : \sum_{i \in I} A_i} \qquad \frac{\Gamma \vdash^{\mathsf{v}} V : \sum_{i \in I} A_i \quad \cdots \quad \Gamma, \mathtt{x} : A_i \vdash^{\mathsf{c}} M_i : \underline{B} \quad \cdots_{\, i \in I}}{\Gamma \vdash^{\mathsf{c}} \mathtt{pm}\ V\ \mathtt{as}\ \{\ldots, (i, \mathtt{x}).M_i, \ldots\} : \underline{B}}$$

$$\frac{\Gamma \vdash^{\mathsf{v}} V : A \quad \Gamma \vdash^{\mathsf{v}} V' : A'}{\Gamma \vdash^{\mathsf{v}} (V, V') : A \times A'} \qquad \frac{\Gamma \vdash^{\mathsf{v}} V : A \times A' \quad \Gamma, \mathtt{x} : A, \mathtt{y} : A' \vdash^{\mathsf{c}} M : \underline{B}}{\Gamma \vdash^{\mathsf{c}} \mathtt{pm}\ V\ \mathtt{as}\ (\mathtt{x}, \mathtt{y}).M : \underline{B}}$$

$$\frac{\cdots \quad \Gamma \vdash^{\mathsf{c}} M_i : \underline{B}_i \quad \cdots_{\, i \in I}}{\Gamma \vdash^{\mathsf{c}} \lambda\{\ldots, i.M_i, \ldots\} : \prod_{i \in I} \underline{B}_i} \qquad \frac{\Gamma \vdash^{\mathsf{c}} M : \prod_{i \in I} \underline{B}_i}{\Gamma \vdash^{\mathsf{c}} \hat{\imath}`M : \underline{B}_{\hat{\imath}}}$$

$$\frac{\Gamma, \mathtt{x} : A \vdash^{\mathsf{c}} M : \underline{B}}{\Gamma \vdash^{\mathsf{c}} \lambda \mathtt{x}.M : A \to \underline{B}} \qquad \frac{\Gamma \vdash^{\mathsf{v}} V : A \quad \Gamma \vdash^{\mathsf{c}} M : A \to \underline{B}}{\Gamma \vdash^{\mathsf{c}} V`M : \underline{B}}$$

Figure 2.1. Terms of Basic Language

Definition 2.2 1 A *returner* is a computation of type FA.

2 A *ground type* is a type of the form $\sum_{i \in I} 1$ (such as $\mathtt{bool} = 1 + 1$).

3 A *ground value* is a value of ground type.

4 A *ground returner* is a computation of type FA, where A is a ground type.

We write `true` and `false` for the closed values of type $\mathtt{bool} = 1 + 1$ and `if` V `then` M `else` M' for the corresponding `pm` construct. □

Notice that CBPV has two forms of product. In the terminology of Sect. 1.3.2, the product of value types is a *pattern-match product* whereas the product of computation types is a *projection product.* The reason we do allow a term πV where V has type $A \times A'$ is that the CBPV operational semantics (presented in Sect. 2.3) exploits the fact that values do not need to be evaluated. So we cannot allow a *complex value* such as $\pi(\texttt{true},\texttt{false})$, which needs to be evaluated to $\texttt{true}$. Complex values are discussed fully in Chap. 3.

2.3 Operational Semantics Without Effects

Although at this stage we give the language without effects, we take care that the definitions, propositions and proofs can easily be adapted to various effects where possible.

2.3.1 Big-Step Semantics

The big-step semantics has the form $M \Downarrow T$, where M and T are closed computations of the same type, and T is *terminal*, meaning that it belongs to the following class:

$$T ::= \ \texttt{return}\, V \mid \lambda\{\ldots, i.M_i, \ldots\} \mid \lambda\texttt{x}.M$$

The big-step rules are given in Fig. 2.2.

Proposition 10 For every closed computation M, there is a unique terminal computation T such that $M \Downarrow T$. □

Proof (in the style of [Tait, 1967]) We define, by mutual induction over types, three families of subsets:

- for each A, a set $\mathsf{red}^{\mathsf{v}}_A$ of closed values of type A
- for each $\underline{B}$, a set $\mathsf{red}^{\mathsf{t}}_{\underline{B}}$ of terminal computations of type $\underline{B}$
- for each $\underline{B}$, a set $\mathsf{red}^{\mathsf{c}}_{\underline{B}}$ of closed computations of type $\underline{B}$.

These definition of these subsets proceeds as follows:

$$\frac{M[V/\mathtt{x}] \Downarrow T}{\mathtt{let}\ V\ \mathtt{be}\ \mathtt{x}.\ M \Downarrow T}$$

$$\frac{}{\mathtt{return}\ V \Downarrow \mathtt{return}\ V} \qquad \frac{M \Downarrow \mathtt{return}\ V \quad N[V/\mathtt{x}] \Downarrow T}{M\ \mathtt{to}\ \mathtt{x}.\ N \Downarrow T}$$

$$\frac{M \Downarrow T}{\mathtt{force}\ \mathtt{thunk}\ M \Downarrow T}$$

$$\frac{M_{\hat{\imath}}[V/\mathtt{x}] \Downarrow T}{\mathtt{pm}\ (\hat{\imath}, V)\ \mathtt{as}\ \{\ldots,(i,\mathtt{x}).M_i,\ldots\} \Downarrow T}$$

$$\frac{M[V/\mathtt{x}, V'/\mathtt{y}] \Downarrow T}{\mathtt{pm}\ (V, V')\ \mathtt{as}\ (\mathtt{x},\mathtt{y}).M \Downarrow T}$$

$$\frac{}{\lambda\{\ldots,i.M_i,\ldots\} \Downarrow \lambda\{\ldots,i.M_i,\ldots\}} \qquad \frac{M \Downarrow \lambda\{\ldots,i.N_i,\ldots\} \quad N_{\hat{\imath}} \Downarrow T}{\hat{\imath}`M \Downarrow T}$$

$$\frac{}{\lambda\mathtt{x}.M \Downarrow \lambda\mathtt{x}.M} \qquad \frac{M \Downarrow \lambda\mathtt{x}.N \quad N[V/\mathtt{x}] \Downarrow T}{V`M \Downarrow T}$$

Note that each of these rules is of the form

$$\frac{M_0 \Downarrow T_0 \quad \cdots \quad M_{r-1} \Downarrow T_{r-1}}{M \Downarrow T} \tag{2.1}$$

for some $r \geqslant 0$.

Figure 2.2. Big-Step Semantics for CBPV

$\mathtt{thunk}\ M \in \mathsf{red}^{\mathsf{v}}_{U\underline{B}}$	iff	$M \in \mathsf{red}^{\mathsf{c}}_{\underline{B}}$
$(\hat{\imath}, V) \in \mathsf{red}^{\mathsf{v}}_{\sum_{i\in I} A_i}$	iff	$V \in \mathsf{red}^{\mathsf{v}}_{A_{\hat{\imath}}}$
$(V, V') \in \mathsf{red}^{\mathsf{v}}_{A\times A'}$	iff	$V \in \mathsf{red}^{\mathsf{v}}_{A}$ and $V' \in \mathsf{red}^{\mathsf{v}}_{A'}$
$\mathtt{return}\ V \in \mathsf{red}^{\mathsf{t}}_{FA}$	iff	$V \in \mathsf{red}^{\mathsf{v}}_{A}$
$\lambda\{\ldots,i.M_i,\ldots\} \in \mathsf{red}^{\mathsf{t}}_{\prod_{i\in I}\underline{B}_i}$	iff	$M_i \in \mathsf{red}^{\mathsf{c}}\underline{B}_i$ for all $i \in I$
$\lambda\mathtt{x}.M \in \mathsf{red}^{\mathsf{t}}_{A\to\underline{B}}$	iff	$M[V/\mathtt{x}] \in \mathsf{red}^{\mathsf{c}}_{\underline{B}}$ for all $V \in \mathsf{red}^{\mathsf{v}}_{A}$
$M \in \mathsf{red}^{\mathsf{c}}_{\underline{B}}$	iff	$M \Downarrow T$ for unique T, and $T \in \mathsf{red}^{\mathsf{t}}_{\underline{B}}$

Notice that if $T \in \mathbb{T}_{\underline{B}}$, then $T \in \mathsf{red}^{\mathsf{c}}_{\underline{B}}$ iff $T \in \mathsf{red}^{\mathsf{t}}_{\underline{B}}$. Finally we show that

- for any computation $A_0, \ldots, A_{n-1} \vdash^{\mathsf{c}} M : \underline{B}$, if $W_i \in \mathsf{red}^{\mathsf{v}}_{A_i}$ for $i = 0, \ldots, n-1$ then $M[\overrightarrow{W_i/\mathtt{x}_i}] \in \mathsf{red}^{\mathsf{c}}_{\underline{B}}$
- for any value $A_0, \ldots, A_{n-1} \vdash^{\mathsf{v}} V : B$, if $W_i \in \mathsf{red}^{\mathsf{v}}_{A_i}$ for $i = 0, \ldots, n-1$ then $V[\overrightarrow{W_i/\mathtt{x}_i}] \in \mathsf{red}^{\mathsf{v}}_{B}$.

This is shown by mutual induction on M and V, and gives the required result. □

2.3.2 CK-Machine

The CK-machine is a general form of operational semantics that can be used for CBV and CBN as well as for CBPV. It was introduced in [Felleisen and Friedman, 1986] as a simplification of Landin's SECD machine[Landin, 1964], and there are many similar machines [Bierman, 1998; Krivine, 1985; Streicher and Reus, 1998]. At any point in time, the machine has configuration M, K when M is the term we are evaluating and K is a stack[1]. There is no need for an environment or for closures, because substitution is used. To incorporate types, we write a configuration of the machine as

$$M \qquad \underline{B} \qquad K \qquad \underline{C}$$

where $\underline{B}$ is the type of the current computation M and $\underline{C}$ is the type of the initial computation (so $\underline{C}$ stays fixed during execution).

The machine is summarized in Fig. 2.3. Let us ignore the types for the moment, and understand the machine by thinking about how we might implement the big-step rules using a stack.

Suppose for example that we are evaluating M `to` $\mathtt{x}.\ N$. The big-step semantics tells us that we must first evaluate M. So we put the context $[\cdot]$ `to` $\mathtt{x}.\ N$ onto the stack, because at present we do not need it. Later, having evaluated M to `return` V, we can remove $[\cdot]$ `to` $\mathtt{x}.\ N$ from the stack and proceed to evaluate $N[V/\mathtt{x}]$, as the big-step semantics suggests.

As another example, suppose we are evaluating $V`M$. The big-step semantics tells us that we must first evaluate M. So we put the operand[2]

[1] When Felleisen and Friedman converted their CK-machine into a small-step semantics, they introduced the term *evaluation context* for stack; but the CK-machine seems easier and more natural to work with. We use the terminology *current stack* rather than *current continuation* because in CBPV (unlike CBV) not all stacks are continuations: for example, the stack $V :: K$ is a pair, not a continuation—see Def. 2.5. We look at this in more detail in Sect. 5.4.5.

[2] If we wanted our usage to be strictly consistent, we would put the context $V`[\cdot]$ rather than just V onto the stack, but this is unnecessarily complicated.

V onto the stack, because at present we do not need it. Later, having evaluated M to $\lambda \mathtt{x}.N$, we can remove the operand V from the stack and proceed to evaluate $N[V/\mathtt{x}]$, as the big-step semantics suggests.

To evaluate a closed computation M, we place it alongside the empty stack `nil` and follow the transitions in Fig. 2.3 until we reach a configuration T, `nil` for a terminal computation T. We shall see in Sect. 2.3.6 that this will happen precisely when $M \Downarrow T$.

Notice that

- the behaviour of $V`M$ is to push V and then evaluate M
- the behaviour of $\lambda \mathtt{x}.M$ is to pop V and then evaluate $M[V/\mathtt{x}]$

From the big-step rules we have thus recovered the push/pop reading described in Sect. I.5.1, although we did not mention there that the stack is used for sequencing as well as for operands.

2.3.3 CK-Machine For Non-Closed Computations

Whereas big-step semantics is suitable only for closed computations, the CK-machine can be quite naturally extended from closed computations to computations on a fixed context Γ. This extension will be useful for studying control effects, where, as we see in Sect. 5.4.3, we will want to treat the empty stack `nil` as a free identifier.

A configuration of the machine now takes the form

$$\Gamma \qquad M \qquad \underline{B} \qquad K \qquad \underline{C} \tag{2.2}$$

and both Γ and $\underline{C}$ stay fixed during execution. The transitions are entirely unchanged, and we need only give some additional terminal configurations, which try to force or pattern-match a free identifier. The details are shown in Fig. 2.4.

2.3.4 Typing the CK-Machine

Given a configuration of the CK-machine

$$\Gamma \qquad M \qquad \underline{B} \qquad K \qquad \underline{C}$$

we know that $\Gamma \vdash^{\mathsf{c}} M : \underline{B}$, but we do not as yet have a way of typing the stack K. We therefore introduce a judgement $\Gamma|\underline{B} \vdash^{\mathsf{k}} K : \underline{C}$ to achieve this task. This judgement means that K is a stack that accompanies a computation of type $\underline{B}$ in the course of executing a computation of type $\underline{C}$, with free identifiers given by Γ. The typing rules for this judgement are immediate from Fig. 2.3 and Fig. 2.4, so the reader is invited to write them out. The solution is given in Fig. 2.5.

Initial Configuration To Evaluate $\vdash^{\mathsf{c}} M : \underline{C}$

	M	$\underline{C}$	$\mathtt{nil}$	$\underline{C}$

Transitions

	$\mathtt{let}\ V\ \mathtt{be}\ \mathtt{x}.\ M$	$\underline{B}$	K	$\underline{C}$
$\leadsto$	$M[V/\mathtt{x}]$	$\underline{B}$	K	$\underline{C}$
	$M\ \mathtt{to}\ \mathtt{x}.\ N$	$\underline{B}$	K	$\underline{C}$
$\leadsto$	M	FA	$[\cdot]\ \mathtt{to}\ \mathtt{x}.\ N :: K$	$\underline{C}$
	$\mathtt{return}\ V$	FA	$[\cdot]\ \mathtt{to}\ \mathtt{x}.\ N :: K$	$\underline{C}$
$\leadsto$	$N[V/\mathtt{x}]$	$\underline{B}$	K	$\underline{C}$
	$\mathtt{force}\ \mathtt{thunk}\ M$	$\underline{B}$	K	$\underline{C}$
$\leadsto$	M	$\underline{B}$	K	$\underline{C}$
	$\mathtt{pm}\ (\hat{\imath}, V)\ \mathtt{as}\ \{\ldots, (i, \mathtt{x}).M_i, \ldots\}$	$\underline{B}$	K	$\underline{C}$
$\leadsto$	$M_{\hat{\imath}}[V/\mathtt{x}]$	$\underline{B}$	K	$\underline{C}$
	$\mathtt{pm}\ (V, V')\ \mathtt{as}\ (\mathtt{x}, \mathtt{y}).M$	$\underline{B}$	K	$\underline{C}$
$\leadsto$	$M[V/\mathtt{x}, V'/\mathtt{y}]$	$\underline{B}$	K	$\underline{C}$
	$\hat{\imath}`M$	$\underline{B}_{\hat{\imath}}$	K	$\underline{C}$
$\leadsto$	M	$\prod_{i \in I} \underline{B}_i$	$\hat{\imath} :: K$	$\underline{C}$
	$\lambda\{\ldots, i.M_i, \ldots\}$	$\prod_{i \in I} \underline{B}_i$	$\hat{\imath} :: K$	$\underline{C}$
$\leadsto$	$M_{\hat{\imath}}$	$\underline{B}_{\hat{\imath}}$	K	$\underline{C}$
	$V`M$	$\underline{B}$	K	$\underline{C}$
$\leadsto$	M	$A \to \underline{B}$	$V :: K$	$\underline{C}$
	$\lambda\mathtt{x}.M$	$A \to \underline{B}$	$V :: K$	$\underline{C}$
$\leadsto$	$M[V/\mathtt{x}]$	$\underline{B}$	K	$\underline{C}$

Terminal Configurations

$\mathtt{return}\ V$	FA	$\mathtt{nil}$	FA
$\lambda\{\ldots, i.M_i, \ldots\}$	$\prod_{i \in I} \underline{B}_i$	$\mathtt{nil}$	$\prod_{i \in I} \underline{B}_i$
$\lambda\mathtt{x}.M$	$A \to \underline{B}$	$\mathtt{nil}$	$A \to \underline{B}$

Figure 2.3. CK-Machine For CBPV, With Types

Initial Configuration To Evaluate $\Gamma \vdash^{\mathsf{c}} M : \underline{C}$

Γ	M	$\underline{C}$	nil	$\underline{C}$

Transitions

As in Fig. 2.3 but with Γ before each line

Terminal Configurations

Γ	$\mathtt{return}\ V$	FA	nil	FA
Γ	$\lambda\{\ldots, i.M_i, \ldots\}$	$\prod_{i \in I} \underline{B}_i$	nil	$\prod_{i \in I} \underline{B}_i$
Γ	$\lambda \mathtt{x}.M$	$A \to \underline{B}$	nil	$A \to \underline{B}$
$\Gamma, \mathtt{z} : U\underline{B}, \Gamma'$	$\mathtt{force\ z}$	$\underline{B}$	K	$\underline{C}$
$\Gamma, \mathtt{z} : \sum_{i \in I} A_i, \Gamma'$	$\mathtt{pm\ z\ as}\ \{\ldots, i.M_i, \ldots\}$	$\underline{B}$	K	$\underline{C}$
$\Gamma, \mathtt{z} : A \times A', \Gamma'$	$\mathtt{pm\ z\ as}\ (\mathtt{x}, \mathtt{y}).M$	$\underline{B}$	K	$\underline{C}$

Figure 2.4. CK-Machine For Non-Closed Computations

$$\frac{}{\Gamma | \underline{C} \vdash^{\mathsf{k}} \mathtt{nil} : \underline{C}} \qquad \frac{\Gamma, \mathtt{x} : A \vdash^{\mathsf{c}} M : \underline{B} \quad \Gamma | \underline{B} \vdash^{\mathsf{k}} K : \underline{C}}{\Gamma | FA \vdash^{\mathsf{k}} [\cdot]\ \mathtt{to\ x}.\ M :: K : \underline{C}}$$

$$\frac{\Gamma | \underline{B}_{\hat{\imath}} \vdash^{\mathsf{k}} K : \underline{C}}{\Gamma | \prod_{i \in I} \underline{B}_i \vdash^{\mathsf{k}} \hat{\imath} :: K : \underline{C}} \qquad \frac{\Gamma \vdash^{\mathsf{v}} V : A \quad \Gamma | \underline{B} \vdash^{\mathsf{k}} K : \underline{C}}{\Gamma | A \to \underline{B} \vdash^{\mathsf{k}} V :: K : \underline{C}}$$

Figure 2.5. Typing Stacks

Definition 2.3 A configuration from Γ to $\underline{C}$ consists of

- a computation type $\underline{B}$
- a computation $\Gamma \vdash^{\mathsf{c}} M : \underline{B}$
- a stack $\Gamma | \underline{B} \vdash^{\mathsf{k}} K : \underline{C}$

□

Proposition 11 (determinism) For every configuration Q from Γ to $\underline{C}$, precisely one of the following holds.

1 Q is not terminal, and $Q \leadsto Q'$ for unique configuration Q' from Γ to $\underline{C}$.

2 Q is terminal, and there does not exist Q' such that $Q \leadsto Q'$.

□

To complete the CK-machine semantics, we write $\leadsto^*$ for the transitive closure of $\leadsto$. More technically:

Definition 2.4 We define inductively the relation $\leadsto^*$ on configurations:

$$\frac{}{M, K \leadsto^* M, K} \qquad \frac{M', K' \leadsto^* N, L}{M, K \leadsto^* N, L}\ (M, K \leadsto M', K')$$

□

By analogy with Prop. 10, we can now formulate the following.

Proposition 12 For every configuration Q from Γ to $\underline{C}$ there is a unique terminal T (which is also from Γ to $\underline{C}$) such that $Q \leadsto^* T$, and there is no infinite sequence of transitions from M, K. □

The only nontrivial part is the last part, and we defer the proof of to Sect. 7.5.3.

Definition 2.5 (complementary to Def. 2.2(1)) A *continuation* is a stack from a type FA to any type. □

Such a stack uses the value that a returner returns. For example, $[\cdot]$ `to x.` $N :: K$ is a continuation. (This usage of "continuation" is consistent with the CBV literature, because in CBV, *every* stack is a continuation, just as every computation is a returner.)

2.3.5 Operations On Stacks

There are 3 operations on stacks which are of great importance in the theory of CBPV. (Of course, they are just as important in the theory of CBV and CBN, when one considers stack terms for these paradigms.)

substitution We can substitute values for free identifiers in a stack K, just like in a computation or in a value.

concatenation Given a stack $\Gamma|\underline{A} \vdash^{\mathsf{k}} K : \underline{B}$ and a stack $\Gamma|\underline{B} \vdash^{\mathsf{k}} L : \underline{C}$, we can concatenate them to obtain $\Gamma|\underline{A} \vdash^{\mathsf{k}} K +\!\!+ L : \underline{C}$.

dismantling A stack $\Gamma|\underline{B} \vdash^{\mathsf{k}} K : \underline{C}$ can be dismantled onto a computation $\Gamma \vdash^{\mathsf{c}} M : \underline{B}$ giving a computation $\Gamma \vdash^{\mathsf{c}} M \bullet K : \underline{B}$. Informally, this is obtained by running the CK-machine in reverse on the configuration M, K, reducing the stack until it is empty.

Formally, each of these is defined by induction on K. The reader can easily write out all the clauses, so we do not present it until Fig. 10.2.

2.3.6 Agreement Of Big-Step and CK-Machine Semantics

The sole aim of this section is to prove

Proposition 13 For closed computations M, T of type $\underline{B}$, the following are equivalent:

1 $M \Downarrow T$;

2 $M, K \leadsto^* T, K$ for every stack K such that $\underline{B} \vdash^{\mathsf{k}} K : \underline{C}$ for some $\underline{C}$;

3 $M, \mathtt{nil} \leadsto^* T, \mathtt{nil}$.

□

(1) ⇒ (2) is a straightforward induction. (2) ⇒ (3) is trivial.

Lemma 14 For all M and N of type $\underline{B}$, if, for all T, $M \Downarrow T$ implies $N \Downarrow T$, then, for every $\underline{B} \vdash^{\mathsf{k}} K : \underline{C}$, we have that $\underline{C}$, $M \bullet K \Downarrow T$ implies $N \bullet K \Downarrow T$. □

Proof Induct on K. □

Lemma 15 If $M, K \leadsto^* T, \mathtt{nil}$ then $M \bullet K \Downarrow T$. □

Proof We induct on the antecedent. As an example clause, suppose that the antecedent is given by

$$\mathtt{return}\ V, [\cdot]\ \mathtt{to}\ \mathtt{x}.\ N :: K \leadsto N[V/\mathtt{x}], K \leadsto^* T, \mathtt{nil}$$

We know by the inductive hypothesis that $(N[V/\mathtt{x}]) \bullet K \Downarrow T$, so by Lemma 14 we know that $(\mathtt{return}\ V\ \mathtt{to}\ \mathtt{x}.\ N) \bullet K \Downarrow T$. □

Prop. 13((3)⇒(1)) is an immediate consequence of Lemma 15.

2.4 Operational Semantics for `print`

We now add printing: more precisely, we add to the syntax of CBPV the typing rule

$$\frac{\Gamma \vdash^{\mathsf{c}} M : \underline{B}}{\Gamma \vdash^{\mathsf{c}} \mathtt{print}\ c.\ M : \underline{B}}$$

and we must adapt the operational semantics accordingly.

2.4.1 Big-Step Semantics for `print`

Since we have added only one term constructor to the basic language, we might expect that we need only add one rule to Fig. 2.2. However, this is clearly not possible: whereas for the basic language the big-step relation has the form $M \Downarrow T$, it now has the form $M \Downarrow m, T$. It seems therefore that we must present the big-step semantics from scratch.

Fortunately this is not the case. We simply replace each rule in Fig. 2.2 of the form (2.1) by

$$\frac{M_0 \Downarrow m_0, T_0 \quad \cdots \quad M_{r-1} \Downarrow m_{r-1}, T_{r-1}}{M \Downarrow m_0 * \ldots * m_{r-1}, T}$$

Then we add the big-step rule

$$\frac{M \Downarrow m, T}{\texttt{print}\ c.\ M \Downarrow c * m, T}$$

Proposition 16 For every computation M, there exists unique m, T such that $M \Downarrow m, T$. □

The proof is easily adapted from the proof of Prop. 10.

2.4.2 CK-Machine For `print`

In Fig. 2.3 a transition has the form (omitting the types)

$$M \qquad K \qquad \rightsquigarrow \qquad M' \qquad K' \tag{2.3}$$

With printing, we want a transition to have the form

$$M \qquad K \qquad \rightsquigarrow \qquad m \qquad M' \qquad K'$$

where $m \in \mathcal{A}^*$. So we replace each transition (2.3) in Fig. 2.3 by

$$M \qquad K \qquad \rightsquigarrow \qquad "" \qquad M' \qquad K'$$

and we add the transition

$$\texttt{print}\ c.\ M \quad K \qquad \rightsquigarrow \qquad c \qquad M \qquad K$$

We replace Def. 2.4 by the following.

Definition 2.6 We define the relation $\rightsquigarrow^*$, whose form is $M, K \rightsquigarrow^* m, M', K'$ inductively:

$$\frac{}{M, K \rightsquigarrow^* "", M, K} \qquad\qquad \frac{M', K' \rightsquigarrow^* n, N, L}{M, K \rightsquigarrow^* m * n, N, L}\ (M, K \rightsquigarrow m, M', K')$$

□

We can state and prove analogues of all the results in Sect. 2.3.4–2.3.2. In particular we have

Proposition 17 $M \Downarrow m, T$ iff $M, \texttt{nil} \rightsquigarrow^* m, T, \texttt{nil}$. □

2.5 Observational Equivalence

As with CBV and CBN, we want to define a notion of observational equivalence.

Definition 2.7 1 A *ground context* is a closed ground returner with zero or more occurrences of a hole which might be a computation or a value.

2 Given two computations $\Gamma \vdash^{\mathsf{c}} M, N : \underline{B}$, we say that $M \simeq N$ when for every ground context $\mathcal{C}[\cdot]$ we have that $\mathcal{C}[M] \Downarrow T$ iff $\mathcal{C}[N] \Downarrow T$ (for every T). We similarly define $\simeq$ for values.

□

We modify this to suit the effect being considered. For printing (without divergence) we say that $M \simeq N$ when for every ground context $\mathcal{C}[\cdot]$, we have that $\mathcal{C}[M] \Downarrow m, T$ iff $\mathcal{C}[N] \Downarrow m, T$ (for every T).

2.6 Denotational Semantics

2.6.1 Values and Computations

In the printing semantics, a value type (and hence a context) denotes a set, and a computation type denotes an $\mathcal{A}$-set. The semantics of types is given by

$$\begin{aligned}
[\![U\underline{B}]\!] &= \text{the carrier of } [\![\underline{B}]\!] \\
[\![\textstyle\sum_{i\in I} A_i]\!] &= \textstyle\sum_{i\in I} [\![A_i]\!] \\
[\![A \times A']\!] &= [\![A]\!] \times [\![A']\!] \\
\\
[\![FA]\!] &= \text{the free } \mathcal{A}\text{-set on } [\![A]\!] \\
[\![\textstyle\prod_{i\in I} \underline{B}_i]\!] &= \textstyle\prod_{i\in I} [\![\underline{B}_i]\!] \\
[\![A \to \underline{B}]\!] &= [\![A]\!] \to [\![\underline{B}]\!]
\end{aligned}$$

Similarly we define the denotation of a context Γ: if Γ is the sequence $A_0, \ldots, A_{n-1}$, then we set $[\![\Gamma]\!]$ to be the set $[\![A_0]\!] \times \cdots \times [\![A_{n-1}]\!]$.

Next we define the semantics of terms. A value $\Gamma \vdash^{\mathsf{v}} V : A$ denotes a function $[\![V]\!]$ from the set $[\![\Gamma]\!]$ to the set $[\![A]\!]$, and a computation $\Gamma \vdash^{\mathsf{c}} M : \underline{B}$ denotes a function from the set $[\![\Gamma]\!]$ to the carrier of the $\mathcal{A}$-set

$[\![\underline{B}]\!]$. Here are some example clauses:

$$\begin{array}{rcl} [\![\mathtt{return}\ V]\!]\rho & = & (1, [\![V]\!]\rho) \\ [\![M\ \mathtt{to}\ \mathtt{x}.\ N]\!]\rho & = & m * [\![N]\!](\rho, \mathtt{x} \mapsto a) \text{ where } [\![M]\!]\rho = (m, a) \\ [\![\mathtt{thunk}\ M]\!]\rho & = & [\![M]\!]\rho \\ [\![\mathtt{force}\ V]\!]\rho & = & [\![V]\!]\rho \\ [\![\lambda \mathtt{x}.M]\!]\rho & = & \lambda x.[\![M]\!](\rho, \mathtt{x} \mapsto x) \\ [\![V`M]\!]\rho & = & ([\![V]\!]\rho)`([\![M]\!]\rho) \\ [\![\mathtt{print}\ m.\ M]\!]\rho & = & m * ([\![M]\!]\rho) \end{array}$$

In the Scott semantics, a value type denotes a cpo and a computation type denotes a pointed cpo. Some example clauses:

$$\begin{array}{rcl} [\![U\underline{B}]\!] & = & [\![\underline{B}]\!] \\ [\![FA]\!] & = & \text{lift of } [\![A]\!] \\ [\![A \to \underline{B}]\!] & = & [\![A]\!] \to [\![\underline{B}]\!] \end{array}$$

Like a value type, a context denotes a cpo, given by $\times$. Then a value $\Gamma \vdash^{\mathsf{v}} V : A$ denotes a continuous function $[\![V]\!]$ from the cpo $[\![\Gamma]\!]$ to the cpo $[\![A]\!]$, and a computation $\Gamma \vdash^{\mathsf{c}} M : \underline{B}$ denotes a continuous function from the cpo $[\![\Gamma]\!]$ to the pointed cpo $[\![\underline{B}]\!]$.

Notice that, in both printing and Scott semantics, the constructs `thunk` and `force` are *invisible* in the sense that

$$\begin{array}{rcl} [\![\mathtt{thunk}\ M]\!] & = & [\![M]\!] \\ [\![\mathtt{force}\ V]\!] & = & [\![V]\!] \end{array}$$

In the Scott semantics, U too is invisible, meaning $[\![U\underline{B}]\!] = [\![\underline{B}]\!]$. By contrast, in many of the semantics in Chap. 5 `thunk` , `force` and U are all visible.

Proposition 18 (Soundness of Denotational Semantics) For any closed computation M, if $M \Downarrow m, T$ then $[\![M]\!] = m * [\![T]\!]$. □

Corollary 19 (by Prop. 10) For any closed ground returner M, we have $M \Downarrow m, \mathtt{return}\ n$ iff $[\![M]\!] = (m, n)$. Hence terms with the same denotation are observationally equivalent. □

If we are dealing with non-closed computations as in Sect. 2.3.3, then Prop. 18 extends to

Proposition 20 For any computation $\Gamma \vdash^{\mathsf{c}} M : \underline{B}$ and any environment $\rho \in [\![\Gamma]\!]$ if $M \Downarrow m, T$ then $[\![M]\!]\rho = m * ([\![T]\!]\rho)$. □

2.6.2 Denotational Semantics Of Stacks

All of the denotational semantics for CBPV considered in this book interpret not just values and computations, but stacks too. (It is this that makes the categorical semantics in Part. III so simple.) For example, let us consider the 2 models we have studied so far. Suppose we have a stack $\Gamma|\underline{B} \vdash^{\mathsf{k}} \underline{C}$.

- In the Scott model, K denotes a continuous function f from $[\![\Gamma]\!] \times [\![\underline{B}]\!]$ to $[\![\underline{C}]\!]$ satisfying $f(\rho, \bot) = \bot$ i.e. it is strict in its second argument.
- In the printing model, K denotes a function f from $[\![\Gamma]\!] \times [\![\underline{B}]\!]$ to $[\![\underline{C}]\!]$ satisfying $f(\rho, c * x) = c * f(\rho, x)$ i.e. it is homomorphic in its second argument.

We omit the equations giving the interpretation of each term constructor; they are straightforward.

2.6.3 Monads and Algebras

We now return to the monad semantics of Sect. 1.6.4 (CBV) and Sect. 1.7.6 (CBN); this section can be omitted by the non-categorical reader. We are going to interpret stacks as algebra homomorphisms, defined as follows.

Definition 2.8 Let (T, η, μ, t) be a strong monad. A *T-algebra homomorphism* from a T-algebra (Y, θ) to a T-algebra (Z, ϕ) over a $\mathcal{C}$-object X is a $\mathcal{C}$-morphism $X \times Y \xrightarrow{f} Z$ such that

$$\begin{array}{ccccc} X \times TY & \xrightarrow{t(\Gamma,Y)} & T(\Gamma \times Y) & \xrightarrow{Tf} & TZ \\ {\scriptstyle X\times\theta}\downarrow & & & & \downarrow{\scriptstyle \phi} \\ X \times Y & & \xrightarrow[f]{} & & Z \end{array} \quad \text{commutes.}$$

□

As in Sect. 1.7.6, suppose we have a strong monad T on a distributive category $\mathcal{C}$, with an exponential from A to B whenever B is the carrier of a T-algebra. We then interpret

- a value type (and hence a context) by a $\mathcal{C}$-object
- a computation type by a T-algebra
- a value $\Gamma \vdash^{\mathsf{v}} V : A$ by a $\mathcal{C}$-morphism from $[\![\Gamma]\!]$ to $[\![A]\!]$

- a computation $\Gamma \vdash^{\mathsf{c}} M : \underline{B}$ by a $\mathcal{C}$-morphism from $[\![\Gamma]\!]$ to the carrier of $[\![\underline{B}]\!]$ (again, this makes `thunk` and `force` invisible)
- a stack $\Gamma|\underline{B} \vdash^{\mathsf{k}} \underline{C}$ by a T-algebra homomorphism from $[\![\underline{B}]\!]$ to $[\![\underline{C}]\!]$ over $[\![\Gamma]\!]$.

It is easy to see that this generalizes the Scott semantics and printing semantics. In particular, a homomorphism between cppos is a strict continuous function.

2.7 Subsuming CBV and CBN

We give translations from the CBV and CBN fragments described in Chap. 1. The full translations and their technical properties (adequacy, full abstraction etc.) are given in Appendix A.

If the reader bears in mind the denotational semantics of CBV and CBN, the translations into CBPV are obvious.

2.7.1 From CBV to CBPV

The big difference between CBV and CBPV is that in CBV $\lambda\mathtt{x}.M$ is a value whereas in CBPV $\lambda\mathtt{x}.M$ is a computation. Thus $\lambda\mathtt{x}$ in CBV decomposes into `thunk` $\lambda\mathtt{x}$ in CBPV.

More generally, a CBV function from A to B is, from a CBPV perspective, a thunk of a computation that pops a value of type A and returns a value of type B. So we have a decomposition of $\to_{\mathbf{CBV}}$ into CBPV given by

$$A \to_{\mathbf{CBV}} B = U(A \to FB) \tag{2.4}$$

It is important to see that this decomposition respects both of our denotational semantics i.e. the two sides of (2.4) have the same denotation. This is essential if the translation is to be regarded as subsumptive.

- In the printing semantics, $A \to FB$ denotes an $\mathcal{A}$-set whose carrier is the set of functions from $[\![A]\!]$ to $\mathcal{A}^* \times [\![B]\!]$. So the thunk type $U(A \to FB)$ denotes this set, as does $A \to_{\mathbf{CBV}} B$.
- In the Scott semantics $A \to FB$ denotes the cppo of continuous functions from $[\![A]\!]$ to $[\![B]\!]_\perp$. So the thunk type $U(A \to FB)$ denotes this cpo, as does $A \to_{\mathbf{CBV}} B$.

The translation from CBV to CBPV is presented in Fig. 2.6. Just as a CBV value V has two denotations $[\![V]\!]^{\mathsf{val}}$ and $[\![V]\!]^{\mathsf{ret}}$, so it has two translations V^{val} and V^{prod} related by

$$V^{\mathsf{prod}} = \mathtt{return}\ V^{\mathsf{val}}$$

C	C^{v} (a value type)
bool	bool i.e. $1+1$
$A+B$	$A^{\mathsf{v}}+B^{\mathsf{v}}$
$A \to B$	$U(A^{\mathsf{v}} \to FB^{\mathsf{v}})$

$A_0,\ldots,A_{n-1} \vdash V : C$	$A_0{}^{\mathsf{v}},\ldots,A_{n-1}{}^{\mathsf{v}} \vdash^{\mathsf{v}} V^{\mathsf{val}} : C^{\mathsf{v}}$
x	x
true	true
false	false
inl V	inl V^{val}
inr V	inr V^{val}
λx.M	thunk λx.M^{prod}

$A_0,\ldots,A_{n-1} \vdash M : C$	$A_0{}^{\mathsf{v}},\ldots,A_{n-1}{}^{\mathsf{v}} \vdash^{\mathsf{c}} M^{\mathsf{prod}} : FC^{\mathsf{v}}$
x	return x
let M be x. N	M^{prod} to x. N^{prod}
true	return true
false	return false
if M then N else N'	M^{prod} to z. if z then N^{prod} else N'^{prod}
inl M	M^{prod} to z. return inl z
inr M	M^{prod} to z. return inr z
pm M as {inl x.N, inr x.N'}	M^{prod} to z. pm z as {inl x.N^{prod}, inr x.N'^{prod}}
λx.M	return thunk λx.M^{prod}
M‘N	M^{prod} to x. N^{prod} to f. x‘(force f)
print c. M	print c. M^{prod}

Figure 2.6. Translation of CBV types, values and returners

in the CBPV equational theory of Chap. 3.

As we explained for $\to_{\mathbf{CBV}}$, the translation preserves denotational semantics. Thus, for both printing and Scott semantics, we have the following:

Proposition 21 1 For any CBV type A, $[\![A]\!]_{\mathbf{CBV}} = [\![A^{\mathsf{v}}]\!]_{\mathbf{CBPV}}$.

2 For any CBV value $\Gamma \vdash V : A$, $[\![V]\!]^{\mathsf{val}}_{\mathbf{CBV}} = [\![V^{\mathsf{val}}]\!]_{\mathbf{CBPV}}$.

3 For any CBV returner $\Gamma \vdash M : A$, $[\![M]\!]^{\mathsf{ret}}_{\mathbf{CBV}} = [\![M^{\mathsf{prod}}]\!]_{\mathbf{CBPV}}$.

□

That the translation respects operational semantics (to a certain degree of intensionality) is proved in Appendix A.

2.7.2 From CBN to CBPV

The translation from CBN to CBPV is motivated as follows.

- Identifiers in CBN are bound to unevaluated terms, so we regard them (from a CBPV perspective) as bound to thunks.

- Similarly, tuple-components and operands in CBN are unevaluated terms, so we regard them as thunks.
- Consequently, identifiers, tuple-components and operands all have type of the form $U\underline{B}$—a thunk type.

We thus have a decomposition of $\to_{\mathbf{CBN}}$ into CBPV.

$$\underline{A} \to_{\mathbf{CBN}} \underline{B} = (U\underline{A}) \to \underline{B} \tag{2.5}$$

It is important to see that this decomposition respects both of our denotational semantics i.e. both sides of (2.5) have the same denotation. This is essential if the translation is to be regarded as subsumptive.

- In the printing semantics, if $\underline{A}$ denotes the $\mathcal{A}$-set $(X, *)$ and $\underline{B}$ denotes the $\mathcal{A}$-set $(Y, *)$ then $U\underline{A}$ denotes the set X and $(U\underline{A}) \to \underline{B}$ denotes $X \to (Y, *)$, as does $\underline{A} \to_{\mathbf{CBN}} \underline{B}$.
- In the Scott semantics, $(U\underline{A}) \to \underline{B}$ denotes the cppo of continuous functions from $[\![\underline{A}]\!]$ to $[\![\underline{B}]\!]$, as does $\underline{A} \to_{\mathbf{CBN}} \underline{B}$.

Similarly we have decompositions

$$\begin{array}{rcl} \mathtt{bool}_{\mathbf{CBN}} & = & F\mathtt{bool} \\ \underline{A} +_{\mathbf{CBN}} \underline{B} & = & F((U\underline{A}) + (U\underline{B})) \end{array}$$

It is easily seen that these also respect denotational semantics, e.g. the Scott semantics for $F((U\underline{A}) + (U\underline{B}))$ is the lifted sum of $[\![\underline{A}]\!]$ and $[\![\underline{B}]\!]$.

C	C^{n} (a computation type)
`bool`	F`bool` i.e. $F(1+1)$
$A + B$	$F(UA^{\mathsf{n}} + UB^{\mathsf{n}})$
$A \to B$	$(UA^{\mathsf{n}}) \to B^{\mathsf{n}}$

$A_0, \ldots, A_{n-1} \vdash M : C$	$UA_0^{\mathsf{n}}, \ldots, UA_{n-1}^{\mathsf{n}} \vdash^{\mathsf{c}} M^{\mathsf{n}} : C^{\mathsf{n}}$
`x`	`force x`
`let` M `be x.` N	`let thunk` M^{n} `be x.` M^{n}
`true`	`return true`
`false`	`return false`
`if` M `then` N `else` N'	M^{n} `to z. if z then` N^{n} `else` N'^{n}
`inl` M	`return inl thunk` M^{n}
`pm` M `as` {`inl x.`N, `inr x.`N'}	M^{n} `to z. pm z as` {`inl x.`N^{n}, `inr x.`N'^{n}}
λ`x.`M	λ`x.`M^{n}
N‘M	(`thunk` N^{n})‘M^{n}
`print` c. M	`print` c. M^{n}

Figure 2.7. Translation of CBN types and terms

As we explained for $\to_{\mathbf{CBN}}$, the translation preserves denotational semantics. Thus, for both printing and Scott semantics, we have the following:

Proposition 22 1 For any CBN type A, $[\![A]\!]_{\mathbf{CBN}} = [\![A^{\mathsf{n}}]\!]_{\mathbf{CBPV}}$.

2 For any CBN term $\Gamma \vdash M : A$, $[\![M]\!]_{\mathbf{CBN}} = [\![M^{\mathsf{n}}]\!]_{\mathbf{CBPV}}$.

□

That the translation respects operational semantics (to a certain degree of intensionality) is proved in Appendix A.

2.8 CBPV As A Metalanguage

We can use CBPV not just as an object language but as a metalanguage. When we do this, we have to specify which CBPV model the metalanguage is referring to. For example, when talking about the printing model, we use FA to refer to the free $\mathcal{A}$-set on the set A, and $U\underline{B}$ to refer to the carrier of the $\mathcal{A}$-set $\underline{B}$.

Another example is the Scott model.

- If A is a cpo we write FA for its lift, a cppo.
- If $a \in A$ we write $\mathsf{return}\ a$ for the corresponding element of FA.
- If $b \in FA$ and f is a function from A to a cppo $\underline{B}$, we write b to $x.\ f(x)$ to mean $\bot$ if $b = \bot$ and $f(a)$ if $b = \mathsf{return}\ a$.
- If $\underline{B}$ is a cppo, we write $U\underline{B}$ to mean the cpo $\underline{B}$.
- If $b \in \underline{B}$ we write $\mathsf{thunk}\ b$ for b regarded as an element of $U\underline{B}$.
- If $a \in U\underline{B}$ we write $\mathsf{force}\ a$ for a regarded as an element of $\underline{B}$.

It may seem superfluous to write U, thunk and force when referring to the Scott model, because they are invisible. But the advantage of doing so is that everything we write in this notation is meaningful not just in the Scott model but in any CBPV model.

2.9 Useful Syntactic Sugar

2.9.1 Pattern-Matching

It is convenient to extend all constructs that bind identifiers to allow pattern-matching. We give some examples.

sugar	unsugared
M $\mathtt{to}$ $\{\ldots,(i,\mathtt{x}).N_i,\ldots\}$	M $\mathtt{to}$ $\mathtt{z}.$ $\mathtt{pm}$ $\mathtt{z}$ $\mathtt{as}$ $\{\ldots,(i,\mathtt{x}).N_i,\ldots\}$
$\lambda(\mathtt{x},\mathtt{y}).M$	$\lambda\mathtt{z}.(\mathtt{pm}\ \mathtt{z}\ \mathtt{as}\ (\mathtt{x},\mathtt{y}).M)$
$\mathtt{pm}\ M\ \mathtt{as}\ (\mathtt{w},(\mathtt{x},\mathtt{y})).N$	$\mathtt{pm}\ M\ \mathtt{as}\ (\mathtt{w},\mathtt{z}).(\mathtt{pm}\ \mathtt{z}\ \mathtt{as}\ (\mathtt{x},\mathtt{y}).N)$

We use such abbreviations only informally, as it would be complicated to give a precise, general description.

2.9.2 Commands

The types $F1$ and $F0$ have a special significance in CBPV, because they can be seen as types of commands, as we now explain.

Computations of type $F1$ are commands such as printing, assignment, divergence and so on. They correspond to terms of type `comm` in Idealized Algol [Reynolds, 1981], and to returners of type `unit` in ML. Such a command can be prefixed to a computation of any type. We write

$$\begin{array}{lll} \texttt{return} & \text{for} & \texttt{return}\ () \\ \texttt{print}\ c & \text{for} & \texttt{print}\ c.\ \texttt{return}\ () \\ M; N & \text{for} & M\ \texttt{to}\ ().\ N \end{array}$$

(The command `return` is written `skip` in Idealized Algol, CSP etc.) The typing rules for these constructs are as follows.

$$\frac{}{\Gamma \vdash^{\mathsf{c}} \texttt{return} : F1} \qquad \frac{\Gamma \vdash^{\mathsf{c}} M : F1 \quad \Gamma \vdash^{\mathsf{c}} N : \underline{B}}{\Gamma \vdash^{\mathsf{c}} M; N : \underline{B}}$$

$$\frac{}{\Gamma \vdash^{\mathsf{c}} \texttt{print}\ c : F1}$$

Computations of type $F0$ constitute a more restricted class of commands, the *nonreturning commands* such as divergence or raising an error. Such a command can be coerced into a computation of any type. Since a nonreturning command ignores its continuation[3], we can say intuitively that any two stacks from $F0$ to $\underline{C}$ are equivalent, for any $\underline{C}$. (Indeed, they will be provably equal in the equational theory of Chap. 3.) We write

$$\begin{array}{lll} \texttt{coerce}_{\underline{B}} M & \text{for} & M\ \texttt{to}\ \{\} \\ \texttt{ignored} & \text{for} & [\cdot]\ \texttt{to}\ \{\} :: \texttt{nil} \end{array}$$

We will make use of $F0$ for technical purposes in Sect. 7.8.2, so we take the trouble here to describe explicitly the typing rules for its constructs.

$$\frac{\Gamma \vdash^{\mathsf{c}} M : F0}{\Gamma \vdash^{\mathsf{c}} \texttt{coerce}_{\underline{B}}\ M : \underline{B}} \qquad \frac{}{\Gamma | F0 \vdash^{\mathsf{k}} \texttt{ignored} : \underline{C}}$$

[3] This would not be the case if we considered exception *handling*, but we do not.

2.10 Computations Matter Most

We have seen that CBV terms and CBN terms are translated into CBPV computations. One generally finds that the computation judgement is the most important in CBPV—the role of values and stacks is an auxiliary one. Thus many results, especially definability results, are really results about computations.

Consider, for example, a definability result stating that every computable element of, say, a domain model or a game model is the denotation of an appropriately typed term [Hyland and Ong, 2000; Plotkin, 1977; Sazonov, 1976]. Such a result can be true only for the computation judgement, because it is impossible for every computable total function $f : \mathbb{N} \longrightarrow \mathbb{N}$ to be the denotation of a value $\mathtt{x} : \mathtt{nat} \vdash^{\mathsf{v}} V : \mathtt{nat}$, written in a finitary syntax.

Other examples in this book of a result which concerns only computations (actually for the Jump-With-Argument language rather than CBPV) is Prop. 79, and the domain enrichment in Sect. 9.5.5.

Sometimes we shall extend a language by adding more value and stack constructs. But we prove that every computation M in the larger language is equal (in some appropriate equational theory) to a computation N in the smaller language. We shall call this a *computation-unaffecting* extension. Because the two languages have the same computations, we shall regard them as essentially the same.

The principal example of this procedure occurs in the next chapter, where we add *complex values* and *complex stacks* to CBPV, so that the equational theory we describe can be shown (in Chap. 10) to have a simple categorical semantics. Unfortunately, the simplicity of the *operational* semantics is lost as a consequence. At that point, the reader may protest:

> Surely this is cheating! You claim that CBPV has all these good properties. But *which* CBPV? For the CBPV (with complex values) that has a simple categorical semantics is different from the CBPV (without complex values) that has a simple operational semantics.

This objection is answered when we show that the extension is computation-unaffecting, so that the languages are essentially the same.

Chapter 3

COMPLEX VALUES AND EQUATIONAL THEORY

3.1 Introduction

In this chapter we look at 2 equational theories, as shown in Fig. 3.1. All the extensions are computation-unaffecting in the sense of Sect. 2.10.

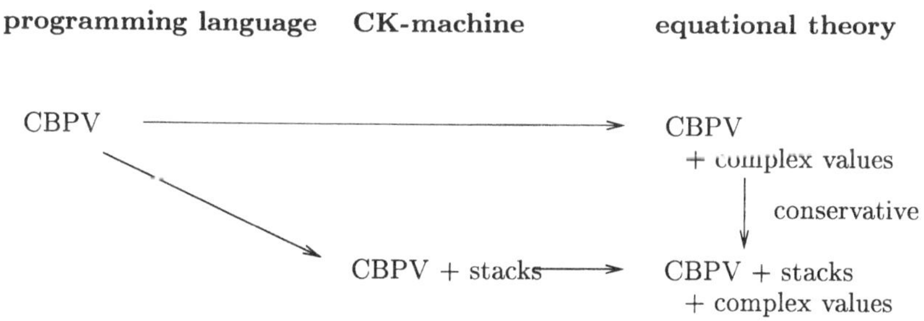

Figure 3.1. Varieties of CBPV

We first see that, in order for the equational theory to have reasonable mathematical properties (in particular, categorical semantics), we need to add in Sect. 3.2 some extra terms called *complex values* for the $\sum$ and $\times$ types.

We present an equational theory for CBPV with complex values in Sect. 3.3. Unfortunately, there is no reversible derivation for F. We use this theory in Sect. 3.5 to show that adding complex values is computation-unaffecting.

We then look present an equational theory for CBPV with stacks. As well as complex values, this requires *complex stacks* for the $\prod$ and $\to$ types (these are roughly dual to the complex values). This equational theory is much better behaved: it has reversible derivations for all the connectives. We use it to construct some isomorphisms between types.

Finally, we will look at how this theory is related to effect-free equational theories.

3.2 Complex Values

Suppose we want to form a value $\mathtt{x} : \mathtt{bool} \vdash^{\mathsf{v}} \mathtt{not}\ \mathtt{x} : \mathtt{bool}$. This seems reasonable: it makes sense denotationally, and a natural way to code it is `if x then false else true`. But the CBPV syntax as defined in Fig. 2.1 will not allow this, because the rules allow pattern-matching into computations only, not into values. (Recall from Sect. 2.2 that `if` is the pattern-match construct for $\mathtt{bool} = 1 + 1$.) A similar problem arises with values such as $\mathtt{w} + 5$ and $\mathtt{x} + \mathtt{y}$, which we used in our example program in Sect. I.5.2.

For another example—and one which which will be indispensable when we come to equational/ categorical issues, because it gives us a cartesian category of values—try to form a value $\mathtt{x} : \mathtt{bool} \times \mathtt{nat} \vdash^{\mathsf{v}} \pi\mathtt{x} : \mathtt{bool}$. Again, this makes sense denotationally, and a natural way to code it is $\mathtt{pm}\ \mathtt{x}\ \mathtt{as}\ (\mathtt{y}, \mathtt{z}).\ \mathtt{y}$. But this too involves pattern-matching into a value.

All these are examples of *complex values*. We incorporate them into the CBPV language by adding the rules in Fig. 3.2.

$$\frac{\Gamma \vdash^{\mathsf{v}} V : A \quad \Gamma, \mathtt{x} : A \vdash^{\mathsf{v}} W : B}{\Gamma \vdash^{\mathsf{v}} \mathtt{let}\ V\ \mathtt{be}\ \mathtt{x}.\ W : B}$$

$$\frac{\Gamma \vdash^{\mathsf{v}} V : \sum_{i \in I} A_i \quad \cdots \quad \Gamma, \mathtt{x} : A_i \vdash^{\mathsf{v}} W_i : B \quad \cdots_{\ i \in I}}{\Gamma \vdash^{\mathsf{v}} \mathtt{pm}\ V\ \mathtt{as}\ \{\ldots, (i, \mathtt{x}).W_i, \ldots\} : B}$$

$$\frac{\Gamma \vdash^{\mathsf{v}} V : A \times A' \quad \Gamma, \mathtt{x} : A, \mathtt{y} : A' \vdash^{\mathsf{v}} W : B}{\Gamma \vdash^{\mathsf{v}} \mathtt{pm}\ V\ \mathtt{as}\ (\mathtt{x}, \mathtt{y}).W : B}$$

Figure 3.2. Complex Values

All these rules make sense in every denotational model, and, as we mentioned above, they are indispensable when we study equational/categorical issues. The reader may therefore wonder: why did we not include these rules from the outset? The answer is that excluding complex values keeps the operational semantics simple: both our big-step and our machine semantics exploit the fact that values do not need to be evaluated.

Furthermore, the range of the transforms from CBN and (coarse-grain) CBV into CBPV does not involve complex values.

It would certainly be possible to extend the operational semantics to include complex values, but at the cost of canonicity. We would have to make an arbitrary decision as to when to evaluate complex values. Since the evaluation of complex values causes no effects, the decision has no semantic significance. We will therefore continue to exclude complex values when treating operational issues, but otherwise we will include them.

We shall see in Sect. 3.5 that adding complex values is *computation-unaffecting* in the sense of Sect. 2.10: a computation containing complex values can be converted into one without. For example,

```
return (if x then false else true)
```

can be converted into

```
if x then return true else return false
```

Less importantly, complex value constructs can also be removed from a closed value.

3.3 Equations

In Sect. 2.1 we looked at equational properties of CBV and CBN, and the informal reasons for them:

- In CBV, the η-law for sum types (and boolean type) holds because identifiers are bound to values.
- In CBN, the η-law for function types holds because a term of function type can be made to evaluate only by applying it.

We stipulated that a subsuming language should have both of these properties. We can now see that CBPV indeed meets these requirements.

- Identifiers are bound to values, so the η-laws for sum types holds.
- A term of function type can be made to evaluate only by applying it, so the η-law for function types holds.

The equational issue is more immediately apparent from a denotational perspective, whether we look at the printing semantics or the Scott semantics:

- In CBPV, $[\![A + A']\!]$ is precisely the disjoint union of $[\![A]\!]$ and $[\![A']\!]$—like in CBV but unlike in CBN.

- In CBPV $[\![A]\!] \to [\![\underline{B}]\!]$ is precisely the set of functions (or the cpo of continuous functions) from $[\![A]\!]$ to $[\![\underline{B}]\!]$—like in CBN but unlike in CBV.

We thus formulate an equational theory for CBPV (with complex values), given in Fig 3.3. M, N and P range over computations, while V and W range over values. Q ranges over all terms, both computations and values.

The equations for `print` and `diverge` are of course specific to our example effects, but there are directly analogous equations for many other effects. We call this theory (without the `print` and `diverge` equations) the *CBPV equational theory*.

For both printing and Scott models we have (as a consequence of a substitution lemma, which we omit)

Proposition 23 Provably equal terms have the same denotation. □

3.4 CK-Machine Illuminates → Equations

We recall from Sect. I.5.2 and Sect. 2.3.2 that $\lambda\mathtt{x}$ and V‘ can be read as commands:

- $\lambda\mathtt{x}$ means "pop x"
- V‘ means "push V".

This reading illuminates many equations involving $\to$ (as well as the analogous equations involving $\prod$). Here are some examples.

1 Consider the equation

$$\mathtt{print}\ \texttt{"hello"}.\ \lambda\mathtt{x}.M = \lambda\mathtt{x}.(\mathtt{print}\ \texttt{"hello"}.\ M) \qquad (3.1)$$

In the CK-machine reading, the LHS means

```
print "hello".
pop x.
M
```

while the RHS means

```
pop x.
print "hello".
M
```

Provided the stack is non-empty (as it must be if the initial configuration was ground), these two behaviours are the same, because popping and printing do not interfere.

$$
\begin{array}{rcl}
 & \beta\text{-}\mathbf{laws} & \\
\texttt{let } V \texttt{ be x. } Q & = & Q[V/\texttt{x}] \\
\texttt{pm } (\hat{\imath}, V) \texttt{ as } \{\ldots,(i,\texttt{x}).Q_i,\ldots\} & = & Q_{\hat{\imath}}[V/\texttt{x}] \\
\texttt{pm } (V,V') \texttt{ as } (\texttt{x},\texttt{y}).\ Q & = & Q[V/\texttt{x},V'/\texttt{y}] \\
\texttt{force thunk } M & = & M \\
(\texttt{return } V) \texttt{ to x. } M & = & M[V/\texttt{x}] \\
\hat{\imath}`\lambda\{\ldots,i.M_i,\ldots\} & = & M_{\hat{\imath}} \\
V`\lambda\texttt{x}.M & = & M[V/\texttt{x}]
\end{array}
$$

$$
\begin{array}{rcl}
 & \eta\text{-}\mathbf{laws} & \\
M & = & M \texttt{ to x. return x} \\
V & = & \texttt{thunk force } V \\
Q[V/\texttt{z}] & = & \texttt{pm } V \texttt{ as } \{\ldots,(i,\texttt{x}).\ {}^{\texttt{x}}Q[(i,\texttt{x})/\texttt{z}],\ldots\} \\
Q[V/\texttt{z}] & = & \texttt{pm } V \texttt{ as } (\texttt{x},\texttt{y}).\ {}^{\texttt{xy}}Q[(\texttt{x},\texttt{y})/\texttt{z}] \\
M & = & \lambda\{\ldots,i.i`M,\ldots\} \\
M & = & \lambda\texttt{x}.(\texttt{x}`\ {}^{\texttt{x}}M)
\end{array}
$$

$$
\begin{array}{rcl}
 & \mathbf{sequencing\ laws} & \\
(M \texttt{ to x. } N) \texttt{ to y. } P & = & M \texttt{ to x. } (N \texttt{ to y. } {}^{\texttt{x}}P) \\
M \texttt{ to x. } \lambda\{\ldots,i.N_i,\ldots\} & = & \lambda\{\ldots,i.(M \texttt{ to x. } N_i),\ldots\} \\
M \texttt{ to x. } \lambda\texttt{y}.N & = & \lambda\texttt{y}.({}^{\texttt{x}}M \texttt{ to x. } N)
\end{array}
$$

$$
\begin{array}{rcl}
 & \texttt{print}\ \mathbf{laws} & \\
(\texttt{print } c.\ M) \texttt{ to x. } N & = & \texttt{print } c.\ (M \texttt{ to x. } N) \\
\texttt{print } c.\ \lambda\{\ldots,i.M_i,\ldots\} & = & \lambda\{\ldots,i.(\texttt{print } c.\ M),\ldots\} \\
\texttt{print } c.\ \lambda\texttt{x}.M & = & \lambda\texttt{x}.(\texttt{print } c.\ M)
\end{array}
$$

$$
\begin{array}{rcl}
 & \texttt{diverge}\ \mathbf{laws} & \\
\texttt{diverge to x. } N & = & \texttt{diverge} \\
\texttt{diverge} & = & \lambda\{\ldots,i.\texttt{diverge},\ldots\} \\
\texttt{diverge} & = & \lambda\texttt{x}.\texttt{diverge}
\end{array}
$$

Figure 3.3. CBPV equations, using conventions of Sect. I.4.2

2 In the β-law

$$V`\lambda\texttt{x}.M \simeq M[V/\texttt{x}]$$

the LHS means

$$
\begin{array}{l}
\texttt{push } V. \\
\texttt{pop x.} \\
M
\end{array}
$$

The first two lines have the effect of binding x to V and leaving the stack unchanged, so the overall effect is to obey M with x bound to V.

3 In the η-law

$$M \simeq \lambda\mathtt{x}.(\mathtt{x}\text{` }^{\mathtt{x}}M)$$

the RHS means

```
pop x.
push x.
M —which does not mention x
```

Assuming again that the stack is non-empty, the first two lines have the effect of leaving the stack unchanged and binding x to the top entry in the stack. But we are assuming that x is not used in M, so the binding is of no consequence.

4 The analogue of (3.1) for divergence is

$$\mathtt{diverge} \simeq \lambda\mathtt{x}.\mathtt{diverge}$$

The RHS means

```
pop x;
diverge
```

Assuming again that the stack is non-empty, this diverges.

As an aside, we notice that all these 4 equations, and the push/pop reading that explains them, hold in CBN as well as in CBPV. Indeed, Krivine at one point adopted the operand-first notation, for just this reason [Krivine, 2001]. Therefore, "call-by-push" would have been a good name for CBN. The name "call-by-push-value" for our new language reflects the fact that it is a value that gets pushed during application, not a general term as in CBN.

3.5 Adding Complex Values is Computation-Unaffecting

Lemma 24 If $M = N$ is provable and M and N do not contain complex values then $M \simeq N$. □

Proof This follows from Prop. 23 and Cor. 19. □

We are now in a position to see that complex values can be eliminated from a computation, as required. As a bonus, we can also eliminate complex values from a closed value.

Proposition 25 1 There is an effective procedure that, given a computation $\Gamma \vdash^{\mathsf{c}} M : \underline{B}$, possibly containing complex values, returns a computation $\Gamma \vdash^{\mathsf{c}} \tilde{M} : \underline{B}$ without complex values, such that $M = \tilde{M}$ is provable.

2 There is an effective procedure that, given a closed value $\vdash^{\mathsf{v}} V : A$, possibly containing complex values, returns a closed value $\vdash^{\mathsf{v}} \tilde{V} : A$ without complex values, such that $V = \tilde{V}$ is provable.

□

Proof

1 We will define one such procedure, and simultaneously, for each value $\Gamma \vdash^{\mathsf{v}} V : A$, possibly containing complex values, and each complex-value-free computation $\Gamma, \mathtt{v} : A \vdash^{\mathsf{c}} N : \underline{B}$, we will define a complex-value-free computation $\Gamma \vdash^{\mathsf{c}} N[V /\!\!/ \mathtt{v}] : \underline{B}$ such that $N[V /\!\!/ \mathtt{v}] = N[V/\mathtt{v}]$ is provable.

By mutual induction on M and V, we define $\tilde{M}$ and $N[V /\!\!/ \mathtt{v}]$ in Fig. 3.4 and we prove the equations

$$\tilde{M} = M$$
$$N[V /\!\!/ \mathtt{v}] \quad = \quad N[V/\mathtt{v}]$$

2 For a value $A_0, \ldots, A_{n-1} \vdash^{\mathsf{v}} V : A$ possibly containing complex values, we will define $\tilde{V}$ to be a function that, when applied to a sequence $W_0, \ldots, W_{n-1}$ of complex-value-free closed values $\vdash^{\mathsf{v}} W_i : A_i$, returns a complex-value-free closed value $\tilde{V}(W_0, \ldots, W_{n-1})$ such that

$$\tilde{V}(W_0, \ldots, W_{n-1}) = V[\overrightarrow{W_i/\mathtt{x}_i}]$$

is provable. This is defined by induction on V in Fig. 3.4. The special case $n = 0$ gives the desired result.

□

By choosing some such procedure, we can extend the operational semantics to include ground returners with complex values: if M is such a returner, we say that $M \Downarrow m, i$ iff $\tilde{M} \Downarrow m, i$. Because of Lemma 24, this relation does not depend on the choice of procedure $\tilde{-}$.

We can adapt the proof of Prop. 25 to show that complex values can be removed from a context. So it is immaterial whether, in our definition of observational equivalence, we allow contexts to contain complex values.

We can extend Prop, 23 and Cor. 19 as follows, for the printing language:

V	$N[V/\!\!/\mathtt{v}]$
$\mathtt{x}$	$N[\mathtt{x}/\mathtt{v}]$
$\mathtt{let}\ W\ \mathtt{be}\ \mathtt{x}.\ U$	$(\mathtt{let}\ \mathtt{w}\ \mathtt{be}\ \mathtt{x}.\ N[U/\!\!/\mathtt{v}])[W/\!\!/\mathtt{w}]$
$(\hat{\imath}, W)$	$(\mathtt{let}\ (\hat{\imath}, \mathtt{w})\ \mathtt{be}\ \mathtt{v}.\ N)[W/\!\!/\mathtt{w}]$
$\mathtt{pm}\ W\ \mathtt{as}\ \{\ldots, (i, \mathtt{x}).U_i, \ldots\}$	$(\mathtt{pm}\ \mathtt{w}\ \mathtt{as}\ \{\ldots, (i, \mathtt{x}).N[U_i/\!\!/\mathtt{v}], \ldots\})[W/\!\!/\mathtt{w}]$
(W, W')	$(\mathtt{let}\ (\mathtt{w}, \mathtt{x})\ \mathtt{be}\ \mathtt{v}.\ N)[W/\!\!/\mathtt{w}][W'/\!\!/\mathtt{x}]$
$\mathtt{pm}\ W\ \mathtt{as}\ (\mathtt{x}, \mathtt{y}).U$	$(\mathtt{pm}\ \mathtt{w}\ \mathtt{as}\ (\mathtt{x}, \mathtt{y}).N[U/\!\!/\mathtt{v}])[W/\!\!/\mathtt{w}]$
$\mathtt{thunk}\ M$	$N[\mathtt{thunk}\ \tilde{M}/\mathtt{v}]$

M	$\tilde{M}$
$\mathtt{let}\ W\ \mathtt{be}\ \mathtt{x}.\ N$	$(\mathtt{let}\ \mathtt{w}\ \mathtt{be}\ \mathtt{x}.\ \tilde{N})[W/\!\!/\mathtt{w}]$
$\mathtt{pm}\ W\ \mathtt{as}\ \{\ldots, (i, \mathtt{x}).N_i, \ldots\}$	$(\mathtt{pm}\ \mathtt{w}\ \mathtt{as}\ \{\ldots, (i, \mathtt{x}).\tilde{N}_i, \ldots\})[W/\!\!/\mathtt{w}]$
$\mathtt{pm}\ W\ \mathtt{as}\ (\mathtt{x}, \mathtt{y}).N$	$(\mathtt{pm}\ \mathtt{w}\ \mathtt{as}\ (\mathtt{x}, \mathtt{y}).\tilde{N})[W/\!\!/\mathtt{w}]$
$\lambda\{\ldots, i.N_i, \ldots\}$	$\lambda\{\ldots, i.\tilde{N}_i, \ldots\}$
$\hat{\imath}`N$	$\hat{\imath}`\tilde{N}$
$\lambda\mathtt{x}.N$	$\lambda\mathtt{x}.\tilde{N}$
$V`N$	$(\mathtt{v}`\tilde{N})[V/\!\!/\mathtt{v}]$
$\mathtt{return}\ V$	$(\mathtt{return}\ \mathtt{v})[V/\!\!/\mathtt{v}]$
$N\ \mathtt{to}\ \mathtt{x}.\ P$	$\tilde{N}\ \mathtt{to}\ \mathtt{x}.\ \tilde{P}$
$\mathtt{force}\ V$	$(\mathtt{force}\ \mathtt{v})[V/\!\!/\mathtt{v}]$
$\mathtt{print}\ c.\ N$	$\mathtt{print}\ c.\ \tilde{N}$

V	$\tilde{V}(\overrightarrow{W_i})$
$\mathtt{x}_i$	W_i
$\mathtt{let}\ W\ \mathtt{be}\ \mathtt{x}.\ U$	$\tilde{U}(\overrightarrow{W_i}, \tilde{W}(\overrightarrow{W_i}))$
$(\hat{\imath}, W)$	$(\hat{\imath}, \tilde{W}(\overrightarrow{W_i}))$
$\mathtt{pm}\ W\ \mathtt{as}\ \{\ldots, (i, \mathtt{x}).U_i, \ldots\}$	$\tilde{U}_{\hat{\imath}}(\overrightarrow{W_i}, V')$ where $\tilde{W}(\overrightarrow{W_i}) = (\hat{\imath}, V')$
(W, W')	$(\tilde{W}(\overrightarrow{W_i}), \tilde{W'}(\overrightarrow{W_i}))$
$\mathtt{pm}\ W\ \mathtt{as}\ (\mathtt{x}, \mathtt{y}).U$	$\tilde{U}(\overrightarrow{W_i}, V', V'')$ where $\tilde{W}(\overrightarrow{W_i}) = (V', V'')$
$\mathtt{thunk}\ M$	$\mathtt{thunk}\ \tilde{M}[\overrightarrow{W_i/\mathtt{x}_i}]$

Figure 3.4. Definitions used in proof of Prop. 25

Proposition 26 Provable equality implies denotational equality, which implies observational equivalence. □

This adapts to the various effects and denotational models we will look at.

3.6 The Problem With the CBPV Equational Theory

The equational theory of Fig. 3.3 is convenient for reasoning about equivalence of programs, and we have seen that it fits well with the CK-machine reading. But it is flawed in two ways.

1 There is no reversible derivation (in the sense of Sect. 1.3.4) for F.

2 As we saw in Sect. 2.6.2, the denotational models we study are typically able to interpret stacks as well as values and computations. So we would like an equational theory that can reason about equality for stacks.

Fortunately, these problems are solved together. We will formulate an equational theory for CBPV with stacks, conservatively extending that of Fig. 3.3, and see that it has reversible derivations for all the connectives, including F.

3.7 Complex Stacks

$$\frac{\Gamma \vdash^{\mathsf{v}} V : A \quad \Gamma, \mathtt{x} : A \vdash^{\mathsf{v}} W : B}{\Gamma \vdash^{\mathsf{v}} \mathtt{let}\ V\ \mathtt{be}\ \mathtt{x}.\ W : B} \qquad \frac{\Gamma \vdash^{\mathsf{v}} V : A \quad \Gamma, \mathtt{x}.A|\underline{B} \vdash^{\mathsf{k}} K : \underline{C}}{\Gamma|\underline{B} \vdash^{\mathsf{k}} \mathtt{let}\ V\ \mathtt{be}\ \mathtt{x}.\ K : \underline{C}}$$

$$\frac{\Gamma \vdash^{\mathsf{v}} V : \sum_{i \in I} A_i \quad \cdots \ \Gamma, \mathtt{x} : A_i \vdash^{\mathsf{v}} W_i : B \ \cdots_{i \in I}}{\Gamma \vdash^{\mathsf{v}} \mathtt{pm}\ V\ \mathtt{as}\ \{\ldots, (i, \mathtt{x}).W_i, \ldots\} : B}$$

$$\frac{\Gamma \vdash^{\mathsf{v}} V : \sum_{i \in I} A_i \quad \cdots \ \Gamma, \mathtt{x} : A_i|\underline{B} \vdash^{\mathsf{k}} K_i : \underline{C} \ \cdots_{i \in I}}{\Gamma|\underline{B} \vdash^{\mathsf{k}} \mathtt{pm}\ V\ \mathtt{as}\ \{\ldots, (i, \mathtt{x}).K_i, \ldots\} : \underline{C}}$$

$$\frac{\Gamma \vdash^{\mathsf{v}} V : A \times A' \quad \Gamma, \mathtt{x} : A, \mathtt{y} : A' \vdash^{\mathsf{v}} W : B}{\Gamma \vdash^{\mathsf{v}} \mathtt{pm}\ V\ \mathtt{as}\ (\mathtt{x}, \mathtt{y}).W : B}$$

$$\frac{\Gamma \vdash^{\mathsf{v}} V : A \times A' \quad \Gamma, \mathtt{x} : A, \mathtt{y} : A'|\underline{B} \vdash^{\mathsf{k}} K : \underline{C}}{\Gamma|\underline{B} \vdash^{\mathsf{k}} \mathtt{pm}\ V\ \mathtt{as}\ (\mathtt{x}, \mathtt{y}).K : \underline{C}}$$

$$\frac{\Gamma|\underline{C} \vdash^{\mathsf{k}} K : \underline{B} \quad \Gamma|\underline{B} \vdash^{\mathsf{k}} L : \underline{D}}{\Gamma|\underline{C} \vdash^{\mathsf{k}} K\ \mathtt{where}\ \mathtt{nil}\ \mathtt{is}\ L : \underline{D}}$$

$$\frac{\cdots \ \Gamma|\underline{C} \vdash^{\mathsf{k}} K_i : \underline{B}_i \ \cdots_{i \in I} \quad \Gamma|\prod_{i \in I} \underline{B}_i \vdash^{\mathsf{k}} L : \underline{D}}{\Gamma|\underline{C} \vdash^{\mathsf{k}} \{\ldots, K_i\ \mathtt{where}\ i :: \mathtt{nil}, \ldots\}\ \mathtt{is}\ L : \underline{D}}$$

$$\frac{\Gamma, \mathtt{x} : A|\underline{C} \vdash^{\mathsf{k}} K : \underline{B} \quad \Gamma|A \to \underline{B} \vdash^{\mathsf{k}} L : \underline{D}}{\Gamma|\underline{C} \vdash^{\mathsf{k}} K\ \mathtt{where}\ \mathtt{x} :: \mathtt{nil}\ \mathtt{is}\ L : \underline{D}}$$

Figure 3.5. Complex Values and Stacks

Just as we have to add complex values to get a well-behaved equational theory, we also have to add *complex stacks*, shown in Fig. 3.5. These new constructs are of two kinds. Firstly we want to be able to pattern-match (or bind) a value V into a stack, just as we can pattern-match V into a computation or a value. These are the `let` and `pm` rules.

Secondly we want to be able to pattern-match or bind a stack, and this requires some explanation. The idea is that if we have a stack K to $\underline{B}$, the symbol `nil` that appears on the right is rather like an identifier, because it can be replaced by a stack L from $\underline{B}$, giving the concatenated stack $K +\!\!+ L$. So it makes sense to be able to bind `nil` to L. We could perhaps write such as binding as `let nil be` L. K. But we have chosen instead the syntax K `where nil is` L, as this agrees with the order of concatenation.

Now suppose we have a stack L from $A \to \underline{B}$. Typically, this is of the form $V :: L'$, where V is a value of type A and L' is a stack from $\underline{B}$. So we can pattern-match L, suggesting the syntax `pm` L `as` $\mathtt{x} :: \mathtt{nil}.\ K$. Again, we choose the more convenient syntax K `where` $\mathtt{x} :: \mathtt{nil}$ `is` L.

Similarly, a stack L from $\prod_{i \in I} \underline{B}_i$ is typically of the form $\hat{\imath} :: L'$, where L' is a value from $\underline{B}_{\hat{\imath}}$. So we can pattern-match L, suggesting the syntax `pm` L `as` $\{i :: \mathtt{nil}.\ K_i\}$. Again, we choose the more convenient syntax $\{\ldots, K_i\ \mathtt{where}\ i :: \mathtt{nil}, \ldots\}$ `is` L, although unfortunately this is difficult to read as English.

Since a stack is never a subterm of a computation or a value, we do not require an eliminability theorem, such as we had for complex values (Sect. 3.5). But we do need to extend to complex stacks the definition of dismantling and concatenation; this is given explicitly in Fig. 10.2.

3.8 Equational Theory For CBPV With Stacks

Now that we have introduced both complex values and complex stacks, we are in a position to give the equational theory for CBPV with stacks.

The equations for `print` and `diverge` are of course specific to our example effects, but there are directly analogous equations for many other effects.

One noteworthy equation of Fig. 3.6 is the η-law

$$M \bullet K = M\ \mathtt{to}\ \mathtt{x}.\ ((\mathtt{return}\ \mathtt{x}) \bullet K) \tag{3.2}$$

This is equivalent to the two equations

$$M = M\ \mathtt{to}\ \mathtt{x}.\ \mathtt{return}\ \mathtt{x} \tag{3.3}$$

$$(M\ \mathtt{to}\ \mathtt{x}.\ N) \bullet K = M\ \mathtt{to}\ \mathtt{x}.\ (N \bullet K) \tag{3.4}$$

$$
\begin{array}{rcl}
 & \beta\textbf{-laws} & \\
\texttt{let } V \texttt{ be x. } Q & = & Q[V/\texttt{x}] \\
K \texttt{ where nil is } L & = & K +\!\!+ L \\
\texttt{pm } (\hat{\imath}, V) \texttt{ as } \{\ldots, (i, \texttt{x}).Q_i, \ldots\} & = & Q_{\hat{\imath}}[V/\texttt{x}] \\
\texttt{pm } (V, V') \texttt{ as } (\texttt{x}, \texttt{y}).Q & = & Q[V/\texttt{x}, V'/\texttt{y}] \\
\texttt{force thunk } M & = & M \\
(\texttt{return } V) \texttt{ to x. } M & = & M[V/\texttt{x}] \\
\hat{\imath}`\lambda\{\ldots, i.M_i, \ldots\} & = & M_{\hat{\imath}} \\
\{\ldots, K_i \texttt{ where } i :: \texttt{nil}, \ldots\} \texttt{ is } \hat{\imath} :: L & = & K_{\hat{\imath}} +\!\!+ L \\
V`\lambda\texttt{x}.M & = & M[V/\texttt{x}] \\
K \texttt{ where x} :: \texttt{nil is } V :: L & = & K[V/\texttt{x}] +\!\!+ L
\end{array}
$$

$$
\begin{array}{rcl}
 & \eta\textbf{-laws} & \\
Q[V/\texttt{z}] & = & \texttt{pm } V \texttt{ as } \{\ldots, (i, \texttt{x}).\ {}^{\texttt{x}}Q[(i, \texttt{x})/\texttt{z}], \ldots\} \\
Q[V/\texttt{z}] & = & \texttt{pm } V \texttt{ as } (\texttt{x}, \texttt{y}).\ {}^{\texttt{xy}}Q[(\texttt{x}, \texttt{y})/\texttt{z}] \\
V & = & \texttt{thunk force } V \\
M \bullet K & = & M \texttt{ to x. } ((\texttt{return x}) \bullet {}^{\texttt{x}}K) \\
K +\!\!+ L & = & [\cdot] \texttt{ to x. } ((\texttt{return x}) \bullet {}^{\texttt{x}}K) :: L \\
M & = & \lambda\{\ldots, i.i`M, \ldots\} \\
K +\!\!+ L & = & \{\ldots, (K +\!\!+ i :: \texttt{nil}) \texttt{ where } i :: \texttt{nil}, \ldots\} \texttt{ is } L \\
M & = & \lambda\texttt{x}.(\texttt{x}`\ {}^{\texttt{x}}M) \\
K +\!\!+ L & = & ({}^{\texttt{x}}K +\!\!+ \texttt{x} :: \texttt{nil}) \texttt{ where x} :: \texttt{nil is } L
\end{array}
$$

$$
\begin{array}{rcl}
 & \texttt{print}\ \textbf{law} & \\
(\texttt{print } c.\ M) \bullet K & = & \texttt{print } c.\ (M \bullet K)
\end{array}
$$

$$
\begin{array}{rcl}
 & \texttt{diverge}\ \textbf{law} & \\
\texttt{diverge} \bullet K & = & \texttt{diverge}
\end{array}
$$

Figure 3.6. Equational laws for CBPV + stacks, using operations of Sect. 2.3.5

As instances of (3.4) we have the sequencing laws of Fig. 3.3. Similarly the `print` and `diverge` laws of Fig. 3.3 are instances of the `print` and `diverge` laws of Fig. 3.6.

It is thus evident that our equational theory for CBPV+stacks extends our CBPV theory. What is not apparent at this stage is that the extension is conservative. We prove this in Appendix B.

3.9 Reversible Derivations

Now that we have an equational theory for CBPV + stacks, we can ask what reversible derivations it possesses, and we see in Fig. 3.7 that it possesses rules for each connective, including F.

Rules for $\sum$

$$\frac{\cdots \quad \Gamma, A_i \vdash^{\mathsf{v}} B \quad \cdots \;{}_{i \in I}}{\Gamma, \sum_{i \in I} A_i \vdash^{\mathsf{v}} B}$$

$$\frac{\cdots \quad \Gamma, A_i \vdash^{\mathsf{c}} \underline{B} \quad \cdots \;{}_{i \in I}}{\Gamma, \sum_{i \in I} A_i \vdash^{\mathsf{c}} \underline{B}}$$

$$\frac{\cdots \quad \Gamma, A_i | \underline{B} \vdash^{\mathsf{k}} \underline{C} \quad \cdots \;{}_{i \in I}}{\Gamma, \sum_{i \in I} A_i | \underline{B} \vdash^{\mathsf{k}} \underline{C}}$$

Rules for U

$$\frac{\Gamma \vdash^{\mathsf{c}} \underline{B}}{\Gamma \vdash^{\mathsf{v}} U\underline{B}}$$

Rules for $\prod$

$$\frac{\cdots \quad \Gamma \vdash^{\mathsf{c}} \underline{B}_i \quad \cdots \;{}_{i \in I}}{\Gamma \vdash^{\mathsf{c}} \prod_{i \in I} \underline{B}_i}$$

$$\frac{\cdots \quad \Gamma | \underline{C} \vdash^{\mathsf{k}} \underline{B}_i \quad \cdots \;{}_{i \in I}}{\Gamma | \underline{C} \vdash^{\mathsf{k}} \prod_{i \in I} \underline{B}_i}$$

Rules for $\times$

$$\frac{\Gamma \vdash^{\mathsf{v}} A \quad \Gamma \vdash^{\mathsf{v}} A'}{\Gamma \vdash^{\mathsf{v}} A \times A'}$$

$$\frac{\Gamma, A, A' \vdash^{\mathsf{v}} B}{\Gamma, A \times A' \vdash^{\mathsf{v}} B}$$

$$\frac{\Gamma, A, A' \vdash^{\mathsf{c}} \underline{B}}{\Gamma, A \times A' \vdash^{\mathsf{c}} \underline{B}}$$

$$\frac{\Gamma, A, A' | \underline{B} \vdash^{\mathsf{k}} \underline{C}}{\Gamma, A \times A' | \underline{B} \vdash^{\mathsf{k}} \underline{C}}$$

Rules for F

$$\frac{\Gamma, A \vdash^{\mathsf{c}} \underline{B}}{\Gamma | FA \vdash^{\mathsf{k}} \underline{B}}$$

Rules for $\to$

$$\frac{\Gamma, A \vdash^{\mathsf{c}} \underline{B}}{\Gamma \vdash^{\mathsf{c}} A \to \underline{B}}$$

$$\frac{\Gamma, A | \underline{C} \vdash^{\mathsf{k}} \underline{B}}{\Gamma | \underline{C} \vdash^{\mathsf{k}} A \to \underline{B}}$$

Figure 3.7. Reversible derivations for CBPV with stacks

3.10 Example Isomorphisms Of Types

We frequently want to say that two CBPV types are "isomorphic". Here are some examples:

$$A \to (B \to \underline{C}) \cong (A \times B) \to \underline{C} \tag{3.5}$$

$$U\underline{A} \times U\underline{A}' \cong U(\underline{A} \sqcap \underline{A}') \tag{3.6}$$

(3.5) is the currying isomorphism. Operationally, this says that popping two operands successively is essentially the same as popping a pair of operands.

(3.6) says that a pair of thunks can be coalesced into a single thunk, of a computation that first pops a binary tag and proceeds accordingly.

These isomorphisms are obviously valid in both the printing model and the Scott model. Here is a syntactic definition:

Definition 3.1 value type An isomorphism from A to B is a pair of values $\mathtt{x} : A \vdash^{\mathsf{v}} V : B$ and $\mathtt{y} : B \vdash W : A$ such that $V[W/\mathtt{x}] = \mathtt{y}$ and $W[V/\mathtt{y}] = \mathtt{x}$ are provable.

computation type An isomorphism from $\underline{A}$ to $\underline{B}$ is a pair of stacks $|\underline{A} \vdash^{\mathsf{k}} K : \underline{B}$ and $|\underline{B} \vdash^{\mathsf{k}} L : \underline{A}$ such that $K +\!\!+ L = \mathtt{nil}$ and $L +\!\!+ K = \mathtt{nil}$ are provable.

□

It is then easy to construct the desired isomorphisms (3.5)–(3.6). As a consequence we have

Proposition 27 Every CBPV type is isomorphic to a *type canonical form*, meaning a type in the following class:

$$\begin{array}{rl} A ::= & \sum_{i \in I} U\underline{B}_i \\ \underline{B} ::= & \prod_{i \in I}(U\underline{B}_i \to FA_i) \end{array}$$

□

3.11 Trivialization

The effect-free language analogous to CBPV is the $\times\sum\prod\to$-calculus shown in Fig. 3.8; essentially this is the $\lambda\mathtt{bool}+$-calculus of Sect. 1.3 extended with both pattern-match products and projection products. As this calculus is effect-free, it is not necessary to have both kinds of product, just convenient.

We use the term "$\times$-calculus" for the fragment of this calculus using just $\times$; similarly "$\times\sum$-calculus" and so forth.

It is easy to see that we can collapse effect-free CBPV + stacks, with complex values and complex stacks, into the $\times\sum\prod\to$-calculus. This

Types

$A ::= \ 1 \mid A \times A \mid \sum_{i \in I} A_i \mid \prod_{i \in I} A_i \mid A \to A$
where each set I is finite (or countable, for *infinitely wide* $\times\sum\prod \to$-calculus)

Terms

$$\frac{}{\Gamma, \mathtt{x} : A, \Gamma' \vdash \mathtt{x} : A} \qquad \frac{\Gamma \vdash M : A \quad \Gamma, \mathtt{x} : A \vdash N : B}{\Gamma \vdash \mathtt{let}\ M\ \mathtt{be}\ \mathtt{x}.\ N : B}$$

$$\frac{\Gamma \vdash M : A \quad \Gamma \vdash M' : A'}{\Gamma \vdash (M, M') : A \times A'} \qquad \frac{\Gamma \vdash M : A \times A' \quad \Gamma, \mathtt{x} : A, \mathtt{y} : A' \vdash N : B}{\Gamma \vdash \mathtt{pm}\ M\ \mathtt{as}\ (\mathtt{x}, \mathtt{y}).N : B}$$

$$\frac{\Gamma \vdash M : A_{\hat{\imath}}}{\Gamma \vdash (\hat{\imath}, M) : \sum_{i \in I} A_i} \qquad \frac{\Gamma \vdash M : \sum_{i \in I} A_i \quad \cdots \ \Gamma, \mathtt{x} : A_i \vdash N_i : B \ \cdots_{\ i \in I}}{\Gamma \vdash \mathtt{pm}\ M\ \mathtt{as}\ \{\ldots, (i, \mathtt{x}).N_i, \ldots\} : B}$$

$$\frac{\cdots \ \Gamma \vdash M_i : B_i \ \cdots_{\ i \in I}}{\Gamma \vdash \lambda\{\ldots, i.M_i, \ldots\} : \prod_{i \in I} B_i} \qquad \frac{\Gamma \vdash M : \prod_{i \in I} B_i}{\Gamma \vdash \hat{\imath}`M : B_{\hat{\imath}}}$$

$$\frac{\Gamma, \mathtt{x} : A \vdash M : B}{\Gamma \vdash \lambda \mathtt{x}.M : A \to B} \qquad \frac{\Gamma \vdash M : A \quad \Gamma \vdash N : A \to B}{\Gamma \vdash M`N : B}$$

Equations, using conventions of Sect. I.4.2

β**-laws**

$$\begin{array}{rcl}
\mathtt{let}\ M\ \mathtt{be}\ \mathtt{x}.\ N & = & N[M/\mathtt{x}] \\
\mathtt{pm}\ (M, M')\ \mathtt{as}\ (\mathtt{x}, \mathtt{y}).N & = & N[M/\mathtt{x}, M'/\mathtt{y}] \\
\mathtt{pm}\ (\hat{\imath}, M)\ \mathtt{as}\ \{\ldots, (i, \mathtt{x}).N_i, \ldots\} & = & N_{\hat{\imath}}[M/\mathtt{x}] \\
\hat{\imath}`\lambda\{\ldots, i.M_i, \ldots\} & = & M_{\hat{\imath}} \\
M`\lambda \mathtt{x}.N & = & N[M/\mathtt{x}]
\end{array}$$

η**-laws**

$$\begin{array}{rcl}
N[M/\mathtt{z}] & = & \mathtt{pm}\ M\ \mathtt{as}\ (\mathtt{x}, \mathtt{y}).N[(\mathtt{x}, \mathtt{y})/\mathtt{z}] \\
N[M/\mathtt{z}] & = & \mathtt{pm}\ M\ \mathtt{as}\ \{\ldots, (i, \mathtt{x}).N[(i, \mathtt{x})/\mathtt{z}], \ldots\} \\
M & = & \lambda\{\ldots, i.i`M, \ldots\} \\
M & = & \lambda \mathtt{x}.(\mathtt{x}`M)
\end{array}$$

Figure 3.8. The $\times\sum\prod \to$-Calculus

transform of equational theories is called *trivialization*. It discards U and F and leaves all other type constructors unchanged. It translates

- a value $\Gamma \vdash^{\mathsf{v}} V : A$ to a term $\Gamma^{\mathsf{tr}} \vdash V^{\mathsf{tr}} : A^{\mathsf{tr}}$
- a computation $\Gamma \vdash^{\mathsf{c}} M : \underline{B}$ to a term $\Gamma^{\mathsf{tr}} \vdash M^{\mathsf{tr}} : \underline{B}^{\mathsf{tr}}$
- a stack $\Gamma|\underline{A} \vdash^{\mathsf{k}} K : \underline{B}$ to a term $\Gamma^{\mathsf{tr}}, \mathtt{x} : \underline{A}^{\mathsf{tr}} \vdash K^{\mathsf{tr}} : \underline{B}^{\mathsf{tr}}$.

We omit the actual translation of terms, which is entirely straightforward.

It follows that every model for $\times\sum\prod\rightarrow$-calculus (in categorical terms, a *bicartesian closed category*—Sect. 1.6.4) gives a model for CBPV (a *trivial model*).

In the opposite direction, we can translate $\times\sum$-calculus into CBPV with complex values—just regard everything as a value. Consequently, every CBPV model must include a model for $\times\sum$-calculus (in categorical terms, a *distributive category*—Def. 1.7) for interpreting values. We call it the *value category* of the CBPV model.

Chapter 4

RECURSION AND INFINITELY DEEP CBPV

This chapter does not depend on Chap. 3.

4.1 Introduction

This chapter is about the computational effect of divergence. There is little original work here; our aim is to recall and adapt well-known material, which is not specific to CBPV, for use in subsequent chapters, notably our treatment of thunk storage in Sect. 6.10.

First, we add recursion to CBPV—both recursive terms and recursive types—and look at the denotational semantics.

Secondly, we look at a language feature that is, in a sense, equivalent to recursion—infinitely deep syntax. This means that the parse tree of a term or type can be infinitely deep, i.e. branches can be infinitely long. Böhm trees are a well-known example of such terms. Our only subsequent use of infinitely deep CBPV will be to state definability results for game semantics in Chap. 8.

It is important not to confuse this infinitely deep syntax with the weaker feature of *infinitely wide* syntax , which is used throughout the book. The latter simply means that we allow types $\sum_{i\in I} A_i$ and $\prod_{i\in I} \underline{B}_i$, where I is countably infinite. (Therefore, parse trees of types and terms are infinitely wide.) Infinitely wide syntax does not introduce divergence into the language, and all the CBPV semantics in the book can interpret it. It provides, for example, an easy way of giving a type of natural numbers: as $\sum_{i\in\mathbb{N}} 1$. We use it also as a metalanguage while describing the possible world semantics in Chap. 6.

It is easy to see that once we allow infinitely deep syntax, we can encode infinitely wide syntax—for example, the infinitely wide type $\sum_{i\in\mathbb{N}} 1$

can be written as the infinitely deep type $1+(1+(1+\cdots))$. In summary:

$$\begin{array}{ccccc} \text{finitary} & \subset & \text{infinitely wide} & \subset & \text{infinitely deep} \\ \text{CBPV} & & \text{CBPV} & & \text{CBPV} \end{array}$$

Of course, all of this applies equally to CBV or to CBN.

There is a serious problem that arises from both infinitely wide and infinitely deep syntax—non-computability. For example, when we define $\mathtt{nat}$ to be $\sum_{i \in \mathbb{N}} 1$, *every* function f from $\mathbb{N}$ to $\mathbb{N}$ is definable, even if f is not computable:

$$\vdash^{\mathsf{c}} \lambda\mathtt{x}.\mathtt{pm}\ \mathtt{x}\ \mathtt{as}\ \{\ldots i.\mathtt{return}\ f(i), \ldots\} : \mathtt{nat} \to F\mathtt{nat}$$

Consequently realizability models of CBPV (which we do not treat in this book) are not models of either infinitely wide CBPV or infinitely deep CBPV. This can probably be remedied by restricting infinitely deep CBPV to "computable terms" and "effectively presented types", as is done for strategies and arenas in game semantics (e.g. [Hyland and Ong, 2000]). However, we have not investigated this.

In the final section of this chapter, Sect. 4.8, we see how we can see how to restrict our interpretation of CBPV in cpo and cppos to a smaller class of predomains and domains. We emphasize that this would not be possible for a language providing a $\to$ operation on value types.

The material in Chap. 3 on complex values and the equational theory is not, at present, applicable to infinitely deep CBPV. We do not have an appropriate notion of congruence needed to define an equational theory, and we need an equational theory to justify the addition of complex values, by showing the extension to be computation-unaffecting. Creating such an equational theory remains a challenge for future work. In particular, it would be desirable to have a form of pattern-matching in complex values sufficiently powerful to construct a syntactic isomorphism between $\mu\mathtt{X}.(\mathtt{bool} \times \mathtt{X})$ and 0.

4.2 Divergence and Recursion

Although we have already described the Scott model of CBPV in Sect. 2.6, the model is rather pointless if we do not add recursion to the language. We now look at this addition.

4.2.1 Divergent and Recursive Terms

We add to the basic CBPV language the constructs

$$\frac{}{\Gamma \vdash^{\mathsf{c}} \mathtt{diverge} : \underline{B}} \qquad \frac{\Gamma, \mathtt{x} : U\underline{B} \vdash^{\mathsf{c}} M : \underline{B}}{\Gamma \vdash^{\mathsf{c}} \mu\mathtt{x}.M : \underline{B}}$$

While `diverge` is not strictly necessary (e.g. it can be desugared as $\mu\mathtt{x}.\mathtt{force}\ \mathtt{x}$) it is convenient to include it as a primitive.

We add the big-step rules

$$\frac{\mathtt{diverge} \Downarrow T}{\mathtt{diverge} \Downarrow T} \qquad\qquad \frac{M[\mathtt{thunk}\ \mu\mathtt{x}.M/\mathtt{x}] \Downarrow T}{\mu\mathtt{x}.M \Downarrow T}$$

The rule for `diverge` can of course never be applied, so it is technically dispensable. We have included it to reflect the operational idea of divergence: to evaluate `diverge`, one evaluates `diverge`, and so forth. We say that M *diverges* iff there does not exist T such that $M \Downarrow T$. (This definition is acceptable only in a deterministic setting.)

To the CK-machine we add transitions:

$$\begin{array}{lll} & \mathtt{diverge} & K \\ \leadsto & \mathtt{diverge} & K \\ \\ & \mu\mathtt{x}.M & K \\ \leadsto & M[\mathtt{thunk}\ \mu\mathtt{x}.M/\mathtt{x}] & K \end{array}$$

We extend the cpo/cppo denotational semantics in Sect. 2.6 by interpreting `diverge` as $\bot$ and interpreting μ as a least pre-fixed point, in the usual way.

Proposition 28 Let M be a closed computation.

Soundness If $M \Downarrow T$ then $[\![M]\!] = [\![T]\!]$.

Adequacy If M diverges then $[\![M]\!] = \bot$.

□

Proof (in the style of [Tait, 1967]) Soundness is straightforward. For our adequacy proof, we say that a subset R of a cpo B is *admissible* when R is closed under directed joins. We say that R is *lift-reflecting* when $a, b \in R$ and $\mathsf{return}\ a \leqslant \mathsf{return}\ b$ imply $a \leqslant b$. Of course, in the Scott model, *every* subset is lift-reflecting, but we use this condition so that our proof generalizes to other (**Cpo**-enriched) models where that may not be the case.

We define, by mutual induction over types, three families of relations:

- for each A, a relation $\leqslant^{\mathsf{v}}_A$ between $[\![A]\!]$ and closed values of type A, such that $\{a \in [\![A]\!] | a \leqslant^{\mathsf{v}}_A V\}$ is admissible and lift-reflecting
- for each $\underline{B}$ a relation $\leqslant^{\mathsf{t}}_{\underline{B}}$ between $[\![\underline{B}]\!]$ and *terminal* computations of type $\underline{B}$, such that $\{b \in [\![\underline{B}]\!] | b \leqslant^{\mathsf{t}} T\}$ is admissible and contains $\bot$.

- for each $\underline{B}$ a relation $\leqslant^{\mathsf{c}}_{\underline{B}}$ between $[\![\underline{B}]\!]$ and closed computations of type $\underline{B}$, such that $\{b \in [\![\underline{B}]\!] | b \leqslant^{\mathsf{c}} M\}$ is admissible and contains $\bot$

The definition of these relations proceeds as follows:

$$\begin{array}{lll}
a \leqslant^{\mathsf{v}}_{U\underline{B}} \mathtt{thunk}\ M & \text{iff} & \mathsf{force}\ a \leqslant^{\mathsf{c}}_{\underline{B}} M \\
a \leqslant^{\mathsf{v}}_{\sum_{i \in I} A_i} (\hat{\imath}, V) & \text{iff} & a = (\hat{\imath}, b) \text{ for some } b \leqslant^{\mathsf{v}}_{A_{\hat{\imath}}} V \\
a \leqslant^{\mathsf{v}}_{A \times A'} (V, V') & \text{iff} & a = (b, b') \text{ for some } b \leqslant^{\mathsf{v}}_{A} V \text{ and } b' \leqslant^{\mathsf{v}}_{A'} V' \\
 & & \\
b \leqslant^{\mathsf{t}}_{FA} \mathtt{return}\ V & \text{iff} & b = \bot \text{ or } b = \mathsf{return}\ a \text{ for some } a \leqslant^{\mathsf{v}}_{A} V \\
f \leqslant^{\mathsf{t}}_{\prod_{i \in I} \underline{B}_i} \lambda\{\ldots, i.M_i, \ldots\} & \text{iff} & \hat{\imath} \in I \text{ implies } \hat{\imath}`f \leqslant^{\mathsf{c}}_{\underline{B}_{\hat{\imath}}} M_{\hat{\imath}} \\
f \leqslant^{\mathsf{t}}_{A \to \underline{B}} \lambda \mathtt{x}.M & \text{iff} & a \leqslant^{\mathsf{v}}_{A} V \text{ implies } a`f \leqslant^{\mathsf{c}}_{\underline{B}} M[V/\mathtt{x}] \\
 & & \\
b \leqslant^{\mathsf{c}}_{\underline{B}} M & \text{iff} & b = \bot \text{ or, for some } T,\ M \Downarrow T \text{ and } b \leqslant^{\mathsf{t}}_{\underline{B}} T
\end{array}$$

Notice that for terminal T, $b \leqslant^{\mathsf{c}}_{\underline{B}} T$ iff $b \leqslant^{\mathsf{t}}_{\underline{B}} T$. Finally, we show that

- for any computation $A_0, \ldots, A_{n-1} \vdash^{\mathsf{c}} M : \underline{B}$, if $a_i \leqslant^{\mathsf{v}}_{A_i} W_i$ for $i = 0, \ldots, n-1$ then $[\![M]\!]\overrightarrow{\mathtt{x}_i \mapsto a_i} \leqslant^{\mathsf{c}}_{\underline{B}} M[\overrightarrow{W_i/\mathtt{x}_i}]$
- for any value $A_0, \ldots, A_{n-1} \vdash^{\mathsf{v}} V : B$, if $a_i \leqslant^{\mathsf{v}}_{A_i} W_i$ for $i = 0, \ldots, n-1$ then $[\![V]\!]\overrightarrow{\mathtt{x}_i \mapsto a_i} \leqslant^{\mathsf{v}}_{B} V[\overrightarrow{W_i/\mathtt{x}_i}]$.

This is shown by mutual induction on M and V, and gives the required result. □

Corollary 29 For any closed ground returner M, we have $M \Downarrow \mathtt{return}\,n$ iff $[\![M]\!] = \mathsf{return}\ n$, and M diverges iff $[\![M]\!] = \bot$. Hence terms with the same denotation are observationally equivalent. □

4.2.2 Type Recursion

Both value types and computation types can be defined recursively, so we extend the type expressions as follows:

$$\begin{array}{rl}
A ::= & \cdots \mid \mathtt{X} \mid \mu\mathtt{X}.A \\
\underline{B} ::= & \cdots \mid \underline{\mathtt{X}} \mid \mu\underline{\mathtt{X}}.\underline{B}
\end{array} \tag{4.1}$$

Here are some examples of CBPV recursive types:

$\mu\mathtt{X}.(1 + \mathtt{bool} \times \mathtt{X})$	finite lists of booleans
$\mu\underline{\mathtt{X}}.F(1 + \mathtt{bool} \times U\underline{\mathtt{X}})$	finite or infinite lazy lists of booleans

Notice again the flexibility of CBPV; it can describe both eager and lazy recursive types.

There is a syntactic difference between the two kinds of recursive type:

$$\frac{\Gamma \vdash^{\mathsf{v}} V : A[\mu\mathtt{X}.A/\mathtt{X}]}{\Gamma \vdash^{\mathsf{v}} \mathtt{fold}\ V : \mu\mathtt{X}.A} \qquad \frac{\Gamma \vdash^{\mathsf{v}} V : \mu\mathtt{X}\,A \quad \Gamma, \mathtt{x} : A[\mu\mathtt{X}.A/\mathtt{X}] \vdash^{\mathsf{c}} M : \underline{B}}{\Gamma \vdash^{\mathsf{c}} \mathtt{pm}\ V\ \mathtt{as}\ \mathtt{fold}\ \mathtt{x}.M : \underline{B}}$$

$$\frac{\Gamma \vdash^{\mathsf{c}} M : \underline{B}[\mu\underline{\mathtt{X}}.\underline{B}/\underline{\mathtt{X}}]}{\Gamma \vdash^{\mathsf{c}} \mathtt{fold}\ M : \mu\underline{\mathtt{X}}.\underline{B}} \qquad \frac{\Gamma \vdash^{\mathsf{c}} M : \mu\underline{\mathtt{X}}.\underline{B}}{\Gamma \vdash^{\mathsf{c}} \mathtt{unfold}\ M : \underline{B}[\mu\underline{\mathtt{X}}.\underline{B}/\underline{\mathtt{X}}]}$$

The reason we have not included a construct $\texttt{unfold}\ V$,where V has type $\mu\mathtt{X}.A$, is that it is a complex value, and so would complicate the operational semantics, as explained in Sect. 3.2. If we allowed pattern-matching into values, as we do in Sect. 3.2, then $\texttt{unfold}\ V$ could be coded as $\texttt{pm}\ V\ \texttt{as fold x. x}$.

For big-step semantics, we first extend the class of terminal computations

$$T ::= \quad \cdots \mid \texttt{fold}\ M$$

and then add the following rules:

$$\frac{M[V/\mathtt{x}] \Downarrow T}{\texttt{pm fold}\ V\ \texttt{as fold x}.M \Downarrow T}$$

$$\frac{}{\texttt{fold}\ M \Downarrow \texttt{fold}\ M} \qquad \frac{M \Downarrow \texttt{fold}\ N \quad N \Downarrow T}{\texttt{unfold}\ M \Downarrow T}$$

To the CK-machine for CBPV we add

$$\begin{array}{llrl}
 & \texttt{pm fold}\ V\ \texttt{as fold x}.M & \underline{B} & K\ \ \underline{C} \\
\leadsto & \underline{B} & M[V/\mathtt{x}] & K\ \ \underline{C} \\
\\
 & \texttt{unfold}\ M & \underline{B}[\mu\underline{\mathtt{X}}.\underline{B}/\underline{\mathtt{X}}] & K\ \ \underline{C} \\
\leadsto & M & \mu\underline{\mathtt{X}}.\underline{B} & \texttt{unfold}\ [\cdot] :: K\ \ \underline{C} \\
\\
 & \texttt{fold}\ M & \mu\underline{\mathtt{X}}.\underline{B} & \texttt{unfold}\ [\cdot] :: K\ \ \underline{C} \\
\leadsto & M & \underline{B}[\mu\underline{\mathtt{X}}.\underline{B}/\underline{\mathtt{X}}] & K\ \ \underline{C}
\end{array}$$

and the additional terminal configuration

$$\texttt{fold}\ M \qquad \underline{C} \qquad \texttt{nil}\ \ \underline{C}$$

This necessitates an additional typing rule for stacks:

$$\frac{\Gamma|\underline{B}[\mu\underline{\mathtt{X}}.\underline{B}/\underline{\mathtt{X}}] \vdash^{\mathsf{k}} K : \underline{C}}{\Gamma|\mu\underline{\mathtt{X}}.\underline{B} \vdash^{\mathsf{k}} \texttt{unfold}\ [\cdot] :: K : \underline{C}}$$

4.3 Denotational Semantics Of Type Recursion

4.3.1 Bilimit-Compact Categories

In order to give denotational semantics for recursive types, we first have to review some key results about solution of domain equations from [Pitts, 1996; Smyth and Plotkin, 1982]. Whilst those papers work with the category $\mathbf{Cpo}^{\perp}$ of pointed cpos and strict continuous functions, everything generalizes[1] to bilimit-compact categories, as we now describe. Much of this material is implicit in [Stark, 1996].

Definition 4.1 Let $\mathcal{C}$ be a **Cpo**-enriched category (i.e. a category whose homsets are cpos and in which composition is continuous).

An *(e, p)-pair* from A to B consists of morphisms

$$A \underset{p}{\overset{e}{\rightleftarrows}} B$$

such that $e; p = \mathsf{id}_A$ and $p; e \leqslant \mathsf{id}_B$. We call e an *embedding* and p a *projection*—they determine each other. We write $\mathcal{C}^{\mathsf{ep}}$ for the category with the same objects as $\mathcal{C}$, and (e, p)-pairs as morphisms.

2 Let D be a directed diagram in $\mathcal{C}^{\mathsf{ep}}$ i.e. a functor from a directed poset $\mathbb{D}$ to $\mathcal{C}^{\mathsf{ep}}$. A *cocone* from D consists of an object V (the *vertex*) together with, for each $d \in \mathbb{D}$, an (e, p)-pair (e_d, p_d) from Dd to V, such that

for $d \leqslant d'$

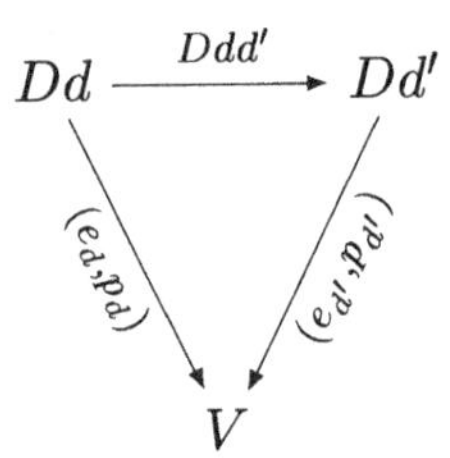

commutes in $\mathcal{C}^{\mathsf{ep}}$.

Such a cocone is an *bilimit* when

$$\bigvee_{d \in \mathbb{D}} (p_d; e_d) = \mathsf{id}_V \tag{4.2}$$

(It is clear that $p_d; e_d$ is monotone in d, so the join has to exist.) □

Many properties and alternative definitions of bilimits are given in [Smyth and Plotkin, 1982].

Definition 4.2 An *bilimit-compact category* $\mathcal{C}$ is a **Cpo**-enriched category with the following properties.

- Each hom-cpo $\mathcal{C}(A, B)$ has a least element $\perp$.

[1]Another generalization is to the "rational categories" of [Abramsky et al., 1994], but they are for call-by-name.

- Composition is bi-strict.

$$\begin{aligned} \bot; g &= \bot \\ f; \bot &= \bot \end{aligned}$$

- $\mathcal{C}$ has a zero object i.e. an object which is both initial and terminal. (Because of bi-strictness, just one of these properties is sufficient.)
- every diagram D of (e,p)-pairs indexed by a countable directed poset $\mathbb{D}$ has an bilimit.

□

As we shall see in Sect. 4.3.3, this is sufficient structure to solve domain equations. Here are some examples of bilimit-compact categories.

Proposition 30 1 The category $\mathbf{Cpo}^{\bot}$ is bilimit-compact.

2 If $\mathcal{C}$ is bilimit-compact then so is $\mathcal{C}^{\mathsf{op}}$.

3 Any small product $\prod_{i \in I} \mathcal{C}_i$ of bilimit-compact categories is bilimit-compact.

4 Let $\mathfrak{I}$ be a small **Cpo**-enriched category. (In particular, it could be a small ordinary category—just regard the homsets as flat cpos.) If $\mathcal{C}$ is bilimit-compact then so is the (locally continuous) functor category $[\mathfrak{I}, \mathcal{C}]$.

□

Proof

(1) standard.

(2) follows from the isomorphism $\mathcal{C}^{\mathsf{op}\,\mathsf{ep}} \simeq \mathcal{C}^{\mathsf{ep}}$.

(3) trivial.

(4) Given a countable directed diagram D in $[\mathfrak{I}, \mathcal{C}]$, set $(Vi, \{(e_{di}, p_{di})\}_{d \in \mathbb{D}})$ to be the bilimit in $\mathcal{C}$ of Di and set $Vi \xrightarrow{Vf} Vj$ to be

$$\bigvee_{d \in \mathbb{D}} (p_{di}; D_{df}; e_{dj}) \tag{4.3}$$

for $i \xrightarrow{f} j$ The required properties are trivial.

□

4.3.2 Sub-Bilimit-Compact Categories

Recursively defined computation types are interpreted as isomorphisms in $\mathbf{Cpo}^{\perp}$, which, as we have said, is bilimit-compact. But recursively defined value types should be interpreted as isomorphisms in **Cpo**, which is not bilimit-compact. What we can do is to find a supercategory (viz. the category **pCpo** of cpos and partial continuous maps) which *is* bilimit-compact, and whose isomorphisms are all guaranteed to be in **Cpo**. We then say that **Cpo** is *sub-bilimit-compact.* However, we require some additional conditions to ensure that sub-bilimit-compactness is closed under $-^{\mathsf{op}}$, under arbitrary product and under $[\mathfrak{I}, -]$ for every small category $\mathfrak{I}$. In particular, in Sect. 6.10 we will need the fact that $[\mathcal{W}, \mathbf{Cpo}]$ is sub-bilimit-compact.

First some preliminaries.

Definition 4.3 1 A subcategory $\mathcal{B}$ of a category $\mathcal{C}$ is *lluf* [Freyd, 1991] when $\mathsf{ob}\ \mathcal{B} = \mathsf{ob}\ \mathcal{C}$.

2 A subcategory $\mathcal{B}$ of a **Cpo**-enriched category $\mathcal{C}$ is *admissible* when it is closed under directed joins.

3 If $\mathcal{B}$ is a subcategory of $\mathcal{D}$ we write $\mathcal{B} \bullet\!\!\rightarrow \mathcal{D}$ for the category with the objects of $\mathcal{B}$ and the morphisms of $\mathcal{D}$, i.e. the unique category $\mathcal{C}$ such that

$$\mathcal{B} \subset_{\mathsf{lluf}} \mathcal{C} \subset_{\mathsf{full}} \mathcal{D}$$

4 A **Cpo**-enriched category $\mathcal{B}$ is *sub-bilimit-compact* within $\mathcal{C}$ when the following conditions hold.

- $\mathcal{C}$ is bilimit-compact category containing $\mathcal{B}$ as a lluf admissible subcategory.
- Suppose D and D' are countable directed diagrams in $\mathcal{C}^{\mathsf{ep}}$, with the same shape $\mathbb{D}$. Suppose $(V, \{(e_d, p_d)\}_{d \in \mathbb{D}})$ is a bilimit of D and $(V', \{(e'_d, p'_d)\}_{d \in \mathbb{D}})$ is a bilimit of D'. Suppose α is a diagram morphism (i.e. natural transformation) from D to D', and $\alpha_d \in \mathcal{B}(Dd, D'd)$ for each d. Then the join

$$\bigvee_{d \in \mathbb{D}} (p_d; \alpha_d; e'_d) \in \mathcal{B}(V, V') \tag{4.4}$$

(It is clear that $p_d; \alpha_d; e'_d$ is monotone in d, so the join has to exist.)

□

Proposition 31 If $\mathcal{B}$ is sub-bilimit-compact within $\mathcal{C}$, then it contains all the isomorphisms in $\mathcal{C}$. □

Proof Let $A \xrightarrow{f} B$ be an isomorphism in $\mathcal{C}$. Let D and D' be the singleton diagram A, and let α be the identity natural transformation. Clearly $(A, (\mathsf{id}_A, \mathsf{id}_A))$ is a bilimit of D and $(B, (f, f^{-1}))$ is a bilimit of D'. For these bilimits, the join (4.4) is f. □

Proposition 32 1 The category **Cpo** is sub-bilimit-compact within **pCpo**.

2 If $\mathcal{C}$ is bilimit-compact, then $\mathcal{C}$ is sub-bilimit-compact within itself.

3 If $\mathcal{B}$ is sub-bilimit-compact within $\mathcal{C}$, then $\mathcal{B}^{\mathsf{op}}$ is sub-bilimit-compact within $\mathcal{C}^{\mathsf{op}}$.

4 If $\mathcal{B}_i$ is sub-bilimit-compact within $\mathcal{C}_i$ for all $i \in I$, then $\prod_{i \in I} \mathcal{B}_i$ is sub-enriched compact within $\prod_{i \in I} \mathcal{C}_i$.

5 Let $\mathfrak{I}$ be a small **Cpo**-enriched category. (In particular, it could be a small ordinary category—just regard the homsets as flat cpos.) If $\mathcal{B}$ is sub-bilimit-compact within $\mathcal{C}$, then the (locally continuous) functor category $[\mathfrak{I}, \mathcal{B}]$ is sub-bilimit-compact within $[\mathfrak{I}, \mathcal{B}] \bullet\!\!\rightarrow [\mathfrak{I}, \mathcal{C}]$.

□

Proof

(1) That **pCpo** is bilimit-compact is well-known. To prove (4.4), take $x \in V$. For sufficiently large d, $p_d(x)$ must be defined, because of (4.2). We know α_d is total, and e_d is total like any embedding. So $e_d(\alpha_d(p_d(x)))$ is defined for sufficiently large d.

(5) To show that $[\mathfrak{I}, \mathcal{B}] \bullet\!\!\rightarrow [\mathfrak{I}, \mathcal{C}]$ is bilimit-compact, recall that the a bilimit in $[\mathfrak{I}, \mathcal{C}]$ are given by (4.3). It must therefore lie in $[\mathfrak{I}, \mathcal{B}]$, because $\mathcal{B}$ is sub-bilimit-compact in $\mathcal{C}$.

The other parts are straightforward. □

4.3.3 Minimal Invariants

Now we recapitulate Pitts' theory, generalized in a straightforward manner to bilimit-compact categories.

Definition 4.4 Let G be a locally continuous functor (i.e. a functor which is continuous on homsets) from $\mathcal{C}^{\mathsf{op}} \times \mathcal{C}$ to $\mathcal{C}$. Then an *invariant* for G is an object D together with an isomorphism $i : G(D, D) \cong D$. It is a *minimal invariant* when the least pre-fixed point of the continuous endofunction on $\mathcal{C}(D, D)$ taking e to $i^{-1}; G(e, e); i$ is the identity. □

Proposition 33 Suppose $\mathcal{C}$ is bilimit-compact. Let G be a locally continuous functor from $\mathcal{C}^{\mathsf{op}} \times \mathcal{C}$ to $\mathcal{C}$. Then G has a minimal invariant, and it is unique up to unique isomorphism. □

Proof This is proved as in [Pitts, 1996]. □

Following [Pitts, 1996], we can show that any bilimit-compact category is **Cpo**-enrichedly *algebraically compact* [Freyd, 1991]. But algebraic compactness, although preserved by $-^{\mathsf{op}}$ and by finite product, does not appear to be preserved by countable product. This is why we work with bilimit-compactness.

Minimal invariants are suitable for interpreting *closed* fixpoint types, but when we have nested type recursion, we need a more general theory, again taken from [Pitts, 1996].

Definition 4.5 Suppose $\mathcal{C}$ and $\mathcal{B}$ are **Cpo**-enriched categories and we have a locally continuous functor

$$G : \mathcal{B}^{\mathsf{op}} \times \mathcal{B} \times \mathcal{C}^{\mathsf{op}} \times \mathcal{C} \longrightarrow \mathcal{C}$$

Then a *parametrized invariant* for G is a locally continuous functor

$$F : \mathcal{B}^{\mathsf{op}} \times \mathcal{B} \qquad\qquad \longrightarrow \mathcal{C}$$

together with an isomorphism

$$i_{D^-,D_+} : G(D^-, D^+, F(D^+, D^-), F(D^-, D^+)) \cong F(D^-, D^+)$$

natural in D^- and D^+. It is a *parametrized minimal invariant* when, for each D^- and D^+, the least pre-fixed point of the continuous endofunction on

$$\mathcal{C}(F(D^+, D^-), F(D^+, D^-)) \times \mathcal{C}(F(D^-, D^+), F(D^-, D^+))$$

taking (e^-, e^+) to

$$((i^{-1}_{D_+,D_-}; G(D^+, D^-, e^+, e^-); i_{D^+,D^-}), (i^{-1}_{D_-,D_+}; G(D^-, D^+, e^-, e^+); i_{D^+,D^-}))$$

is $(\mathsf{id}_{F(D^+,D_-)}, \mathsf{id}_{F(D^-,D^+)})$. □

We have seen that minimal invariants always exist, and the same is true for parametrized minimal invariants.

Proposition 34 Suppose $\mathcal{C}$ is a bilimit-compact (or merely algebraically compact) category, and $\mathcal{B}$ is any **Cpo**-enriched category. Then any locally continuous functor

$$G : \mathcal{B}^{\mathsf{op}} \times \mathcal{B} \times \mathcal{C}^{\mathsf{op}} \times \mathcal{C} \longrightarrow \mathcal{C}$$

has a parametrized minimal invariant, unique up to unique isomorphism. □

We will need to be able to relate minimal invariants computed in different categories. We first recall the following result, sometimes called "Plotkin's axiom".

Proposition 35 Suppose we have a commutative square of continuous functions

$$\begin{array}{ccc} C & \xrightarrow{f} & C \\ {\scriptstyle h}\downarrow & & \downarrow{\scriptstyle h} \\ D & \xrightarrow[g]{} & D \end{array}$$

where C and D are pointed and h is strict. Let a be the least prefixed point of f. Then fa is the least prefixed point of g. □

The result we require is a categorical analogue to Prop. 35.

Proposition 36 Suppose we have a commuting square of locally continuous functors commuting up to natural isomorphism α

$$\begin{array}{ccc} \mathcal{C}^{\mathsf{op}} \times \mathcal{C} & \xrightarrow{F} & \mathcal{C} \\ {\scriptstyle H^{\mathsf{op}} \times H}\downarrow & \cong & \downarrow{\scriptstyle H} \\ \mathcal{D}^{\mathsf{op}} \times \mathcal{D} & \xrightarrow[G]{} & \mathcal{D} \end{array}$$

where $\mathcal{C}$ and $\mathcal{D}$ are bilimit-compact and H is locally *strictly* continuous. Let (D, i) be a minimal invariant for F. Then $(HD, (\alpha_{D\,D}; Hi))$ is a minimal invariant for G. □

This is proved using Prop. 35. Prop. 36 can be generalized to parametrized minimal invariants.

4.3.4 Interpreting Recursive Types

We know from the previous section that products of **pCpo** and $\mathbf{Cpo}^{\perp}$ are bilimit-compact categories, and we wish to use them to interpret recursive types. To enable this, the cpo-operations that we have used to interpret type constructors must be extended to locally continuous functors between these categories.

Proposition 37
- U extends to a locally continuous functor from $\mathbf{Cpo}^{\perp}$ to **pCpo**, when we set $(Ug)(\mathsf{thunk}\ a)$ to be $\mathsf{thunk}\ (ga)$. (Recall

that the thunk can be ignored here; we write this only so that the metalanguage conforms to CBPV.)

- $\sum_{i \in I}$ extends canonically to a locally continuous functor from $\mathbf{pCpo}^I$ to $\mathbf{pCpo}$, when we set $(\sum_{i \in I} f_i)(\hat{\imath}, a)$ to be $(\hat{\imath}, f_{\hat{\imath}} a)$ if $f_{\hat{\imath}} a$ is defined, and to be undefined otherwise.
- $\times$ extends to a locally continuous functor from $\mathbf{pCpo} \times \mathbf{pCpo}$ to $\mathbf{pCpo}$, when we set $(f \times f')(a, a')$ to be $(fa, f'a')$ if fa and $f'a'$ are both defined, and to be undefined otherwise.
- F extends to a locally continuous functor from $\mathbf{pCpo}$ to $\mathbf{Cpo}^{\perp}$, when we set $(Ff)(\mathsf{return}\ a)$ to be $\mathsf{return}\ (fa)$ if fa is defined, and to be $\perp$ otherwise, and we set $(Ff)\perp$ to be $\perp$.
- $\prod_{i \in I}$ extends to a locally continuous functor from $(\mathbf{Cpo}^{\perp})^I$ to $\mathbf{Cpo}^{\perp}$, when we set $(\prod_{i \in I} g_i)b$ to be $\lambda i.(g_i(i`b))$.
- $\to$ extends to a locally continuous functor from $\mathbf{pCpo}^{\mathsf{op}} \times \mathbf{Cpo}^{\perp}$ to $\mathbf{Cpo}^{\perp}$, when we set $(f \to g)b$ to be

$$\lambda x. \begin{cases} g((fx)`b) & \text{if } fx \text{ is defined} \\ \perp & \text{otherwise.} \end{cases}$$

□

Using Prop. 37 together with Prop. 30 we can interpret

- a value type A with m free value type identifiers and n free value type identifiers by a locally continuous functor

$$(\mathbf{pCpo}^{\mathsf{op}} \times \mathbf{pCpo})^m \times (\mathbf{Cpo}^{\perp\mathsf{op}} \times \mathbf{Cpo}^{\perp})^n \longrightarrow \mathbf{pCpo}$$

- a computation type $\underline{B}$ with m free value type identifiers and n free value type identifiers by a locally continuous functor

$$(\mathbf{pCpo}^{\mathsf{op}} \times \mathbf{pCpo})^m \times (\mathbf{Cpo}^{\perp\mathsf{op}} \times \mathbf{Cpo}^{\perp})^n \longrightarrow \mathbf{Cpo}^{\perp}$$

In particular, a closed value type denotes a cpo and a closed computation type denotes a cppo. We interpret $\mu\mathtt{X}.A$ and $\mu\mathtt{x}.\underline{B}$ as parametrized minimal invariants. Where these types are closed, we have isomorphisms

$$\begin{array}{lcll} [\![\mu\mathtt{X}.A]\!] & \cong & [\![A[\mu\mathtt{X}.A/\mathtt{X}]]\!] & \text{in } \mathbf{Cpo} \\ [\![\mu\underline{\mathtt{X}}.\underline{B}]\!] & \cong & [\![\underline{B}[\mu\underline{\mathtt{X}}.\underline{B}/\underline{\mathtt{X}}]]\!] & \text{in } \mathbf{Cpo}^{\perp} \end{array}$$

We thus obtain a semantics for CBPV with recursive types. It is obviously sound. To prove adequacy, we use Pitts' methods [Pitts, 1996], which work for any bilimit-compact category.

4.4 Infinitely Deep Terms

4.4.1 Syntax

We now want to allow the branches of a term's parse tree to be infinite branches, as in a Böhm tree. However, this is not always acceptable. To see this, consider the following infinitely deep terms:

1

$$\dfrac{\dfrac{\dfrac{\dfrac{\vdots}{\vdash^{\mathsf{v}} \texttt{thunk force thunk} \ldots : U\underline{B}}}{\vdash^{\mathsf{c}} \texttt{force thunk force thunk} \ldots : \underline{B}}}{\vdash^{\mathsf{v}} \texttt{thunk force thunk force thunk} \ldots : U\underline{B}}}{\vdash^{\mathsf{c}} \texttt{force thunk force thunk force thunk} \ldots : \underline{B}}$$

This is acceptable. It is a divergent computation and hence should denote $\bot$.

2

$$\dfrac{\vdash^{\mathsf{v}} \texttt{true} : \texttt{bool} \qquad \dfrac{\vdash^{\mathsf{v}} \texttt{true} : \texttt{bool} \qquad \dfrac{\vdots}{\texttt{x}_0 : \texttt{bool}, \texttt{x}_1 : \texttt{bool} \vdash^{\mathsf{c}} \texttt{let true be x}_2.\ \ldots : \underline{B}}}{\texttt{x}_0 : \texttt{bool} \vdash^{\mathsf{c}} \texttt{let true be x}_1.\ \texttt{let true be x}_2.\ \ldots : \underline{B}}}{\vdash^{\mathsf{c}} \texttt{let true be x}_0.\ \texttt{let true be x}_1.\ \texttt{let true be x}_2.\ \ldots : \underline{B}}$$

This is acceptable. It is a divergent computation and hence should denote $\bot$.

3

$$\dfrac{\vdash^{\mathsf{v}} \texttt{true} : \texttt{bool} \qquad \dfrac{\vdash^{\mathsf{v}} \texttt{true} : \texttt{bool} \qquad \dfrac{\vdots}{\vdash^{\mathsf{v}} \texttt{fold (true,} \ldots\texttt{)} : \mu X.(\texttt{bool} \times X)}}{\vdash^{\mathsf{v}} \texttt{fold (true,fold (true,} \ldots\texttt{))} : \mu X.(\texttt{bool} \times X)}}{\vdash^{\mathsf{v}} \texttt{fold (true,fold (true,fold (true,} \ldots\texttt{)))} : \mu X.(\texttt{bool} \times X)}$$

This is not acceptable. It should denote an element of $[\![\mu \texttt{X}.(\texttt{bool} \times \texttt{X})]\!]$—but this is the empty cpo.

Each of these trees has a single infinite branch. We need a condition on branches that is satisfied by (1)–(2) but not by (3). The correct requirement is as follows.

Definition 4.6 An infinitely deep parse-tree for a term is *acceptable* when it has no infinite branch that eventually (i.e. from some point upwards) consists only of value-forming rules. □

We stress that a branch in an acceptable tree *is* allowed to contain infinitely many value-forming rules—this happens in (1).

The branch condition in Def. 4.6 appears inelegant. A more abstract characterization using induction and coinduction is given in Sect. 4.6.

For stack terms, we classify stack-forming rules together with value-forming rules, so we do not allow to an infinite branch to eventually consist of value-forming and stack-forming rules. The reason for this is not apparent at this stage, but when we study control effects in Sect. 5.4.2, we treat a stack as a special kind of value.

4.4.2 Partial Terms and Scott Semantics

Now that we have defined CBPV with infinitely deep terms, we want to give describe Scott semantics for it, extending the Scott semantics for finitely deep CBPV. To do this we add to the syntax an additional rule

$$\frac{}{\Gamma \vdash^{\mathsf{c}} \bot : \underline{B}}$$

This larger set of terms that may use $\bot$ are called *partial terms*, whilst terms that do not use $\bot$ are called *total terms*. The reader may wonder what the difference is between $\bot$ and `diverge`, so we explain this as follows. An infinitely deep term can be implemented as lazy tree, essentially a program that repeatedly answers questions about its syntax. For example, the computation λ`x.diverge` can be asked "what is your outermost term-constructor?" and replies "λx". Next, it can be asked "and what term-constructor follows that?" and it replies "`diverge`".

By contrast, the computation λx.$\bot$, when asked the second question, diverges i.e. never replies. That is why a term without $\bot$ is called "total": it always replies to queries about its syntax. The operational behaviour of $\bot$ is to diverge, but this is not because there is a transition that ensures this, as is the case for `diverge`. Rather, the CK-machine must ask the current computation for its outermost term-constructor in order to know which transition to apply. If the current computation is $\bot$, the machine never receives a reply, so execution diverges.

Definition 4.7 We say that $M \lhd_{\mathsf{fin}} N$ ("M is a finite approximant of N") when M is finitely deep and N is obtained by replacing every occurrence of $\bot$ in M by a computation. More abstractly, we can define $\lhd_{\mathsf{fin}}$ to be the least binary relation between terms such that $\lhd_{\mathsf{fin}}$ is compatible (i.e. closed under all term constructors) and $\bot \lhd_{\mathsf{fin}} N$ for all N. □

For every finitely deep term M we define its interpretation $[\![M]\!]^{\mathsf{fin}}$ in the usual way, setting $[\![\bot]\!]\rho$ to be $\bot$. Then, for every term M, we define its interpretation $[\![M]\!]^{\mathsf{inf}}$ to be $\bigsqcup_{M \lhd_{\mathsf{fin}} N}[\![M]\!]$. It is clear that the set $\{[\![M]\!] : M \lhd_{\mathsf{fin}} N\}$ is directed—the acceptability condition on branches ensures its nonemptiness—so the join must exist. Furthermore, if M is finitely deep then $[\![M]\!]^{\mathsf{inf}} = [\![M]\!]^{\mathsf{fin}}$, so $[\![-]\!]^{\mathsf{inf}}$ is an extension of $[\![-]\!]^{\mathsf{fin}}$.

Finally, we need to adapt the adequacy proof in Sect. 2.8 to include infinitely deep terms. We define the various relations and prove their admissibility just as we did there. We then prove that for any computation $A_0, \ldots, A_{n-1} \vdash^{\mathsf{c}} M \lhd_{\mathsf{fin}} N : \underline{B}$, if $a_i \leqslant^{\mathsf{v}}_{A_i} W_i$ for $i = 0, \ldots, n-1$ then $[\![M]\!]\overrightarrow{\mathtt{x}_i \mapsto a_i} \leqslant^{\mathsf{c}}_{\underline{B}} N[\overrightarrow{W_i/\mathtt{x}_i}]$; and similarly for any values $A_0, \ldots, A_{n-1} \vdash^{\mathsf{v}} U \lhd_{\mathsf{fin}} V : A$. This is shown by mutual induction on M and U. By admissibility, we see that $[\![M]\!] \leqslant^{\mathsf{c}}_{\underline{B}} M$ for any closed computation M, and this gives the desired result.

4.5 Infinitely Deep Types

4.5.1 Syntax

Having extended CBPV with infinitely deep terms, we proceed to allow the parse-tree of a *type* to have infinite branches. Fortunately here there is no need for restrictions on branches: any parse tree with finite or infinite branches is acceptable.

For example, we have a value type

$$\mathtt{bool} \times (\mathtt{bool} \times (\mathtt{bool} \times \cdots))$$

This is the unwinding of $\mu\mathtt{X}.(\mathtt{bool} \times \mathtt{X})$. There is no closed value of this type because a putative term such as $(\mathtt{true}, (\mathtt{true}, (\mathtt{true}, \ldots)))$ would violate the branch restriction on terms.

4.5.2 Partial Types and Scott Semantics

Just as in Sect. 4.4 we extended the semantics for recursion to a semantics for infinitely deep terms, so we can extend the cpo semantics for type recursion to a semantics for infinitely deep types. We need to introduce partial types, and we do this by adding a value type $\bot$ and a computation type $\underline{\bot}$. We define $A \lhd_{\mathsf{fin}} B$ just as we did for types.

We first define $[\![-]\!]^{\mathsf{fin}}$, the semantics of finitely deep types. For this, we interpret $\bot$ the same way as 0, viz. by the empty cpo, and we interpret $\underline{\bot}$ the same way as $\underline{1}$ (the nullary $\prod$ type), viz. by the 1-element cppo. Then we define $[\![B]\!]^{\mathsf{inf}}$ to be the bilimit of $[\![A]\!]^{\mathsf{fin}}$ over all $A \lhd_{\mathsf{fin}} B$, using the fact that the set of finite approximants of B is directed and countable. Soundness is trivial and adequacy is proved as for recursive types.

We obtain isomorphisms such as

$$\begin{array}{rcll} [\![A]\!]^{\mathsf{inf}} & \cong & [\![A]\!]^{\mathsf{fin}} & (A \text{ finitely deep}) \\ [\![\underline{B}]\!]^{\mathsf{inf}} & \cong & [\![\underline{B}]\!]^{\mathsf{fin}} & (\underline{B} \text{ finitely deep}) \\ [\![A \times A']\!]^{\mathsf{inf}} & \cong & [\![A]\!]^{\mathsf{inf}} \times [\![A']\!]^{\mathsf{inf}} & \\ [\![U\underline{B}]\!]^{\mathsf{inf}} & \cong & U[\![\underline{B}]\!]^{\mathsf{inf}} & \end{array}$$

It would be desirable for these isomorphisms to be identities, but in the cpo model they are not. If it were possible to make them into identities, then it would also be possible to make the isomorphisms

$$\begin{array}{rcl} [\![\mu \mathtt{X}.A]\!]^{\mathsf{inf}} & \cong & [\![A[\mu \mathtt{X}.A/\mathtt{X}]]\!]^{\mathsf{inf}} \\ [\![\mu \underline{X}.\underline{B}]\!]^{\mathsf{fin}} & \cong & [\![\underline{B}[\mu \underline{X}.\underline{B}/\mathtt{x}]]\!]^{\mathsf{inf}} \end{array}$$

into identities, and this is clearly not possible in the cpo model because of the Axiom of Foundation. By contrast, in models such as information systems [Scott, 1982], games [McCusker, 1996] and non-well-founded cpos (replacing the Axiom of Foundation by the Anti-Foundation Axiom [Aczel, 1988]) all these isomorphisms can be made into identities—we omit details.

4.6 The Inductive/Coinductive Formulation of Infinitely Deep Syntax

The branch condition on infinitely deep terms appears inelegant. In this section (which the reader may omit as it is not used in the remainder of the book) we give a more abstract characterization of infinitely deep syntax. We write

- valtypes for the set of value types
- comptypes for the set of computation types
- contexts for the set of contexts
- valterms for the object in the category $\mathbf{Set}^{\mathsf{contexts}\times\mathsf{valtypes}}$ such that $\mathsf{valterms}_{\Gamma,A}$ is the set of values $\Gamma \vdash^{\mathsf{v}} V : A$
- compterms for the object in the category $\mathbf{Set}^{\mathsf{contexts}\times\mathsf{comptypes}}$ such that $\mathsf{compterms}_{\Gamma,\underline{B}}$ is the set of computations $\Gamma \vdash^{\mathsf{c}} M : \underline{B}$.

We discuss only the case of countable tag sets, but an analogous discussion applies to finite tag sets.

To see the issues, consider first the definition of (infinitely wide) CBPV types:

$$\begin{array}{rcl} A & ::= & U\underline{B} \mid \sum_{i \in I} A_i \mid 1 \mid A \times A \\ \underline{B} & ::= & FA \mid \prod_{i \in I} \underline{B}_i \mid A \to \underline{B} \end{array}$$

where I must be countable. This defines an endofunctor on **Set** × **Set** given by:

$$(X,Y) \mapsto (Y + \sum_{I \text{ countable}} X^I + 1 + X^2, X + \sum_{I \text{ countable}} Y^I + XY)$$

The reader may worry that taking a sum over all countable sets is problematic. But there are two straightforward solutions.

- We can take the sum over all countable *small* sets, relative to a Grothendieck universe.
- Less drastically, we can take the sum over all initial segments of $\mathbb{N}$. This requires us to fix bijections that allow us to regard the set of such segments as closed under finite product and countable sum (indexed over an initial segment of $\mathbb{N}$).

Thus size is not a serious problem, and we will not address it further.

For finitely deep syntax, the definition of types is inductive i.e. the pair (valtypes, comptypes) is the carrier of an initial algebra for this endofunctor. When we allow infinitely deep types, the definition of types becomes *coinductive* i.e. (valtypes, comptypes) is the carrier of a terminal coalgebra for this endofunctor.

The definition of CBPV terms given in Fig. 2.1 similarly describes an endofunctor[2] on $\mathbf{Set}^{\mathsf{contexts}\times\mathsf{valtypes}} \times \mathbf{Set}^{\mathsf{contexts}\times\mathsf{comptypes}}$. It must be of the form (F, G) where

- $F : \mathbf{Set}^{\mathsf{contexts}\times\mathsf{valtypes}} \times \mathbf{Set}^{\mathsf{contexts}\times\mathsf{comptypes}} \longrightarrow \mathbf{Set}^{\mathsf{contexts}\times\mathsf{valtypes}}$
- $G : \mathbf{Set}^{\mathsf{contexts}\times\mathsf{valtypes}} \times \mathbf{Set}^{\mathsf{contexts}\times\mathsf{comptypes}} \longrightarrow \mathbf{Set}^{\mathsf{contexts}\times\mathsf{comptypes}}$

but we will not write F and G explicitly.

We see that (valterms, compterms) is the carrier of an initial algebra for this endofunctor. But when we allow infinitely deep terms, it is not the case that (valterms, compterms) is the carrier of a terminal coalgebra for this endofunctor, because of the branch condition. However, (valterms, compterms) is certainly a fixpoint of the endofunctor in the sense that

$$\begin{aligned} \mathsf{valterms} &\cong F(\mathsf{valterms}, \mathsf{compterms}) \\ \mathsf{compterms} &\cong G(\mathsf{valterms}, \mathsf{compterms}) \end{aligned}$$

How can we characterize this fixpoint? Clearly we require some kind of mixed inductive/coinductive definition: values are inductively defined,

[2]We gloss over issues of identifier binding. See [Altenkirch and Reus, 1999; Fiore et al., 1999] for a fuller discussion.

computations are coinductively defined. But the coinductive part must be (loosely speaking) on the outside, because otherwise we will exclude infinite branches that contain infinitely many value-forming rules, and we have already seen that this is too restrictive.

To express this, we write $\mu\mathtt{X}$ for initial algebra and $\nu\mathtt{X}$ for terminal coalgebra.

Proposition 38 We have the following description of infinitely deep terms:

$$\begin{aligned}\mathsf{compterms} &\cong \nu\mathtt{Y}.G(\mu\mathtt{X}.F(\mathtt{X},\mathtt{Y}),\mathtt{Y})\\ \mathsf{valterms} &\cong \mu\mathtt{X}.F(\mathtt{X},\mathsf{compterms})\end{aligned}$$

□

This is proved by giving an explicit representation of the function

$$\mathtt{Y} \mapsto \mu\mathtt{X}F(\mathtt{X},\mathtt{Y})$$

We omit details.

4.7 Relationship Between Recursion and Infinitely Deep Syntax

We said in Sect. 4.1 that recursion and infinitely deep syntax are, in a sense, equivalent. We now explain this sense, in a very informal way, using examples.

Any recursion can be unwound infinitely often. For example, if we take the recursive type $\mu\mathtt{X}.(1+\mathtt{X})$, this can be unwound to give $1+\mu\mathtt{X}.(1+\mathtt{X})$, then unwound again to give $1+(1+\mu\mathtt{x}.(1+\mathtt{X}))$, and so forth. Ultimately, we obtain the infinitely deep type $1+(1+(1+\cdots))$. Thus if we have a semantics for infinitely deep syntax, it provides a semantics for recursion too.

On the other hand, if we have an infinitely deep term or type, it can be expressed using a countable collection of simultaneously recursive definitions. For example, take the infinitely deep type $A = 0+(1+(2+(\cdots)))$. We name the subexpressions of this as follows:

$$\begin{aligned}A_0 &= 0+(1+(2+\cdots))\\ A_1 &= 1+(2+(3+\cdots))\\ A_2 &= 2+(3+(4+\cdots))\end{aligned}$$

Now we can define these types by mutual recursion:

$$\begin{aligned}A_0 &= 0+A_1\\ A_1 &= 1+A_2\\ A_2 &= 2+A_3\end{aligned}$$

Thus any semantics for recursion that can interpret countable simultaneous recursions gives us a semantics for infinitely deep syntax.

Sometimes it is easier to work with recursion, sometimes with infinitely deep syntax. One advantage of the former is that, by allowing only finitely deep syntax, we have a clear notion of *compositional semantics*. Finitely deep syntax is an initial algebra for an endofunctor, and a compositional semantics is one specified by another algebra for this endofunctor—the interpretation is then given by the unique algebra homomorphism from the syntax to this algebra [Goguen et al., 1979]. By contrast, it is not clear in what sense the semantics for infinitely deep CBPV that we have considered can be said to be compositional.

4.8 Predomain Semantics For CBPV

4.8.1 Domains and Predomains

So far, we have seen how to interpret CBPV in cpos and cppos. But frequently one wishes to restrict attention to a special class of ω-algebraic (or at least ω-continuous) cpos called "predomains". A predomain with a least element is called a "domain".

There are various classes of predomain one could work with, but they share the following properties.

- They all include the flat cpo $\mathbb{N}$.
- They all exclude the uncountable flat cpo $\mathbb{N} \to \mathbb{N}$, because it is not ω-continuous.
- Consequently, they are not cartesian closed categories.

Some languages considered in the literature (e.g. the monadic metalanguage of [Benton and Wadler, 1996]) allow a $\to$ operation on value types. This has two disadvantages, as far as cpos are concerned.

- It rules out a predomain semantics, because the type `nat` $\to$ `nat` cannot be interpreted in predomains.
- It rules out general type recursion, if a cpo semantics is desired, because there is no cpo A such that $A \cong A \to 2$.

We have seen that CBPV, by contrast, allows general type recursion, and we shall soon see that it has a predomain semantics.

Definition 4.8 1 A cppo $\underline{A}$ is a (Scott-Ershov) *domain* when it is ω-algebraic and consistently complete i.e. every consistent (upper-bounded) finite subset has a join. We write **DomStr** for the full subcategory of $\mathbf{Cpo}^{\perp}$ on domains.

2 A cpo A is a (Scott-Ershov/Abramsky-McCusker) *predomain* when it is a countable disjoint union of domains. (This implies A is ω-algebraic.) We write **Predom** for the full subcategory of **Cpo** on predomains.

□

It is clear that a domain is precisely a pointed predomain.

When working with a predomain, one generally wants to consider the poset of compact elements, as explained in [Plotkin, 1983; Stoltenberg-Hansen et al., 1994].

Definition 4.9 1 A *countable conditional upper semilattice* or *ω-cusl* is a consistently complete, countable poset $(A, \leqslant)$ with a least element $\bot$.

2 A poset is an *ω-multicusl*[3] when it is a countable disjoint union of ω-cusls.

□

We have the following constructions, inverse up to poset isomorphism.

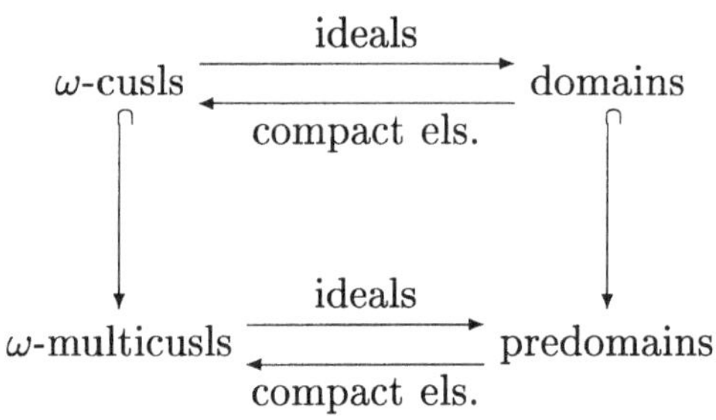

4.8.2 Interpreting Type Recursion In Predomains and Domains

We would like to show that **Predom** is sub-bilimit-compact within some appropriate supercategory. The definition of this supercategory, like our definition of predomain, is based on [Abramsky and McCusker, 1998].

Definition 4.10 A *partial-on-minimals* function from predomain A to predomain B is a partial continuous function such that f is defined on x iff f is defined on the least element below x. We write **Partmin** for the category of predomains and partial-on-minimals functions. □

[3]The term "pre-cusl" has already been used.

Following [Stoltenberg-Hansen et al., 1994], we can characterize in a simple way the (e, p)-pairs in **Partmin** and in **DomStr**.

Proposition 39 1 Let A and B be the ω-multicusls of compact elements of predomains X and Y respectively. Let i be the an injection i from A to B that preserves and reflects $\leqslant$, minimality, inconsistency and joins. Then i determines an (e, p)-pair in **Partmin** from X to Y, where

$$\begin{aligned}
ex &= \bigvee_{a \in A, a \leqslant x} i(a) \\
py &= \begin{cases} \bigvee_{a \in A, i(a) \leqslant y} a & \text{if } \{a \in A | i(a) \leqslant y\} \text{ is non-empty} \\ \text{undefined} & \text{otherwise} \end{cases} \\
\text{so } e(py) &= \begin{cases} \bigvee_{a \in A, i(a) \leqslant y} i(a) & \text{if } \{a \in A | i(a) \leqslant y\} \text{ is non-empty} \\ \text{undefined} & \text{otherwise} \end{cases}
\end{aligned}$$

Furthermore, every (e, p) pair in **Partmin** is obtained in this way, setting i to be the restriction of e to compact elements.

2 Let A and B be the ω-cusls of compact elements of domains $\underline{X}$ and $\underline{Y}$. An injection i from A to B that preserves and reflects $\leqslant$, $\perp$, inconsistency and joins determines an (e, p)-pair in **DomStr** from $\underline{X}$ to $\underline{Y}$, by the same formulas as in (1). Furthermore, every (e, p) pair in **DomStr** is obtained in this way, setting i to be the restriction of e to compact elements.

□

Proposition 40 1 **DomStr** is bilimit-compact.

2 **Predom** is sub-bilimit-compact in **Partmin**.

□

Proof (2) Given a countable directed diagram D of (e, p)-pairs, we use Prop. 39(1) to obtain a diagram of ω-multicusls Ad and injections $Ad \xrightarrow{i_{dd'}} Ad'$ and construct the colimit of this as an ω-multicusl C with injections $Ad \xrightarrow{i_d} C$. (C is constructed as a quotient of the disjoint union of the posets $\{A_d\}_{d \in \mathbb{D}}$, just like colimits in **Set**.) We let V be the predomain corresponding to C. To prove (4.2), it is sufficient to prove it for compact elements of V. If c is such a compact element (i.e. $c \in C$), then, by the construction of C, c is in the range of i_d for sufficiently large d. For such d, the explicit description in Prop. 39(1) shows that $p_d; e_d$ takes a to a. Hence $\bigvee_{d \in \mathbb{D}}(p_d; e_d)$ takes a to a. (1) is proved similarly. □

Of course, Prop. 40 can also be proved directly, as done in [Smyth and Plotkin, 1982], but working with compact elements makes the construction of the bilimit more intelligible.

We want to interpret infinitely deep CBPV in predomains and domains, making sure that this agrees with the interpretation in cpos and cppos. We first note that the embedding of **DomStr** in $\mathbf{Cpo}^{\perp}$ and the embedding of **Partmin** in **pCpo** are locally strictly continuous. So Prop. 36 is applicable.

We then must show that all the functors described in Prop. 37 do indeed restrict from **pCpo** and $\mathbf{Cpo}^{\perp}$ to **Partmin** and **DomStr**. This is straightforward.

II

CONCRETE SEMANTICS

Chapter 5

SIMPLE MODELS OF CBPV

This chapter does not depend on Chap. 3 or Chap. 4. However, all the models in this chapter validate the equations of Chap. 3, and can interpret complex values.

5.1 Introduction

The goal of Part II is to advance the following claim.

> Whenever we are studying effectful higher-order languages (and remember that mere divergence makes a language "effectful"), the semantic primitives are given by CBPV. It is therefore a good choice for our language of study.

While this is certainly a radical claim, we assemble, throughout Part II, extensive evidence for it. In a wide range of fields, we see firstly that CBPV semantics is simpler than the traditional CBV and CBN semantics, and secondly that these traditional semantics can be recovered from the CBPV semantics—so we do not lose out by shifting our focus to CBPV.

Admittedly, we encounter exceptions, where a CBV model does not decompose naturally into CBPV. These exceptions are

- the model for erratic choice with the constraint that choice must be finite (Sect. 5.5.3)
- the possible worlds model for cell generation with the constraint of "parametricity in initializations" (Sect. 6.9)
- the model for input based on Moggi's input monad [Moggi, 1991].

The first two are constrained variants of a simpler CBV model that *does* decompose into a CBPV model, so we do not consider them to be a major objection to our claim. A more optimistic position regarding the finite choice and input effects is argued in Sect. 12.2.

We have already seen the decomposition into CBPV for printing and divergence, as well as for operational semantics, in Chap. 2. In this chapter we look at global store, control effects, erratic choice and errors, and at various combinations of these effects. This range of "simple models" is based on [Moggi, 1991], although we do not treat Moggi's example of "interactive input".

We first look at semantics of values, in Sect. 5.2, as this is common to all the models in the chapter. Then we devote one section to each effect; these sections can be read independently of each other.

As the semantics of values is straightforward, the difficulty lies in the semantics of computations. To invent it, there are two useful heuristics that can be applied. One is to take a known CBV (or CBN) model and look for a decomposition.

- For example, consider the traditional global store interpretation of $A \to_{\mathbf{CBV}} B$ viz. $S \to (\llbracket A \rrbracket \to (S \times \llbracket B \rrbracket))$. We immediately see the decomposition into $U(A \to FB)$; it appears that U will denote $S \to -$, that $\to$ will denote $\to$, and that F will denote $S \times -$. Thus the CBPV type constructors have simpler interpretations than $\to_{\mathbf{CBV}}$.

- As another example, consider the traditional continuation semantics for $A \to_{\mathbf{CBV}} B$ viz. $(\llbracket A \rrbracket \times (\llbracket B \rrbracket \to \mathsf{Ans})) \to \mathsf{Ans}$. We immediately see the decomposition into $U(A \to FB)$; it appears that U and F will both denote $- \to \mathsf{Ans}$, and $\to$ will denote $\times$. Again the CBPV type constructors have simpler interpretations than $\to_{\mathbf{CBV}}$.

- For both of these examples, there are corresponding CBN semantics presented in the literature: $\underline{A} \to_{\mathbf{CBN}} \underline{B}$ is interpreted in [O'Hearn, 1993] as $(S \to \llbracket \underline{A} \rrbracket) \to \llbracket \underline{B} \rrbracket$ (for global store) and in [Streicher and Reus, 1998] as $(\llbracket \underline{A} \rrbracket \to \mathsf{Ans}) \times \llbracket \underline{B} \rrbracket$. Each of these makes apparent the decomposition of $\underline{A} \to_{\mathbf{CBN}} \underline{B}$ into $(U\underline{A}) \to \underline{B}$. Again, the semantics of the CBPV type constructors are simpler. In fact the CBN semantics looks quite strange; CBPV thus provides a rational reconstruction for it.

The other heuristic, which we shall use in this chapter, is to guess in advance the form of the soundness theorem and proceed from there.

5.2 Semantics of Values

We describe the semantics of values at the outset, because it is straightforward and it is the same across the different models in the chapter (except for the new model of Sect. 5.5.4). We will consider models using sets and models using cpos. In all of the set models,

- a value type denotes a set
- in particular, $\sum_{i \in I} A_i$ denotes the set $\sum_{i \in I} [\![A_i]\!]$ and $A \times A'$ denotes the set $[\![A]\!] \times [\![A']\!]$
- a context $A_0, \ldots, A_{n-1}$ denotes the set $[\![A_0]\!] \times \cdots \times [\![A_{n-1}]\!]$
- a value $\Gamma \vdash^{\mathsf{v}} V : A$ denotes a function from $[\![\Gamma]\!]$ to $[\![A]\!]$.

Similarly, in the cpo models,

- a value type denotes a cpo;
- in particular, $\sum_{i \in I} A_i$ denotes the cpo $\sum_{i \in I} [\![A_i]\!]$ and $A \times A'$ denotes the cpo $[\![A]\!] \times [\![A']\!]$;
- a context $A_0, \ldots, A_{n-1}$ denotes the cpo $[\![A_0]\!] \times \cdots \times [\![A_{n-1}]\!]$;
- a value $\Gamma \vdash^{\mathsf{v}} V : A$ denotes a continuous function from $[\![\Gamma]\!]$ to $[\![A]\!]$.

5.3 Global Store

5.3.1 The Language

We take the simplest possible case of global store: we suppose there is just one cell `cell` that stores an element of the countable set S. If we want the syntax to be finitary, we restrict to the case that S is finite.

We thus add to the basic CBPV language constructs for assignment and reading:

$$\frac{\Gamma \vdash^{\mathsf{c}} M : \underline{B}}{\Gamma \vdash^{\mathsf{c}} \texttt{cell} := s.\ M : \underline{B}}\ (s \in S) \qquad \frac{\cdots \quad \Gamma \vdash M_s : A \quad \cdots_{\ s \in S}}{\Gamma \vdash^{\mathsf{c}} \texttt{read-cell-as}\ \{\ldots, s.M_s, \ldots\}}$$

(We continue our practice of treating commands as prefixes.)

For big-step semantics, we define a relation of the form $s, M \Downarrow s', T$, where $s, s' \in S$. We often write s as $\mathsf{cell} \mapsto V$, meaning "cell contains the value V". To define $\Downarrow$, we replace each rule in Fig. 2.2 of the form (2.1) by

$$\frac{s_0, M_0 \Downarrow s_1, T_0 \quad \cdots \quad s_{r-1}, M_{r-1} \Downarrow s_r, T_{r-1}}{s_0, M \Downarrow s_r, T}$$

and add the rules

$$\frac{s', M \Downarrow s'', T}{s, (\texttt{cell} := s'.\ M) \Downarrow s'', T} \qquad \frac{s', M_{s'} \Downarrow s'', T}{s', \texttt{read-cell-as}\ \{\ldots, s.M_s, \ldots\} \Downarrow s'', T}$$

Proposition 41 For every $s \in S$ and closed computation M, there exists unique s', T such that $s, M \Downarrow s', T$. □

This is proved similarly to Prop. 10. We can also adapt the CK-machine to this computational effect, but we omit this.

We adapt the definition of observational equivalence.

Definition 5.1 Given two computations $\Gamma \vdash^{\mathsf{c}} M, N : \underline{B}$, we say $M \simeq N$ when for any ground context $\mathcal{C}[\cdot]$ and any $s \in S$, we have

$$s, \mathcal{C}[M] \Downarrow s', \mathtt{return}\ n \text{ iff } s, \mathcal{C}[N] \Downarrow s', \mathtt{return}\ n$$

Similarly for two values $\Gamma \vdash^{\mathsf{v}} V, W : A$. □

5.3.2 Denotational Semantics

Computations

We seek a denotational semantics for the language of Sect. 5.3.1. Since there is no divergence, we use a set model. The semantics of values is given in Sect. 5.2—the difficulty lies in the interpretation of computations.

While logically we should present the semantics first, and then state the soundness theorem, this makes the interpretation of type constructors appear unintuitive. So we will proceed in reverse order. We will state first the soundness theorem that we are aiming to achieve, even though it is not yet meaningful, and use this to motivate the semantics.

We expect the soundness result to look like this:

Proposition 42 (soundness) Let M be a closed computation. If $s, M \Downarrow s', T$ then $[\![M]\!]s = [\![T]\!]s'$. □

In Prop. 42, the denotation of M takes a store as an argument. More generally, we would expect a computation $\Gamma \vdash^{\mathsf{c}} M : \underline{B}$ to take both store $s \in S$ and environment $\rho \in [\![\Gamma]\!]$ as arguments. Thus M should denote a function from $S \times [\![\Gamma]\!]$ to some set—we call this set $[\![\underline{B}]\!]$. Intuitively, $[\![\underline{B}]\!]$ is the set of *behaviours* for a computation of type $\underline{B}$, in a given store and environment.

We can use this idea that a computation type denotes a set of behaviours to motivate the interpretation of the type constructors.

- The behaviour of a computation of type FA is to terminate in a state $s \in S$ returning a value V of type A. So FA denotes $S \times [\![A]\!]$.
- The behaviour of a computation of type $\prod_{i \in I} \underline{B}_i$ is to pop $i \in I$, and, depending on the i popped, to behave as a computation of type $\underline{B}_i$. So $\prod_{i \in I} \underline{B}_i$ denotes $\prod_{i \in I} [\![\underline{B}_i]\!]$.

- The behaviour of a computation of type $A \to \underline{B}$ is to pop a value of type A, and, depending on the value popped, to behave as a computation of type $\underline{B}$. So $A \to \underline{B}$ denotes $[\![A]\!] \to [\![\underline{B}]\!]$.
- A value of type $U\underline{B}$ can be forced in any store $s \in S$, and depending on this store, will behave as a computation of type $\underline{B}$. So $U\underline{B}$ denotes $S \to [\![\underline{B}]\!]$.

Notice in particular that a returner $\Gamma \vdash^{\mathsf{c}} M : FA$ denotes a function from $S \times [\![\Gamma]\!]$ to $S \times [\![A]\!]$. Notice too the semantics of CBV functions:

$$[\![A \to_{\mathbf{CBV}} B]\!] = [\![U(A \to FB)]\!] = S \to ([\![A]\!] \to (S \times [\![B]\!]))$$

As we said in Sect. 5.1, this is the traditional CBV semantics for global store.

The semantics of terms is straightforward. Here are some example clauses:

$$\begin{array}{rcl}
[\![\mathtt{return}\ V]\!](s,\rho) & = & (s,[\![V]\!]\rho) \\
[\![M\ \mathtt{to}\ \mathtt{x}.\ N]\!](s,\rho) & = & [\![N]\!](s',(\rho,\mathtt{x} \mapsto a)) \\
 & & \text{where } [\![M]\!](s,\rho) = (s',a) \\
[\![\mathtt{thunk}\ M]\!]\rho & = & \lambda s.([\![M]\!](s,\rho)) \\
[\![\mathtt{force}\ V]\!](s,\rho) & = & s`([\![V]\!]\rho) \\
[\![\lambda\mathtt{x}.M]\!](s,\rho) & = & \lambda x.([\![M]\!](s,(\rho,\mathtt{x} \mapsto x))) \\
[\![\mathtt{cell} := s'.\ M]\!](s,\rho) & = & [\![M]\!](s',\rho) \\
[\![\mathtt{read\text{-}cell\text{-}as}\ \{\ldots,s.M_s,\ldots\}]\!](s',\rho) & = & [\![M_{s'}]\!](s',\rho)
\end{array}$$

It is easy to prove Prop. 42.

Corollary 43 (by Prop. 41) If M is a ground returner, then $s, M \Downarrow s', \mathtt{return}\ n$ iff $[\![M]\!]s = (s',n)$. Hence terms with the same denotation are observationally equivalent. □

Stacks

As we stated in Sect. 2.6.2, for each of our models we can also give semantics of stacks. In the case of global store, we can see that a stack $\Gamma|\underline{B} \vdash^{\mathsf{k}} K : \underline{C}$, in a given environment $\rho \in [\![\Gamma]\!]$, transforms a behaviour of a computation M of type $\underline{B}$ into a behaviour of a computation of type $\underline{C}$. This transformation is not dependent on the initial store, although M may set the store (say if $\underline{B} = FA$) and the transformation will then depend on this new store.

Thus K will denote a function from $[\![\Gamma]\!] \times [\![\underline{B}]\!]$ to $[\![\underline{C}]\!]$. We omit the interpretation of stack terms, which is straightforward.

5.3.3 Combining Global Store With Other Effects

The model for global store in Sect. 5.3.2 generalizes. If we have any CBPV model, we can obtain from it a model for global store.

As an example, consider the printing model for CBPV. We seek a model for global store together with `print`. The big-step semantics will have the form $s, M \Downarrow m, s', T$—we omit the details, which are straightforward.

The denotational semantics for global store with `print` is organized as follows:

- a value type (and hence a context) denotes a set
- a computation type denotes an $\mathcal{A}$-set
- a value $\Gamma \vdash^{\mathsf{v}} V : A$ denotes a function from $[\![\Gamma]\!]$ to $[\![A]\!]$
- a computation $\Gamma \vdash^{\mathsf{c}} M : \underline{B}$ denotes a function from $S \times [\![\Gamma]\!]$ to the carrier of $[\![\underline{B}]\!]$
- a stack $\Gamma|\underline{B} \vdash^{\mathsf{k}} K : \underline{C}$ denotes a homomorphism over $[\![\Gamma]\!]$ from $[\![\underline{B}]\!]$ to $[\![\underline{C}]\!]$.

We use CBPV as a metalanguage describing the printing model, as explained in Sect. 2.8. For example,we write FA for the free $\mathcal{A}$-set on the set A, and we write $U\underline{B}$ for the carrier of the $\mathcal{A}$-set $\underline{B}$. With this notation, the semantics of types is given by

$$\begin{array}{rclcrcl}
[\![U\underline{B}]\!] & = & U(S \to [\![\underline{B}]\!]) & & [\![FA]\!] & = & F(S \times [\![A]\!]) \\
[\![\sum_{i \in I} A_i]\!] & = & \sum_{i \in I} [\![A_i]\!] & & [\![\prod_{i \in I} \underline{B}_i]\!] & = & \prod_{i \in I} [\![\underline{B}_i]\!] \\
[\![A \times A']\!] & = & [\![A]\!] \times [\![A']\!] & & [\![A \to \underline{B}]\!] & = & [\![A]\!] \to [\![\underline{B}]\!]
\end{array}$$

Semantics of terms: some example clauses

$$\begin{array}{rcl}
[\![\texttt{return}\ V]\!](s,\rho) & = & \mathsf{return}\ (s, [\![V]\!]\rho) \\
[\![M\ \texttt{to}\ \texttt{x}.\ N]\!](s,\rho) & = & [\![M]\!](s,\rho)\ \mathsf{to}\ (s',a). \\
& & \quad [\![N]\!](s',(\rho, \texttt{x} \mapsto a)) \\
[\![\texttt{thunk}\ M]\!]\rho & = & \mathsf{thunk}\ \lambda s.([\![M]\!](s,\rho)) \\
[\![\texttt{force}\ V]\!](s,\rho) & = & s`\mathsf{force}\ ([\![V]\!]\rho) \\
[\![\lambda \texttt{x}.M]\!](s,\rho) & = & \lambda x.([\![M]\!](s,(\rho, \texttt{x} \mapsto x))) \\
[\![[\cdot]\ \texttt{to}\ \texttt{x}.\ N :: K]\!]\rho & = & [\cdot]\ \mathsf{to}\ (s,x). \\
& & \quad [\![N]\!](\rho, \texttt{x} \mapsto x) :: [\![K]\!]\rho \\
[\![\texttt{cell} := s'.\ M]\!](s,\rho) & = & [\![M]\!](s',\rho) \\
[\![\texttt{read-cell-as}\ \{\ldots, s.M_s, \ldots\}]\!](s',\rho) & = & [\![M_{s'}]\!](s',\rho) \\
[\![\texttt{print}\ c.\ M]\!](s,\rho) & = & c * ([\![M]\!](s,\rho))
\end{array}$$

Proposition 44 (soundness) If $s, M \Downarrow m, s', T$ then $[\![M]\!]s = m * ([\![T]\!]s)$. □

Corollary 45 (by the analogue of Prop. 41) If M is a ground returner, then $s, M \Downarrow m, s', \texttt{return}\ n$ iff $[\![M]\!]s = (m, s', n)$. Hence terms with the same denotation are observationally equivalent. □

In a similar way, we can obtain a model for global store with recursion. The semantics of types and terms are as above except that we now understand the metalinguistic U,F etc. as referring to the Scott model rather than the printing model.

5.4 Control Effects

5.4.1 `letstk` and `changestk`

For our explanation of control effects, we will use the CK-machine for non-closed computations as in Fig. 2.4. Big-step semantics is unsuitable for a language with control effects.

We add to CBPV[1] two commands `letstk x` and `changestk` K.

- `letstk x` means "let `x` be the current stack".
- `changestk` K means "change the current stack to K".

Thus we have two additional machine transitions

	`letstk x`. M	K
$\leadsto$	$M[K/\texttt{x}]$	K
	`changestk` K. M	L
$\leadsto$	M	K

We illustrate these constructs with an example program:

```
let  thunk (
        λx.
        changestk x.
        λz.
```

[1]The CBV control operators (as in ML) are translated into CBPV as follows:

`cont` A	as	`stk` FA
`letcc x`. M	as	`letstk x`. M
`throw` M N	as	M `to x`. N `to y`. (`changestk x. return y`)

We have to use the terminology "current stack" rather than "current continuation" (the phrase used in CBV) because in CBPV not every stack is a continuation.

```
        return 3 + z
     ).  be y
( 7‘
  print "hello".
  letstk a.
  ( ( λ u.
      true‘
      a‘
      force y
    ) to v in
  return v+2
  )
) to w.
return w + 5
```

- By the time we reach the line `letstk a`, we have printed `hello` and the current stack consists of an operand 7 together with the `to w` continuation (i.e. the continuation that begins on the line `to w in`), so `a` is bound to this stack.
- By the time we force `y` the current stack consists of operands `a` and `true` and the `to v` continuation.
- When we force `y`, we pop the top operand, which is `a`, and bind `x` to it. We now change the current stack to `a`.
- We pop the top operand, which is 7 (again) and bind `z` to it.
- We return 10 to the current continuation which is the `to w` continuation. Hence we return 15.

Notice how the stack discipline has been completely lost; in the absence of the control effects we would expect `force y` to return a value to the `to v` continuation, but we have used `changestk` to override this.

5.4.2 Typing Control Effects

The way that we type control effects follows [Duba et al., 1991]—we merely adapt it from CBV to CBPV. The idea is that whereas, in Chap. 2, a stack inhabited a special judgement $\vdash^{\mathsf{k}}$, it will now be regarded as a "first-class citizen", i.e. a value. Since it is a value, it must have a value type. So for every computation type $\underline{B}$, we introduce a value type $\mathtt{stk}\ \underline{B}$. A value of this type is stack from $\underline{B}$.

Thus the two classes of types are now given by

$$\begin{array}{rl} A ::= & U\underline{B} \mid \sum_{i\in I} A_i \mid 1 \mid A \times A \mid \mathtt{stk}\ \underline{B} \\ \underline{B} ::= & FA \mid \prod_{i\in I} \underline{B}_i \mid A \to \underline{B} \end{array}$$

and we add the rules for **letstk** and **changestk** in Fig. 5.1. Notice

$$\frac{\Gamma, \mathtt{x} : \mathtt{stk}\ \underline{B} \vdash^{\mathsf{c}} M : \underline{B}}{\Gamma \vdash^{\mathsf{c}} \mathtt{letstk}\ \mathtt{x}.\ M : \underline{B}} \qquad \frac{\Gamma \vdash^{\mathsf{v}} K : \mathtt{stk}\ \underline{B} \quad \Gamma \vdash^{\mathsf{c}} M : \underline{B}}{\Gamma \vdash^{\mathsf{c}} \mathtt{changestk}\ K.\ M : \underline{B}'}$$

Figure 5.1. Typing Rules for Control Operators

that **changestk** K. M can be given any type, like **diverge**. It is a *nonreturning command* in the sense of Sect. 2.9.2.

The language that we have presented so far is called *CBPV + control*. Notice that our transition rule for **letstk** takes us outside this language, because it substitutes a stack K for an identifier x. Thus, if we wish to typecheck configurations, we are going to need a way of incorporating stacks into computations and values. This is what we do in the next section.

The various forms of CBPV+control are shown in Fig. 5.2, which extends Fig. 3.1. We defer the equational theory, and the proof that the extension is computation-unaffecting, until Sect. 7.8.3.

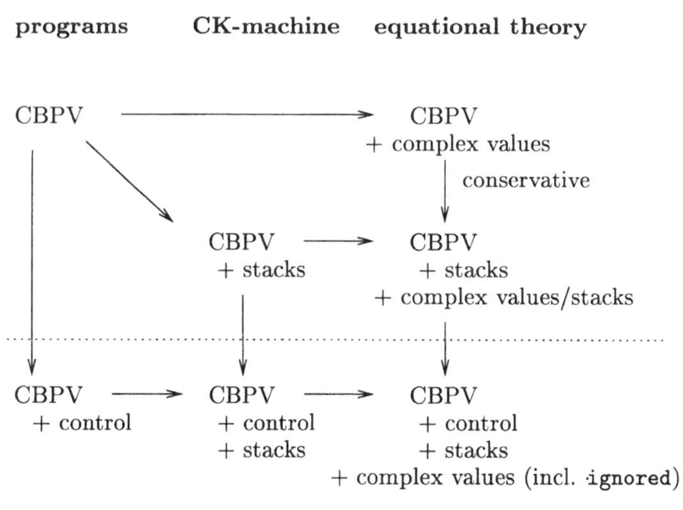

Figure 5.2. Varieties of CBPV with control

5.4.3 Regarding nil As A Free Identifier

Let us consider the following trace.

	(4'λy.return y) to x. return true	Fbool	nil	Fbool
⇝	4'λy.return y	Fnat	[·] to x. return true :: nil	Fbool
⇝	λy.return y	nat → Fnat	4 :: [·] to x. return true :: nil	Fbool
⇝	return 4	Fnat	[·] to x. return true :: nil	Fbool
⇝	return true	Fbool	nil	Fbool

We said in Sect. 5.4.2 that a stack from $\underline{B}$ should be a value of type stk $\underline{B}$. For example, the stack

$$| \ \mathtt{nat} \to F\mathtt{nat} \ \vdash^{\mathsf{k}} \quad 4 :: [\cdot] \ \mathtt{to\ x.\ return\ true} :: \mathtt{nil} \quad : F\mathtt{bool} \tag{5.1}$$

in the above example should be a value of type stk (nat → Fnat). Now this stack contains nil which is a stack from Fbool, and we can substitute for nil any stack from Fbool (that is concatenation). So it makes sense to think of nil as an identifier of type stk Fbool. Thus, when we regard (5.1) as a value, it makes sense to typecheck it as follows.

$$\mathtt{nil} : \mathtt{stk}\ F\mathtt{bool} \vdash^{\mathsf{v}} \quad 4 :: [\cdot] \ \mathtt{to\ x.\ return\ true} :: \mathtt{nil} \quad : \mathtt{stk}\ (\mathtt{nat} \to F\mathtt{nat})$$

More generally, in CBPV + control + stacks, we do not require the $\vdash^{\mathsf{k}}$ judgement, because instead of writing $\Gamma|\underline{B} \vdash^{\mathsf{k}} K : \underline{C}$ we can write

$$\Gamma, \mathtt{nil} : \mathtt{stk}\ \underline{C} \vdash^{\mathsf{v}} K : \mathtt{stk}\ \underline{B}$$

Notice the swapping of $\underline{B}$ and $\underline{C}$ around the $\vdash$. It will lead us to an important duality in Sect. 5.7.

A configuration of the CK-machine will have the form

$$\Gamma \qquad\qquad M \qquad\qquad \underline{B} \qquad\qquad K$$

where $\Gamma \vdash^{\mathsf{c}} M : \underline{B}$ and $\Gamma \vdash^{\mathsf{v}} K : \mathtt{stk}\ \underline{B}$. We call this a *configuration in context* Γ. There is no need for a type $\underline{C}$ on the right, because nil : stk $\underline{C}$ will be included in Γ. Recall from Sect. I.4.2 that, given a term P in context Γ, we write ${}^{\mathtt{nil}}P$ to indicate the weakened term in context Γ, nil.

Looking at Fig. 5.3, we can immediately write down the typing rules for stacks, presented in Fig. 5.4. We have added the symbol · to the stack-forming constructs, which can probably by ignored by all but the most pedantic readers. The difference between :: and ·:: is that :: contains an implicit weakening by nil. This is spelt out in Fig. 5.5.

All the results of Sect. 2.3.4 adapt. In particular we have

Proposition 46 (determinism) (cf. Prop. 11) For every configuration M, K in context Γ, precisely one of the following holds.

1 M, K is not terminal, and $M, K \leadsto N, L$ for a unique configuration N, L in context Γ.

Initial Configuration To Evaluate $\Gamma \vdash^{\mathsf{c}} M : \underline{C}$

$\Gamma, \mathtt{nil} : \mathtt{stk}\ \underline{C}$	${}^{\mathtt{nil}}M$	$\underline{C}$	$\mathtt{nil}$

Transitions

	Γ	$\mathtt{letstk\ x}.\ M$	$\underline{B}$	K
$\rightsquigarrow$	Γ	$M[K/\mathtt{x}]$	$\underline{B}$	K
	Γ	$\mathtt{changestk}\ K.\ M$	$\underline{B}'$	L
$\rightsquigarrow$	Γ	M	$\underline{B}$	K
	Γ	$\mathtt{let}\ V\ \mathtt{be\ x}.\ M$	$\underline{B}$	K
$\rightsquigarrow$	Γ	$M[V/\mathtt{x}]$	$\underline{B}$	K
	Γ	$M\ \mathtt{to\ x}.\ N$	$\underline{B}$	K
$\rightsquigarrow$	Γ	M	FA	$[\cdot]\ \mathtt{to\ x}.\ N :: K$
	Γ	$\mathtt{return}\ V$	FA	$[\cdot]\ \mathtt{to\ x}.\ N :: K$
$\rightsquigarrow$	Γ	$N[V/\mathtt{x}]$	$\underline{B}$	K
	Γ	$\mathtt{force\ thunk}\ M$	$\underline{B}$	K
$\rightsquigarrow$	Γ	M	$\underline{B}$	K
	Γ	$\mathtt{pm}\ (\hat{\imath}, V)\ \mathtt{as}\ \{\ldots, (i, \mathtt{x}).M_i, \ldots\}$	$\underline{B}$	K
$\rightsquigarrow$	Γ	$M_{\hat{\imath}}[V/\mathtt{x}]$	$\underline{B}$	K
	Γ	$\mathtt{pm}\ (V, V')\ \mathtt{as}\ (\mathtt{x}, \mathtt{y}).M$	$\underline{B}$	K
$\rightsquigarrow$	Γ	$M[V/\mathtt{x}, V'/\mathtt{y}]$	$\underline{B}$	K
	Γ	$\hat{\imath}`M$	$\underline{B}_{\hat{\imath}}$	K
$\rightsquigarrow$	Γ	M	$\prod_{i \in I} \underline{B}_i$	$\hat{\imath} :: K$
	Γ	$\lambda\{\ldots, i.M_i, \ldots\}$	$\prod_{i \in I} \underline{B}_i$	$\hat{\imath} :: K$
$\rightsquigarrow$	Γ	$M_{\hat{\imath}}$	$\underline{B}_{\hat{\imath}}$	K
	Γ	$V`M$	$\underline{B}$	K
$\rightsquigarrow$	Γ	M	$A \to \underline{B}$	$V :: K$
	Γ	$\lambda \mathtt{x}.M$	$A \to \underline{B}$	$V :: K$
$\rightsquigarrow$	Γ	$M[V/\mathtt{x}]$	$\underline{B}$	K

Terminal Configurations

$\Gamma, \mathtt{z} : \mathtt{stk}\ FA, \Gamma'$	$\mathtt{return}\ V$	FA	$\mathtt{z}$
$\Gamma, \mathtt{z} : \mathtt{stk} \prod_{i \in I} \underline{B}_i, \Gamma'$	$\lambda\{\ldots, i.M_i, \ldots\}$	$\prod_{i \in I} \underline{B}_i$	$\mathtt{z}$
$\Gamma, \mathtt{z} : \mathtt{stk}\ (A \to \underline{B}), \Gamma'$	$\lambda \mathtt{x}.M$	$A \to \underline{B}$	$\mathtt{z}$
$\Gamma, \mathtt{z} : U\underline{B}, \Gamma'$	$\mathtt{force\ z}$	$\underline{B}$	K
$\Gamma, \mathtt{z} : \sum_{i \in I} A_i, \Gamma'$	$\mathtt{pm\ z\ as}\ \{\ldots, i.M_i, \ldots\}$	$\underline{B}$	K
$\Gamma, \mathtt{z} : A \times A', \Gamma'$	$\mathtt{pm\ z\ as}\ (\mathtt{x}, \mathtt{y}).M$	$\underline{B}$	K

Figure 5.3. CK-Machine With Control

$$\frac{\Gamma, \mathtt{x} : A \vdash^{\mathsf{c}} M : \underline{B} \quad \Gamma \vdash^{\mathsf{v}} K : \mathtt{stk}\ \underline{B}}{\Gamma \vdash^{\mathsf{v}} [\cdot]\ \mathtt{to}\ \mathtt{x}.\ M \cdot\!:: K : \mathtt{stk}\ FA}$$

$$\frac{\Gamma \vdash^{\mathsf{v}} K : \mathtt{stk}\ \underline{B}_{\hat{\imath}}}{\Gamma \vdash^{\mathsf{v}} \hat{\imath} \cdot\!:: K : \mathtt{stk}\ \prod_{i \in I} \underline{B}_i} \qquad \frac{\Gamma \vdash^{\mathsf{v}} V : A \quad \Gamma \vdash^{\mathsf{v}} K : \mathtt{stk}\ \underline{B}}{\Gamma \vdash^{\mathsf{v}} V \cdot\!:: K : \mathtt{stk}\ (A \to \underline{B})}$$

Figure 5.4. Typing Stacks in CBPV + Control

CBPV	CBPV + control
stack $\Gamma \vert \underline{B} \vdash^{\mathsf{k}} K : \underline{C}$	$\Gamma, \mathtt{nil} : \mathtt{stk}\ \underline{C} \vdash^{\mathsf{v}} K : \mathtt{stk}\ \underline{B}$
configuration $\Gamma, M, \underline{B}, K, \underline{C}$	$(\Gamma, \mathtt{nil} : \mathtt{stk}\ \underline{C})$, ${}^{\mathtt{nil}}M, \underline{B}, K$
$[\cdot]\ \mathtt{to}\ \mathtt{x}.\ N :: K$	$[\cdot]\ \mathtt{to}\ \mathtt{x}.\ {}^{\mathtt{nil}}N \cdot\!:: K$
$\hat{\imath} :: K$	$\hat{\imath} \cdot\!:: K$
$V :: K$	${}^{\mathtt{nil}}V \cdot\!:: K$

Figure 5.5. Translating CK-machine for CBPV into CK-machine for CBPV+control

2 M, K is terminal, and there does not exist N, L such that $M, K \leadsto N, L$.

□

Proposition 47 (cf. Prop. 12) For every Γ configuration M, K there is a unique terminal Γ-configuration N, L such that $M, K \leadsto^* N, L$, and there is no infinite sequence of transitions from M, K. □

We defer the proof of this to Sect. 7.5.3.

5.4.4 Observational Equivalence

In the presence of control effects, we replace Def. 2.7(2) by

Definition 5.2 Given two computations $\Gamma \vdash^{\mathsf{c}} M, N : \underline{B}$, we say $M \simeq N$ when for any (closed) ground context $\mathcal{C}[\cdot]$ we have

$${}^{\mathtt{nil}}\mathcal{C}[M], \mathtt{nil} \leadsto^* \mathtt{return}\ n, \mathtt{nil} \text{ iff } {}^{\mathtt{nil}}\mathcal{C}[N], \mathtt{nil} \leadsto^* \mathtt{return}\ n, \mathtt{nil}$$

Similarly for values $\Gamma \vdash^{\mathsf{v}} V, W : A$. □

Since $\mathcal{C}[M]$ and $\mathcal{C}[N]$ are required to be closed (by the definition of ground context), they cannot contain `nil`.

5.4.5 Denotational Semantics

Computations

We seek a denotational semantics for the language described in Sect. 5.4.1. This kind of model is traditionally called *continuation semantics.* We will continue to use this name, even though "stack-passing" would be more appropriate in the CBPV setting.

Since there is no divergence, we work with sets rather than cpos. The semantics of values is given in Sect. 5.2—the difficulty lies in the interpretation of computations. As in Sect. 5.3.2, we will state first the soundness theorem that we are aiming to achieve, even though it is not yet meaningful, and use this to motivate the semantics.

Proposition 48 (soundness) Suppose that $M, K \leadsto N, L$ where

$$\begin{array}{ll} \Gamma \vdash^{\mathsf{c}} M : \underline{B} & \Gamma \vdash^{\mathsf{v}} K : \mathtt{stk}\ \underline{B} \\ \Gamma \vdash^{\mathsf{c}} N : \underline{B}' & \Gamma \vdash^{\mathsf{v}} L : \mathtt{stk}\ \underline{B}' \end{array}$$

Then, for any environment $\rho \in [\![\Gamma]\!]$,

$$[\![M]\!](\rho, [\![K]\!]\rho) = [\![N]\!](\rho, [\![L]\!]\rho) \tag{5.2}$$

□

Notice the similarity between this statement and Prop. 42. There, a computation M can change the store, so $[\![M]\!]$ takes store as an argument. Here, a computation M can change its stack, so $[\![M]\!]$ takes a stack as an argument.

In Prop. 48, we know that $[\![K]\!]\rho \in [\![\mathtt{stk}\ \underline{B}]\!]$. So $[\![M]\!]$ must be a function from $[\![\Gamma]\!] \times [\![\mathtt{stk}\ \underline{B}]\!]$ to some set; similarly, $[\![N]\!]$ must be a function from $[\![\Gamma]\!] \times [\![\mathtt{stk}\ \underline{C}]\!]$ to the same set. This set, which we call Ans (the "set of answers") cannot depend on the type of M, because M and N have different types. It is an arbitrary set which remains fixed throughout the denotational semantics.

We proceed to the semantics of types, which is given in Fig. 5.6(1). At first sight, these equations looks strange, but they make sense once we know that whenever we see $[\![\underline{B}]\!]$, for a computation type $\underline{B}$, we should mentally replace it with $[\![\mathtt{stk}\ \underline{B}]\!]$. To put it another way, the semantic brackets $[\![-]\!]$ for computation types contain a "hidden" **stk**. To explain the semantics of types, we first look at some equations which do not require any mental replacement (because they do not mention denotations of computation types) and so make sense immediately.

- $[\![\mathtt{stk}\ (A \to \underline{B})]\!] = [\![A]\!] \times [\![\mathtt{stk}\ \underline{B}]\!]$ follows from the fact that a stack from $A \to \underline{B}$ consists of a value of type A together with a stack from $\underline{B}$.

1 Official presentation—compositional

Values types	Computation types
$\llbracket U\underline{B}\rrbracket = \llbracket\underline{B}\rrbracket \to \mathsf{Ans}$	$\llbracket FA\rrbracket = \llbracket A\rrbracket \to \mathsf{Ans}$
$\llbracket\sum_{i\in I}A_i\rrbracket = \sum_{i\in I}\llbracket A_i\rrbracket$	$\llbracket\prod_{i\in I}\underline{B}_i\rrbracket = \sum_{i\in I}\llbracket\underline{B}_i\rrbracket$
$\llbracket A\times A'\rrbracket = \llbracket A\rrbracket \times \llbracket A'\rrbracket$	$\llbracket A\to\underline{B}\rrbracket = \llbracket A\rrbracket \times \llbracket\underline{B}\rrbracket$
$\llbracket \mathtt{stk}\ \underline{B}\rrbracket = \llbracket\underline{B}\rrbracket$	

2 Intuitive presentation—not compositional

Value types	
$\llbracket U\underline{B}\rrbracket = \llbracket \mathtt{stk}\ \underline{B}\rrbracket \to \mathsf{Ans}$	$\llbracket \mathtt{stk}\ FA\rrbracket = \llbracket A\rrbracket \to \mathsf{Ans}$
$\llbracket\sum_{i\in I}A_i\rrbracket = \sum_{i\in I}\llbracket A_i\rrbracket$	$\llbracket \mathtt{stk}\ \prod_{i\in I}\underline{B}_i\rrbracket = \sum_{i\in I}\llbracket \mathtt{stk}\ \underline{B}_i\rrbracket$
$\llbracket A\times A'\rrbracket = \llbracket A\rrbracket \times \llbracket A'\rrbracket$	$\llbracket \mathtt{stk}\ (A\to\underline{B})\rrbracket = \llbracket A\rrbracket \times \llbracket \mathtt{stk}\ \underline{B}\rrbracket$
Computation types	
$\llbracket\underline{B}\rrbracket = \llbracket \mathtt{stk}\ \underline{B}\rrbracket$	

Figure 5.6. Semantics of types—two equivalent presentations

- $\llbracket\prod_{i\in I}\underline{B}_i\rrbracket = \sum_{i\in I}\llbracket\underline{B}_i\rrbracket$ follows from the fact that stack from $\prod_{i\in I}\underline{B}_i$ consists of a tag $\hat{\imath} \in I$ together with a stack from $\underline{B}_{\hat{\imath}}$.

- $\llbracket \mathtt{stk}\ FA\rrbracket = \llbracket A\rrbracket \to \mathsf{Ans}$ is plausible, because an A-accepting continuation, when it receives an A-value, executes leading to an answer.

- $\llbracket U\underline{B}\rrbracket = \llbracket \mathtt{stk}\ \underline{B}\rrbracket \to \mathsf{Ans}$ because a value of type $U\underline{B}$ is a thunk that can be forced against any stack from $\underline{B}$, and then executes leading to an answer.

Now these equations, together with the standard equations of $\sum$ and $\times$, completely determine the semantics of value types. In other words, there is a unique function $\llbracket - \rrbracket$ from value types to sets that satisfies them. But this function is not given compositionally. Therefore, in Fig. 5.6, we write $\llbracket\underline{B}\rrbracket$ as shorthand for $\llbracket \mathtt{stk}\ \underline{B}\rrbracket$, for the sole reason that this enables us to rearrange these equations into a compositional semantics.

We can now say that a computation $\Gamma \vdash^{\mathsf{c}} M : \underline{B}$ denotes a function from $\llbracket\Gamma\rrbracket \times \llbracket\underline{B}\rrbracket$ to Ans. Here are some example clauses for semantics of terms:

$$\begin{aligned}
[\![\texttt{return } V]\!](\rho, k) &= ([\![V]\!]\rho)`k \\
[\![M \texttt{ to x. } N]\!](\rho, k) &= [\![M]\!](\rho, \lambda a.([\![N]\!]((\rho, \texttt{x} \mapsto a), k))) \\
[\![\texttt{thunk } M]\!]\rho &= \lambda k.([\![M]\!](\rho, k)) \\
[\![\texttt{force } V]\!](\rho, k) &= k`([\![V]\!]\rho) \\
[\![\lambda\texttt{x}.M]\!](\rho, (a, k)) &= [\![M]\!]((\rho, \texttt{x} \mapsto a), k) \\
[\![V`M]\!](\rho, k) &= [\![M]\!](\rho, ([\![V]\!]\rho, k)) \\
[\![\texttt{letstk x. } M]\!](\rho, k) &= [\![M]\!]((\rho, \texttt{x} \mapsto k), k) \\
[\![\texttt{changestk } K.\ M]\!](\rho, k) &= [\![M]\!](\rho, [\![K]\!]\rho) \\
[\![[\cdot] \texttt{ to x. } N :: K]\!]\rho &= \lambda a.([\![N]\!]((\rho, \texttt{x} \mapsto a), [\![K]\!]\rho)) \\
[\![V :: K]\!]\rho &= ([\![V]\!]\rho, [\![K]\!]\rho)
\end{aligned}$$

These semantic equations are of course very similar to the machine transitions.

We can now prove Prop. 48 straightforwardly.

Corollary 49 (by Prop. 47) Suppose Ans has 2 elements $a \neq b$. For a set N and element $n \in N$, write $I_{N,n}$ for the function from N to Ans that takes n to a and everything else to b. Then for any closed ground returner M of type $F\sum_{i \in N} 1$, we have $M, \texttt{nil} \leadsto^* \texttt{return}\ n, \texttt{nil}$ iff $[\![M]\!]I_{N,n} = a$. Hence denotational equality implies observational equivalence. □

The reason for the following terminology is given in Chap. 7.

Definition 5.3 A *jump-point* is a value that (for a given environment) denotes a function to Ans. □

We note that there are two kinds of jump-point: thunks and continuations.

We emphasize[2] that stacks and jump-points are quite distinct concepts. Only continuations fall into both categories. A stack such as $V :: K$ is not a jump-point because it denotes a pair. The situation is summarized in Fig. 5.7. The reader should beware: other authors sometimes use the word "continuation" to mean "jump-point" or "stack". For example:

- [Thielecke, 1997a] uses "continuation" to mean what we call a jump-point.

[2]especially to readers familiar with the CBV setting, where all stacks are continuations, and so the current stack is usually called the "current continuation"

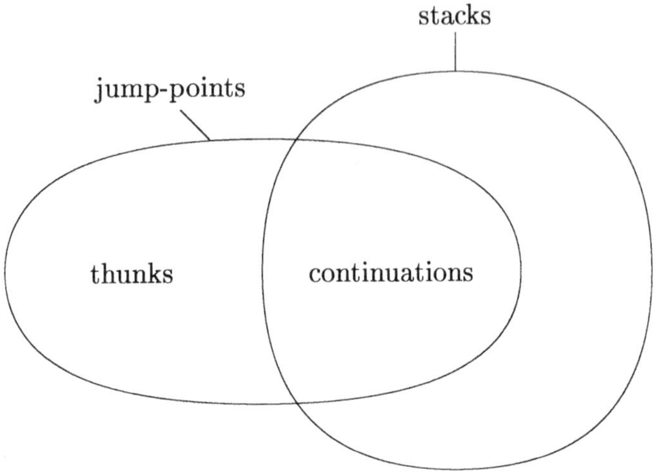

Figure 5.7. Continuations and Stacks

- [Hofmann and Streicher, 1997; Laird, 1998; Streicher and Reus, 1998], which treat CBN as well as CBV, use "continuation" to mean what we call a stack.

However, our definition of "continuation" follows the standard usage in ML and Scheme (e.g. in ML, there is a type $\mathtt{cont}A$, which is a type of continuations).

Notice the semantics of CBV functions:

$$[\![A \to_{\mathbf{CBV}} B]\!] = [\![U(A \to FB)]\!] = ([\![A]\!] \times ([\![B]\!] \to \mathsf{Ans})) \to \mathsf{Ans}$$

As we said in Sect. 5.1, this is precisely the traditional continuation-passing semantics for CBV.

Stacks

As in Sect. 2.6.2, we wish to give a semantics for stacks $\Gamma|\underline{B} \vdash^{\mathsf{k}} K : \underline{C}$. But in this case, the semantics is immediate, for we know from Sect. 5.4.3 that we can regard this as a value $\Gamma, \mathtt{nil} : \mathtt{stk}\ \underline{C} \vdash^{\mathsf{k}} K : \mathtt{stk}\ \underline{B}$, so it denotes a function from $[\![\Gamma]\!] \times [\![\underline{C}]\!]$ to $[\![\underline{B}]\!]$.

This can be summarized by the slogan "in a continuation model, there is a duality between values and stacks".

5.4.6 Combining Control Effects With Other Effects

This section is perhaps the most important in the chapter, as it leads to the Jump-With-Argument language of Chap. 7.

The continuation model in Sect. 5.4.5 generalizes. If we have any CBPV model, we can obtain from it a continuation model for control effects.

As an example, consider the printing model for CBPV. We seek a model for control effects together with `print`. The CK-machine semantics will have the form $M, K \rightsquigarrow m, N, L$ as in Sect. 2.4.2.

We fix an $\mathcal{A}$-set $\underline{\mathsf{Ans}}$. This plays the same role as the set Ans in Sect. 5.4.5. The denotational semantics for control effects with `print` is then organized as follows:

- a value type (and hence a context) denotes a set;
- a computation type denotes[3] a set;
- a value $\Gamma \vdash^{\mathsf{v}} V : A$ denotes a function from $[\![\Gamma]\!]$ to $[\![A]\!]$;
- a computation $\Gamma \vdash^{\mathsf{c}} M : \underline{B}$ denotes a function from $[\![\Gamma]\!] \times [\![\underline{B}]\!]$ to the carrier of $\underline{\mathsf{Ans}}$.

As explained in Sect. 2.8, we use CBPV as a metalanguage describing the printing model. For example, we write $U\underline{B}$ for the carrier of an $\mathcal{A}$-set $\underline{B}$ and FA for the free $\mathcal{A}$-set on a set A. With this notation, the semantics of types is given by

$$\begin{array}{rcl} [\![U\underline{B}]\!] & = & U([\![\underline{B}]\!] \to \underline{\mathsf{Ans}}) \\ [\![\sum_{i\in I} A_i]\!] & = & \sum_{i\in I}[\![A_i]\!] \\ [\![A \times A']\!] & = & [\![A]\!] \times [\![A']\!] \\ [\![\mathtt{stk}\ \underline{B}]\!] & = & [\![\underline{B}]\!] \end{array} \qquad \begin{array}{rcl} [\![FA]\!] & = & U([\![A]\!] \to \underline{\mathsf{Ans}}) \\ [\![\prod_{i\in I} \underline{B}_i]\!] & = & \sum_{i\in I}[\![\underline{B}_i]\!] \\ [\![A \to \underline{B}]\!] & = & [\![A]\!] \times [\![\underline{B}]\!] \end{array}$$

[3]As in Sect. 5.4.5, $[\![\underline{B}]\!]$ should be thought of as shorthand for $[\![\mathtt{stk}\ \underline{B}]\!]$.

Semantics of terms: some example clauses.

$$
\begin{array}{rcl}
[\![\mathtt{return}\ V]\!](\rho,k) & = & ([\![V]\!]\rho)`(\mathsf{force}\ k) \\
[\![M\ \mathtt{to}\ \mathtt{x}.\ N]\!](\rho,k) & = & [\![M]\!](\rho,\mathsf{thunk}\ \lambda a.([\![N]\!]((\rho,\mathtt{x}\mapsto a),k))) \\
[\![\mathtt{thunk}\ M]\!]\rho & = & \mathsf{thunk}\ \lambda k.([\![M]\!](\rho,k)) \\
[\![\mathtt{force}\ V]\!](\rho,k) & = & k`\mathsf{force}\ ([\![V]\!]\rho) \\
[\![\lambda\mathtt{x}.M]\!](\rho,(a,k)) & = & [\![M]\!]((\rho,\mathtt{x}\mapsto a),k) \\
[\![V`M]\!](\rho,k) & = & [\![M]\!](\rho,([\![V]\!]\rho,k)) \\
[\![\mathtt{letstk}\ \mathtt{x}.\ M]\!](\rho,k) & = & [\![M]\!]((\rho,\mathtt{x}\mapsto k),k) \\
[\![\mathtt{changestk}\ K.\ M]\!](\rho,k) & = & [\![M]\!](\rho,[\![K]\!]\rho) \\
[\![\mathtt{print}\ c.\ M]\!](\rho,k) & = & c * ([\![M]\!](\rho,k)) \\
[\![[\cdot]\ \mathtt{to}\ \mathtt{x}.\ N :: K]\!]\rho & = & \mathsf{thunk}\ \lambda a.([\![N]\!]((\rho,\mathtt{x}\mapsto a),[\![K]\!]\rho)) \\
[\![V :: K]\!]\rho & = & ([\![V]\!]\rho,[\![K]\!]\rho)
\end{array}
$$

Proposition 50 (soundness) Suppose that $M, K \rightsquigarrow m, N, L$ where

$$
\begin{array}{ll}
\Gamma \vdash^{\mathsf{c}} M : \underline{B} & \quad \Gamma \vdash^{\mathsf{v}} K : \mathtt{stk}\ \underline{B} \\
\Gamma \vdash^{\mathsf{c}} N : \underline{C} & \quad \Gamma \vdash^{\mathsf{v}} L : \mathtt{stk}\ \underline{C}
\end{array}
$$

Then, for any environment $\rho \in [\![\Gamma]\!]$,

$$[\![M]\!](\rho,[\![K]\!]\rho) = m * [\![N]\!](\rho,[\![L]\!]\rho) \tag{5.3}$$

□

Corollary 51 (by the analogue of Prop. 47) Suppose $\underline{\mathsf{Ans}}$ has two elements a, b with the property that

$$
\begin{array}{rclll}
m * a & \neq & m' * a & \quad \text{for } m \neq m' \in \mathcal{A}^* \\
m * a & \neq & m' * b & \quad \text{for } m, m' \in \mathcal{A}^*
\end{array}
$$

(This property is satisfied by the free $\mathcal{A}$-set on a set of size $\geqslant 2$.) For a countable set N and element $n \in N$, write $I_{N,n}$ for the function from N to $\underline{\mathsf{Ans}}$ that takes n to a and everything else to b.

Then for any closed ground returner M of type $F\sum_{i\in N} 1$, we have $M, \mathtt{nil} \rightsquigarrow^* m, \mathtt{return}\ n, \mathtt{nil}$ iff $[\![M]\!] I_{N,n} = m * a$. Hence terms with the same denotation are observationally equivalent. □

In a similar way, we can obtain a model for control effects with recursion. The semantics of types and terms are as above except that we now understand the metalinguistic U etc. as referring to the Scott model rather than the printing model. We need only replace the semantic equation for `print` by an equation for recursion.

5.5 Erratic Choice

5.5.1 The Language

We add to the language the following erratic choice construct

$$\frac{\cdots \quad \Gamma \vdash^{\mathsf{c}} M_i : \underline{B} \quad \cdots \;_{i \in I}}{\Gamma \vdash^{\mathsf{c}} \mathtt{choose}\ \{\ldots, i.M_i, \ldots\} : \underline{B}}$$

where i ranges over I. According to taste, we can allow I to be finite, countable, nonempty or an arbitrary set (although in the last case the set of terms will not be small). The meaning of $\mathtt{choose}\ \{\ldots, i.M_i, \ldots\}$ is "choose some $i \in I$, then execute M_i". Thus to the big-step semantics we add

$$\frac{M_{\hat{\imath}} \Downarrow T}{\mathtt{choose}\ \{\ldots, i.M_i, \ldots\} \Downarrow T}$$

and to the CK-machine we add the transition

$$\begin{array}{lll} & \mathtt{choose}\ \{\ldots, i.M_i, \ldots\} & K \\ \rightsquigarrow & M_{\hat{\imath}} & K \end{array}$$

It can be proved that computations cannot diverge. Our formulation of the big-step semantics does not allow us to express this fact, but the CK-machine does allow us to express it:

Proposition 52 There is no infinite sequence of transitions from any configuration M, K. □

5.5.2 Denotational Semantics

Computations

We seek a denotational semantics for the language described in Sect. 5.5.1. Since there is no divergence, we use a set model, and the semantics of values is given in Sect. 5.2.

Recall that, in the global store model of Sect. 5.3.2, a computation type $\underline{B}$ denotes the set of *behaviours* that a computation $\Gamma \vdash^{\mathsf{c}} M : \underline{B}$ can exhibit in a given environment and store. Our erratic choice model is somewhat similar, for $\underline{B}$ will denote the set of *possible behaviours* that M can exhibit in a given environment. Thus M will denote a relation from $[\![\Gamma]\!]$ to $[\![\underline{B}]\!]$, and we call this semantics *relation semantics*. The semantics of types is given as follows.

- A possible behaviour of a computation of type FA is to return a value V of type A. So FA denotes $[\![A]\!]$.
- A possible behaviour of a computation of type $\prod_{i \in I} \underline{B}_i$ is to pop a particular $\hat{\imath} \in I$ and then exhibit some possible behaviour of a computation of type $\underline{B}_{\hat{\imath}}$. So $\prod_{i \in I} \underline{B}_i$ denotes $\sum_{i \in I} [\![\underline{B}_i]\!]$.

- A possible behaviour of a computation of type $A \to \underline{B}$ is to pop a particular value V of type A and then exhibit a possible behaviour of a computation of type $\underline{B}$. So $A \to \underline{B}$ denotes $[\![A]\!] \times [\![\underline{B}]\!]$.
- A value of type $U\underline{B}$, when forced, might exhibit any of a range of possible behaviours for a computation of type $\underline{B}$. So $U\underline{B}$ denotes $\mathcal{P}[\![\underline{B}]\!]$.

Semantics of terms—some example clauses:

$$\begin{array}{rcl}
(\rho, a) \in [\![\texttt{return}\ V]\!] & \text{iff} & [\![V]\!]\rho = a \\
(\rho, b) \in [\![M\ \texttt{to x.}\ N]\!] & \text{iff} & \text{for some } a,\ (\rho, a) \in [\![M]\!] \text{ and} \\
 & & \quad ((\rho, \texttt{x} \mapsto a), b) \in [\![N]\!] \\
[\![\texttt{thunk}\ M]\!]\rho & = & \{b \in [\![\underline{B}]\!] : (\rho, b) \in [\![M]\!]\} \\
(\rho, b) \in [\![\texttt{force}\ V]\!] & \text{iff} & b \in [\![V]\!]\rho \\
(\rho, (a, b)) \in [\![\lambda\texttt{x}.M]\!] & \text{iff} & ((\rho, \texttt{x} \mapsto a), b) \in [\![M]\!] \\
(\rho, b) \in [\![V`M]\!] & \text{iff} & (\rho, ([\![V]\!]\rho, b)) \in [\![M]\!] \\
(\rho, b) \in [\![\texttt{choose}\ \{\ldots, i.M_i, \ldots\}]\!] & \text{iff} & \text{for some } i,\ (\rho, b) \in [\![M_i]\!]
\end{array}$$

Proposition 53 (soundness and adequacy) For any closed computation M, we have

$$[\![M]\!] = \bigcup_{M \Downarrow T} [\![T]\!]$$

□

Proof For ($\supseteq$) we induct on $M \Downarrow T$. For ($\subseteq$) we define subsets $\mathsf{red}^{\mathsf{v}}_A$, $\mathsf{red}^{\mathsf{t}}_{\underline{B}}$ and $\mathsf{red}^{\mathsf{c}}_{\underline{B}}$ exactly as in Prop. 10, except that we replace the clause for $M \in \mathsf{red}^{\mathsf{c}}_{\underline{B}}$ by the following:

$M \in \mathsf{red}^{\mathsf{c}}_{\underline{B}}$ iff for all $b \in [\![M]\!]$ there exists $T \in \mathsf{red}^{\mathsf{t}}_{\underline{B}}$ such that $M \Downarrow T$ and $b \in [\![T]\!]$.

We note that if $T \in \mathsf{red}^{\mathsf{t}}_{\underline{B}}$ then $T \in \mathsf{red}^{\mathsf{c}}_{\underline{B}}$ (unlike in the proof of 10, the converse is not apparent at this stage of the proof). The rest of the proof follows that of Prop. 10. □

Corollary 54 For any closed ground returner M, we have $M \Downarrow \texttt{return}\ n$ iff $n \in [\![M]\!]$. □

Notice the semantics of CBV functions:

$$[\![A \to_{\mathbf{CBV}} B]\!] = \mathcal{P}([\![A]\!] \times [\![B]\!])$$

This is the set of relations from $[\![A]\!]$ to $[\![B]\!]$, a traditional nondeterministic semantics for CBV. Notice too that for every computation type $\underline{B}$, the complete lattice $\mathcal{P}[\![\underline{B}]\!]$ is the denotation of $\underline{B}$ in algebra semantics—an algebra for the powerset monad on **Set** is precisely a complete lattice. (This is an instance of Prop. 127 below.)

Stacks

As in Sect. 2.6.2, we wish to give semantics of stacks. Suppose we have a stack $\Gamma|\underline{B} \vdash^{\mathsf{k}} K : \underline{C}$. We can see that, in a given environment $\rho \in [\![\Gamma]\!]$, the stack takes a possible behaviour for a computation of type $\underline{B}$ to a range of behaviours for a computation of type $\underline{C}$. This is apparent in a stack such as

$$|F\texttt{bool} \vdash^{\mathsf{k}} [\cdot]\ \texttt{to} \begin{cases} \texttt{true.} & (\texttt{return}\ 3 \mid \texttt{return}\ 4) \\ \texttt{false.} & \texttt{return}\ 5 \end{cases} :: \texttt{nil} : F\texttt{nat}$$

Thus $\Gamma|\underline{B} \vdash^{\mathsf{k}} K : \underline{C}$ denotes a relation from $[\![\Gamma]\!] \times [\![\underline{B}]\!]$ to $[\![\underline{C}]\!]$.

5.5.3 Finite Choice

We said in Sect. 5.5.1 that in the rule

$$\frac{\cdots\ \Gamma \vdash^{\mathsf{c}} M_i : \underline{B}\ \cdots\ _{i \in I}}{\Gamma \vdash^{\mathsf{c}} \texttt{choose}\ \{\ldots, i.M_i, \ldots\} : \underline{B}}$$

we can, if we like, restrict the set I that i ranges over to be finite (or countable, or nonempty). Now this effect clearly has a CBV model, in which $A \rightarrow_{\mathbf{CBV}} B$ denotes not $[\![A]\!] \rightarrow \mathcal{P}[\![B]\!]$, as it previously did (up to isomorphism), but $[\![A]\!] \rightarrow \mathcal{P}_{\mathsf{fin}}[\![B]\!]$, writing $\mathcal{P}_{\mathsf{fin}}X$ for the set of finite subsets of X.

Our decomposition above of CBV erratic choice semantics into CBPV was based on the isomorphism

$$A \rightarrow \mathcal{P}B \cong \mathcal{P}(A \times B)$$

but there is no analogous isomorphism for $A \rightarrow \mathcal{P}_{\mathsf{fin}}B$. (See Sect. 12.2 for further discussion on modelling CBPV with finite choice.)

This is a situation we shall see again in Sect. 6.9: a "basic" CBV model (in this example, general erratic choice) exhibits a decomposition into CBPV that a constrained model (finite erratic choice) does not.

5.5.4 New Model For May Testing

When we have both erratic choice and divergence, there are various notions of observational equivalence we can consider, An important one is the following.

Definition 5.4 Two computations $\Gamma \vdash^{\mathsf{c}} M, N : \underline{B}$ are *may-testing equivalent* when for any ground context $\mathcal{C}[\cdot]$, we have

$$\mathcal{C}[M] \Downarrow \texttt{return}\ n \quad \text{iff} \quad \mathcal{C}[N] \Downarrow \texttt{return}\ n$$

Similarly, we can define may-testing equivalence on values. □

The following model, closely related to the lower (Hoare) powerdomain and to the linear logic models of [Huth et al., 2000], appears to interpret erratic choice + divergence, using may-testing equivalence. It has just come to light at the time of writing, so the details remain to be worked out. We do not yet have a computational understanding in terms of behaviours, so our treatment will be very brief.

Definition 5.5 Let R and S be posets. A relation $\geqslant$ from R to S is an *opbimodule* when $r' \geqslant_R r \geqslant s \geqslant_S s'$ implies $r' \geqslant s'$. An opbimodule $\geqslant$ is *pointwise ideal* when for each $r \in R$ the lower set $\{s \in S | r \geqslant s\}$ is an ideal i.e. contains an upper bound for every finite subset. □

In the may-testing model of CBPV, each value type (and hence each context) and each computation type denotes a poset.

- A value $\Gamma \vdash^{\mathsf{v}} V : A$ denotes a pointwise ideal opbimodule from $[\![\Gamma]\!]$ to $[\![A]\!]$.
- A computation $\Gamma \vdash^{\mathsf{c}} M : \underline{B}$ denotes an opbimodule from $[\![\Gamma]\!]$ to $[\![\underline{B}]\!]$.
- A stack $\Gamma | \underline{B} \vdash^{\mathsf{k}} \underline{C}$ denotes an opbimodule from $[\![\Gamma]\!] \times [\![\underline{B}]\!]$ to $[\![\underline{C}]\!]$.

We interpret $\sum$ and $\times$ as disjoint union and cartesian product of posets. To describe the semantics of the U connective, we require the following.

Definition 5.6 Let R be a poset. For each element $a \in R$ we write $\downarrow a$ for the set $\{r \in R | r \leqslant_R a\}$. A lower set of a poset R is *finitely generated* when it is of the form

$$\downarrow a_0 \cup \cdots \cup \downarrow a_{n-1}$$

We write $\mathsf{fglow}\, R$ for the set of finitely generated lower sets of R. □

The key equations are

$$\begin{array}{rcl} [\![U\underline{B}]\!] & = & \mathsf{fglow}[\![\underline{B}]\!] \\ [\![FA]\!] & = & [\![A]\!] \\ [\![\prod_{i\in I}\underline{B}_i]\!] & = & \sum_{i\in I}[\![\underline{B}_i]\!] \\ [\![A \to \underline{B}]\!] & = & [\![A]\!]^{\mathsf{op}} \times [\![\underline{B}]\!] \end{array}$$

The semantics of terms is straightforward. In particular, we interpret

- `diverge` as the empty relation
- $\mu\mathtt{x}.M$ as the least prefixed point
- `choose` $\{\ldots, i.M_i, \ldots\}$ as the union of $[\![M_i]\!]$

- an infinitely deep term as the union of denotations of its finite approximants.

To interpret computation type recursion, we notice that the category **OpBimod** of posets and opbimodules is bilimit-compact, because every (e,p) pair from R to S is given by an order-preserving-and-reflecting injection. Furthermore, the category of posets and pointwise ideal opbimodules is sub-bilimit-compact within **OpBimod**, so we can interpret value type recursion.

We expect to be able to show that

$$[\![M]\!] = \bigcup_{M \Downarrow T} [\![T]\!]$$

for any closed computation M, but this has not yet been done at time of writing. This would imply that terms with the same denotation are may-testing equivalent.

Notice the following.

- A value type always denotes a countable disjoint union of join-semilattices. Compare this to our definition of predomain (Def. 4.8(2))
- For each value type A, the cpo of ideals of $[\![A]\!]$ is the denotation of A in usual lower powerdomain semantics
- For each computation type $\underline{B}$, the complete lattice of lower sets of $[\![\underline{B}]\!]$ is the denotation of $\underline{B}$ in usual lower powerdomain semantics.

5.6 Errors

The *errors* feature we consider here is much weaker than the *exceptions* feature of ML and Java. In particular, we do not provide a handling facility.

5.6.1 The Language

We fix a set E of *errors*. We add to CBPV a command `error` e for each $e \in E$. The effect of this command is to halt execution, reporting e as an "error message". Thus we add to the basic CBPV language the rule

$$\frac{}{\Gamma \vdash^{\mathsf{c}} \texttt{error}\ e : \underline{B}}$$

Notice that `error` e can have any type, like `diverge`.

The big-step semantics now has two judgements $M \Downarrow T$ and $M \Downarrow e$. The rules are shown in Fig. 5.8.

$$\frac{M[V/\mathtt{x}] \Downarrow e}{\mathtt{let}\ V\ \mathtt{be}\ \mathtt{x}.\ M \Downarrow e}$$

$$\frac{M \Downarrow e}{M\ \mathtt{to}\ \mathtt{x}.\ N \Downarrow e} \qquad \frac{M \Downarrow \mathtt{return}\ V \quad N[V/\mathtt{x}] \Downarrow e}{M\ \mathtt{to}\ \mathtt{x}.\ N \Downarrow e}$$

$$\frac{M \Downarrow e}{\mathtt{force}\ \mathtt{thunk}\ M \Downarrow e}$$

$$\frac{M_{\hat{\imath}}[V/\mathtt{x}] \Downarrow e}{\mathtt{pm}\ (\hat{\imath}, V)\ \mathtt{as}\ \{\ldots, (i, \mathtt{x}).M_i, \ldots\} \Downarrow e}$$

$$\frac{M[V/\mathtt{x}, V'/\mathtt{y}] \Downarrow e}{\mathtt{pm}\ (V, V')\ \mathtt{as}\ (\mathtt{x}, \mathtt{y}).M \Downarrow e}$$

$$\frac{M \Downarrow e}{\hat{\imath}`M \Downarrow e} \qquad \frac{M \Downarrow \lambda\{\ldots, i.N_i, \ldots\} \quad N_{\hat{\imath}} \Downarrow e}{\hat{\imath}`M \Downarrow e}$$

$$\frac{M \Downarrow e}{V`M \Downarrow e} \qquad \frac{M \Downarrow \lambda\mathtt{x}.N \quad N[V/\mathtt{x}] \Downarrow e}{V`M \Downarrow e}$$

$$\frac{}{\mathtt{error}\ e \Downarrow e}$$

Figure 5.8. Big-Step Rules for Errors in CBPV

Proposition 55 For every closed computation M, precisely one of the following holds

- there exists unique T such that $M \Downarrow T$ and there does not exist e such that $M \Downarrow e$
- there does not exist T such that $M \Downarrow T$ and there exists unique e such that $M \Downarrow e$.

□

For the CK-machine, we first extend the class of configurations: a configuration may be either of the form M, K as in Sect. 2.3.2, or of the form e for $e \in E$. We extend the class of terminal configurations to be the following:

$\texttt{return}\ V$	$\texttt{nil}$
$\lambda\{\ldots, i.M_i, \ldots\}$	$\texttt{nil}$
$\lambda\texttt{x}M$	$\texttt{nil}$
e	

and we add the transition

$$\begin{array}{ll} & \texttt{error}\ e \qquad\qquad K \\ \leadsto & e \end{array}$$

5.6.2 Denotational Semantics

Computations

We seek a denotational semantics for the language described in Sect. 5.6.1. Since there is no divergence, we use a set model. The semantics of values is given in Sect. 5.2—the difficulty lies in the interpretation of computations.

The semantics is very similar to our $\mathcal{A}$-set semantics for printing. (In Sect. 9.7.2, we see that they are all algebra models.)

Definition 5.7 (cf. Def. 1.13) An *E-set* (X, error) consists of a set X together with a function error from E to X. We call X the carrier and error the structure. □

Here are some ways of constructing E-sets (cf. Def. 1.14).

1 For any set X, the *free* E-set on X has carrier $X + E$ and we set $\mathsf{error}\ e$ to be $\mathsf{inr}\ e$.

2 For an $i \in I$-indexed family of E-sets (X_i, error), we set $\prod_{i\in I}(X_i, \mathsf{error})$ to have carrier $\prod_{i\in I} X_i$ and structure given pointwise: $\hat{\imath}`(\mathsf{error}\ e) = \mathsf{error}\ e$.

3 For any set X and E-set (Y, error), we define the E-set $X \to (Y, \mathsf{error})$ to have carrier $X \to Y$ and structure given pointwise: $x`(\mathsf{error}\ e) = \mathsf{error}\ e$.

Computation types denote E-sets and value types denote sets in the evident way: in particular FA denotes the free E-set on $[\![A]\!]$ and $U\underline{B}$ denotes the carrier of $[\![\underline{B}]\!]$. A computation $\Gamma \vdash^{\mathsf{c}} M : \underline{B}$ denotes a function from $[\![\Gamma]\!]$ to the carrier of $[\![\underline{B}]\!]$. We omit semantics of terms; the key clause is

$$[\![\texttt{error}\ e]\!]\rho = \mathsf{error}\ e$$

Proposition 56 (soundness) ▪ If $M \Downarrow T$ then $[\![M]\!] = [\![T]\!]$.

- If $M \Downarrow e$ then $[\![M]\!] = \mathsf{error}\ e$.

□

Corollary 57 (by Prop. 55) For a closed ground returner M, we have $M \Downarrow \mathtt{return}\ n$ iff $[\![M]\!] = \mathsf{inl}\ n$, and $M \Downarrow e$ iff $[\![M]\!] = \mathsf{inr}\ e$. □

Notice that we recover the traditional semantics of CBV functions:

$$[\![A \to B]\!] = [\![A]\!] \to ([\![B]\!] + E)$$

Stacks

Just as for printing, a stack $\Gamma|\underline{B} \vdash^{\mathsf{k}} K : \underline{C}$ denotes a homomorphism from $[\![\underline{B}]\!]$ to $[\![\underline{C}]\!]$ over $[\![\Gamma]\!]$. Explicitly, if $\underline{B}$ denotes (X, error) and $\underline{C}$ denotes (Y, error'), then K denotes a function from $[\![\Gamma]\!] \times X$ to Y such that

$$[\![K]\!](\rho, \mathsf{error} e) = \mathsf{error}' e$$

for all ρ and all e.

5.7 Summary

The easy part of all these models—the semantics of values— was given in Sect. 5.2, except for the may-testing model, where a value type denotes a poset and a value $\Gamma \vdash^{\mathsf{v}} V : A$ denotes a pointwise ideal opbimodule from $[\![\Gamma]\!]$ to $[\![A]\!]$. In Fig. 5.9 we summarize the more difficult part—the semantics of computations and stacks. The set-based models are written above the lines, the cpo- and poset-based models below the line.

Remember that, when giving semantics for global store + printing and the semantics for control + printing, we use CBPV as a metalanguage for the printing model: U means "carrier", F means "free $\mathcal{A}$-set". But we can combine global store or control with any effect that we have a model for, by understanding U and F as referring to this model.

For each effect, we see that the semantics of UF gives a monad in the style of Moggi [Moggi, 1991]. It is remarkable how each monad decomposes into U and F in a specific way that fits the operational semantics. In some ways, we have covered the same ground of Moggi; but the key improvements are that

- we have included CBN as well as CBV;
- our language has operational semantics.

effect	comp. type denotes	computation $\Gamma \vdash^{\mathsf{c}} M : \underline{B}$ denotes
printing	$\mathcal{A}$-set	function from $[\![\Gamma]\!]$ to $[\![\underline{B}]\!]$
global store	set	function from $S \times [\![\Gamma]\!]$ to $[\![\underline{B}]\!]$
global store + printing	$\mathcal{A}$-set	function from $S \times [\![\Gamma]\!]$ to $[\![\underline{B}]\!]$
control	set	function from $[\![\Gamma]\!] \times [\![\underline{B}]\!]$ to Ans
control + printing	set	function from $[\![\Gamma]\!] \times [\![\underline{B}]\!]$ to $\underline{\mathsf{Ans}}$
erratic choice	set	relation from $[\![\Gamma]\!]$ to $[\![\underline{B}]\!]$
errors	E-set	function from $[\![\Gamma]\!]$ to $[\![\underline{B}]\!]$
divergence	cppo	cont. function from $[\![\Gamma]\!]$ to $[\![\underline{B}]\!]$
may-testing	poset	opbimodule from $[\![\Gamma]\!]$ to $[\![\underline{B}]\!]$

effect	stack $\Gamma \vert \underline{B} \vdash^{\mathsf{k}} K : \underline{C}$ denotes
printing	$\mathcal{A}$-set homomorphism over $[\![\Gamma]\!]$ from $[\![\underline{B}]\!]$ to $[\![\underline{C}]\!]$
global store	function from $[\![\Gamma]\!] \times [\![\underline{B}]\!]$ to $[\![\underline{C}]\!]$
global store + printing	$\mathcal{A}$-set homomorphism over $S \times [\![\Gamma]\!]$ from $[\![\underline{B}]\!]$ to $[\![\underline{C}]\!]$
control	function from $[\![\Gamma]\!] \times [\![\underline{C}]\!]$ to $[\![\underline{B}]\!]$
control + printing	function from $[\![\Gamma]\!] \times [\![\underline{C}]\!]$ to $[\![\underline{B}]\!]$
erratic choice	relation from $[\![\Gamma]\!] \times [\![\underline{B}]\!]$ to $[\![\underline{C}]\!]$
errors	E-set homomorphism over $[\![\Gamma]\!]$ from $[\![\underline{B}]\!]$ to $[\![\underline{C}]\!]$
divergence	strict-in-2nd-arg. cont. function from $[\![\Gamma]\!] \times [\![\underline{B}]\!]$ to $[\![\underline{C}]\!]$
may-testing	opbimodule from $[\![\Gamma]\!] \times [\![\underline{B}]\!]$ to $[\![\underline{C}]\!]$

effect	U	F	$UF = T$
printing	carrier	free $\mathcal{A}$-set	$\mathcal{A}^* \times -$
global store	$S \to -$	$S \times -$	$S \to (S \times -)$
global store + printing	$U(S \to -)$	$F(S \times -)$	$U(S \to F(S \times -))$
control	$\to \mathsf{Ans}$	$- \to \mathsf{Ans}$	$(- \to \mathsf{Ans}) \to \mathsf{Ans}$
control + printing	$U(- \to \underline{\mathsf{Ans}})$	$U(- \to \underline{\mathsf{Ans}})$	$U(U(- \to \underline{\mathsf{Ans}}) \to \underline{\mathsf{Ans}})$
erratic choice	$\mathcal{P}$	$-$	$\mathcal{P}$
errors	carrier	free E-set	$- + E$
divergence	$-$	lift	lift
may-testing	fglow	$-$	fglow

effect	$\to$	$\prod_{i \in I}$
printing	$\to$	$\prod_{i \in I}$
global store	$\to$	$\prod_{i \in I}$
global store + printing	$\to$	$\prod_{i \in I}$
control	$\times$	$\sum_{i \in I}$
control + printing	$\times$	$\sum_{i \in I}$
erratic choice	$\times$	$\sum_{i \in I}$
errors	$\to$	$\prod_{i \in I}$
divergence	$\to$	$\prod_{i \in I}$
may-testing	$-^{\mathsf{op}} \times -$	$\sum_{i \in I}$

Figure 5.9. Summary of simple CBPV models

effect	$[\![A \to_{\mathbf{CBV}} B]\!] = [\![U(A \to FB)]\!]$	$[\![\underline{A} \to_{\mathbf{CBN}} \underline{B}]\!] = [\![U\underline{A} \to \underline{B}]\!]$
global store	$S \to ([\![A]\!] \to (S \times [\![B]\!]))$	$(S \to [\![\underline{A}]\!]) \to [\![\underline{B}]\!]$
control	$([\![A]\!] \times ([\![B]\!] \to \mathsf{Ans})) \to \mathsf{Ans}$	$([\![\underline{A}]\!] \to \mathsf{Ans}) \times [\![\underline{B}]\!]$
erratic choice	$\mathcal{P}([\![A]\!] \times [\![B]\!])$	$(\mathcal{P}[\![A]\!]) \times [\![B]\!]$

Figure 5.10. Induced semantics for CBV and CBN function types

Looking at a few examples, we see in Fig. 5.10 that we recover traditional semantics for CBV, and we obtain strange-looking semantics for CBN. As stated in Sect. 5.1, the CBN semantics for global store was presented in [O'Hearn, 1993], while the CBN semantics for control appeared in [Streicher and Reus, 1998]. We can see that CBPV provides an explanation of these apparently mysterious models.

Chapter 6

POSSIBLE WORLD MODEL FOR CELL GENERATION

6.1 Cell Generation

This chapter has two parts. The first part (Sect. 6.1–6.5) presents the language and operational semantics, whereas the second part (Sect. 6.6–6.10) presents a denotational model using possible worlds. The pointer game semantics in Chap. 8 models cell generation too, and therefore Chap. 8 presupposes Sect. 6.1–6.3.

In Sect. 5.3, we looked at semantics of *global* storage cells. But most programming languages provide facilities for generating new cells (i.e. memory locations) during execution. In such a language there may be, at one time, 3 cells storing a boolean and 1 cell storing a number, and at a later time, 5 boolean-storing cells and 6 number-storing cells, because 2 new boolean-storing cells and 5 new number-storing cells have been generated. Consequently, to describe the state of the memory at a given time, we require two pieces of information:

- the *world*, which tells us how many boolean-storing cells, how many number-storing cells, etc., are in existence

- the *store* , which tells us what values these cells are storing.

These two pieces of information together are called a *world-store*.

We stress that the world can only increase: new cells are generated but (at least in principle) none are ever destroyed. Writing w for the earlier world and w' for the later world, in the above example, we say that $w \leqslant w'$. We write $\mathcal{W}$ for the poset of worlds.

6.2 The Language

We add `ref` types to CBPV, giving the following type system:

$$\begin{array}{rl} A ::= & U\underline{B} \mid \sum_{i\in I} A_i \mid 1 \mid A \times A \mid \texttt{ref}\ A \\ \underline{B} ::= & FA \mid \prod_{i\in I} \underline{B}_i \mid A \to \underline{B} \end{array}$$

As in ML, a value of type `ref` A is a cell (i.e. a memory location) that stores a value of type A. For example, a value of type `ref ref bool` is a cell storing a cell storing a boolean.

Maintaining our convention that commands are prefixes, we add the following terms.

Assigning to a cell

$$\frac{\Gamma \vdash^{\mathsf{v}} V : \texttt{ref}\ A \quad \Gamma \vdash^{\mathsf{v}} W : A \quad \Gamma \vdash^{\mathsf{c}} M : \underline{B}}{\Gamma \vdash^{\mathsf{c}} V := W.\ M : \underline{B}}$$

This computation replaces the current contents of the cell V with W, and then executes M.

Reading a cell

$$\frac{\Gamma \vdash^{\mathsf{v}} V : \texttt{ref}\ A \quad \Gamma, \texttt{x} : A \vdash^{\mathsf{c}} M : \underline{B}}{\Gamma \vdash^{\mathsf{c}} \texttt{read}\ V\ \texttt{as}\ \texttt{x}.\ M : \underline{B}}$$

This computation binds `x` to the current contents of the cell V (in other words, substitutes the current contents of V for `x`), and then executes M.

Generating a new cell

$$\frac{\Gamma \vdash^{\mathsf{v}} V : A \quad \Gamma, \texttt{x} : \texttt{ref}\ A \vdash^{\mathsf{c}} M : \underline{B}}{\Gamma \vdash^{\mathsf{c}} \texttt{new}\ \texttt{x} := V.\ M}$$

This computation generates a new cell `x`, initially storing V, and then executes M. We can extend this to *recursive* initialization, where V is allowed to mention `x`, as follows:

$$\frac{\Gamma, \texttt{x} : \texttt{ref}\ A \vdash^{\mathsf{v}} V : A \quad \Gamma, \texttt{x} : \texttt{ref}\ A \vdash^{\mathsf{c}} M : \underline{B}}{\Gamma \vdash^{\mathsf{c}} \texttt{newrec}\ \texttt{x} := V;\ M}$$

More generally still, we can consider *mutually recursive* initialization $\texttt{newrec}\ \texttt{x}_0 := V_0, \ldots, \texttt{x}_{n-1} := V_{n-1}.\ M$. However, we will not treat this.

Equality of cells One might expect the primitive for equality testing to be

$$\frac{\Gamma \vdash^{\mathsf{v}} V : \mathtt{ref}\ A \quad \Gamma \vdash^{\mathsf{v}} V' : \mathtt{ref}\ A}{\Gamma \vdash^{\mathsf{v}} V = V' : \mathtt{bool}}$$

But this is a complex value (in the sense of Sect. 3.2). So, for the sake of operational semantics, we take instead

$$\frac{\Gamma \vdash^{\mathsf{v}} V : \mathtt{ref}\ A \quad \Gamma \vdash^{\mathsf{v}} V' : \mathtt{ref}\ A \quad \Gamma \vdash^{\mathsf{c}} M : \underline{B} \quad \Gamma \vdash^{\mathsf{c}} M' : \underline{B}}{\Gamma \vdash^{\mathsf{c}} \mathtt{if}\ V = V'\ \mathtt{then}\ M\ \mathtt{else}\ M' : \underline{B}}$$

6.3 Operational Semantics

6.3.1 Worlds

As computation proceeds, new cells are generated. The number of cells presently in existence, together with their type, is described by a *world*.

Definition 6.1 1 A *world* w is a finite multiset of value types, i.e. a function from the set $\mathsf{valtypes}$ of value types to $\mathbb{N}$, such that

$$\sum_{A \in \mathsf{valtypes}} w_A < \infty$$

Informally, w_A is the number of A-storing cells in the world w. The condition ensures that the total collection of cells in a world is finite.

2 Using the notation of Sect. I.4.1, we write $\mathsf{cells}\ w$ for the finite set $\sum_{A \in \mathsf{valtypes}} \$ w_A$, the *set of cells in* w.

3 The *empty world* 0 is given by

$$0_A = 0 \text{ for all } A.$$

4 Let w be a world and let A be a value type. We use the phrase w *extended with an* A*-storing cell* l to mean the world w' defined by

$$w'_B = \begin{cases} w_B + 1 & \text{if } B = A \\ w_B & \text{otherwise} \end{cases}$$

The new cell l is then defined to be w_A.

5 Let w and w' be worlds. We say that $w \leqslant w'$ when

$$w_A \leqslant w'_A \text{ for all } A.$$

6 We write $\mathcal{W}$ for the poset of worlds regarded as a category. If $w \leqslant x$, we write $\frac{w}{x}$ for the unique morphism from w to x.

□

The judgements that we use for cell generation are more general than before: we want, for each world w, a class of *w-values* and a class of *w-computations*. These are terms that can explicitly refer to the cells in w. We write these judgements

$$w|\Gamma \vdash^{\mathsf{v}} V : A \qquad\qquad w|\Gamma \vdash^{\mathsf{c}} M : \underline{B}$$

where w is a world. (Where $w = 0$, we sometimes omit it.) We modify each typing rule by adding $w|$ to each premise and to the conclusion. Additionally we introduce the rule

$$\frac{}{w|\Gamma \vdash^{\mathsf{v}} \mathtt{cell}_{A} i : \mathtt{ref}\ A} \quad \text{where } i \in \$w_A \qquad (6.1)$$

for explicit mention of a cell.

Proposition 58 f $w \leqslant w'$ then every w-value is also a w'-value and every w-computation is also a w'-computation. □

6.3.2 Stores

The world tells us only how many cells there are of each type, not what they contain. This latter information is provided by the *store*.

Definition 6.2 1 Let w be a world. A *w-store* is a function associating to each pair (A, i), where A is a value type and $i \in \$w_A$, a closed w-value $w| \vdash^{\mathsf{v}} V_{Ai} : A$. We write it

$$(\ldots, \mathtt{cell}_{A} i \mapsto V_{Ai}, \ldots)$$

2 A *world-store* is a world w together with a w-store s.

□

The operations on stores are similar to the store-handling constructs in the language:

Definition 6.3 Let (w, s) be a world-store and let A be a value type.

reading If l is an A-storing cell in w, we use the phrase *the contents of cell l in s* to mean the function s applied to (A, l).

assignment If l is an A-storing cell in w and $w| \vdash V : A$, we use the phrase *s with cell l assigned V* for the w-store s' which is the same as s except that

$$s'(A, l) = V$$

cell generation, with recursive initialization Supppose w' is w extended with an A-storing cell l, and we have a w-store s is and a w'-value V of type A. (In the nonrecursive special case, V is a w-value.) Then s *extended with a cell* l *storing* V is the w'-store in which

- each w-cell stores the same value as in s (except that in s' it is regarded as a w'-value rather than a w-value)
- the new cell l stores V.

□

6.3.3 Operational Rules

The operational semantics of our dynamically generated store is similar to the operational semantics of global store in Sect. 5.3.1. We define a relation of the form $w, s, M \Downarrow w', s', T$ where

- w, s is a world-store;
- M is a closed w-computation;
- w', s' is a world-store such that $w' \geqslant w$;
- T is a closed terminal w'-computation.

We replace each rule in Fig. 2.2 of the form (2.1) by

$$\frac{w_0, s_0, M_0 \Downarrow w_1, s_1, T_0 \quad \cdots \quad w_{r-1}, s_{r-1}, M_{r-1} \Downarrow w_r, s_r, T_{r-1}}{w_0, s_0, M \Downarrow w_r, s_r, T}$$

and we add the following rules:

$$\frac{w, s, M[V/\mathtt{x}] \Downarrow w', s', T}{w, s, \mathtt{read}\ \mathtt{cell}_A l\ \mathtt{as}\ \mathtt{x}.\ M \Downarrow w', s', T}$$

V is the contents of A-storing cell l in s

$$\frac{w, s', M \Downarrow w', s', T}{w, s, \mathtt{cell}_A l := V.\ M \Downarrow w', s', T}$$

s' is s with A-storing cell l assigned V

$$\frac{w', s', M[\mathtt{cell}_A l/\mathtt{x}] \Downarrow w'', s'', T}{w, s, \mathtt{newrec}\ \mathtt{x} := V.\ M \Downarrow w'', s'', T}$$

(w', s') is (w, s) extended with a cell l storing $V[\mathtt{cell}_A l/\mathtt{x}]$

$$\frac{w, s, M \Downarrow w', s', T}{w, s, \texttt{if cell}_A l = \texttt{cell}_A l \texttt{ then } M \texttt{ else } M' \Downarrow w', s', T}$$

$$\frac{w, s, M' \Downarrow w', s', T}{w, s, \texttt{if cell}_A l = \texttt{cell}_A l' \texttt{ then } M \texttt{ else } M' \Downarrow w', s', T} \ (l \neq l')$$

Similarly, we can adapt the CK-machine to include these constructs. Of course, this requires a class of w-stacks for each world w, written $w|\Gamma|\underline{B} \vdash^{\mathsf{k}} K : \underline{C}$.

6.3.4 Observational Equivalence

Definition 6.4 Given two computations $w|\Gamma \vdash^{\mathsf{c}} M, N : \underline{B}$, we say that $M \simeq N$ when for every ground w-context $\mathcal{C}[\cdot]$ and every w-store s and every n we have

$$\exists w', s'(w, s, \mathcal{C}[M] \Downarrow w', s', \texttt{return}\ n)$$

iff

$$\exists w', s'(w, s, \mathcal{C}[N] \Downarrow w', s', \texttt{return}\ n)$$

We similarly define $\simeq$ for values. □

Some important observational equivalences are

$$\texttt{new x} := V.\ M \quad \simeq \quad {}^{\texttt{x}}M \qquad (6.2)$$

$$\texttt{new x} := V.\ \texttt{new y} := {}^{\texttt{x}}W.\ M \quad \simeq \quad \texttt{new y} := W.\ \texttt{new x} := {}^{\texttt{y}}V.\ M \qquad (6.3)$$

using the conventions of Sect. I.4.2.

6.4 Cell Types As Atomic Types

The definition of "world" depends on the syntax of types. This presents a problem when we move between different languages with different syntax of types, and we wish to show preservation of operational or denotational semantics. Furthermore, there are (in the infinitely wide language) uncountably many `ref` types and hence uncountably many worlds, and we shall see in Sect. 6.8.3 that this is undesirable. In addition, the `ref` type constructor is badly behaved in the sense that $A \cong B$ does not imply `ref` $A \cong$ `ref` B.

A way of rectifying all these problems is to abolish the `ref` type constructor. Instead, we fix a countable set $\mathcal{N}$, whose elements we call *cell types*, together with a function ϕ from $\mathcal{N}$ to the set of value types generated from $\mathcal{N}$. This set is defined by

$$\begin{array}{rl} A ::= & U\underline{B} \mid \sum_{i \in I} A_i \mid 1 \mid A \times A \mid \texttt{ref}_C \\ \underline{B} ::= & FA \mid \prod_{i \in I} \underline{B}_i \mid A \to \underline{B} \end{array}$$

where C ranges over $\mathcal{N}$.

A value of type $\texttt{ref}_C$ is a cell storing values in $\phi(C)$. We do not require ϕ to be injective or surjective, so we might have, for example, two types of `bool`-storing cells and no type of `nat`-storing cells. Only cells in the same cell type can be tested for equality. A world is now defined to be a finite multiset in $\mathcal{N}$. This definition does not involve syntax of types, and it makes the poset of worlds countable because $\mathcal{N}$ is assumed countable.

We modify the typing rules in Sect. 6.2 for reading, assignment, cell generation and equality testing: we replace $\texttt{ref}\ A$ by $\texttt{ref}_C$ and A by $\phi(C)$ throughout. We likewise modify the typing rule (6.1) for explicit mention of a cell:

$$\frac{}{w|\Gamma \vdash^{\mathsf{v}} \texttt{cell}_{C}i : \texttt{ref}_C} \quad \text{where } i \in \$w_C$$

To see the advantage of this formulation when we are dealing with translation between languages, consider the translation from CBV to CBPV. If we take CBV with $(\mathcal{N}, \phi)$-cell generation, we can translate it into CBPV with $(\mathcal{N}, \phi')$-cell generation, where $\phi'(C)$ is defined to be the translation of $\phi(C)$. In both languages, a world is, by definition, a finite multiset in $\mathcal{N}$. So preservation of operational semantics is straightforward to describe. We do not need to consider whether the translation is injective on types, as we would if worked with $\texttt{ref}A$ types.

6.5 Excluding Thunk Storage

In Chap. 5, we studied various effects in the absence of divergence. This enabled us to present simple set-based semantics and to avoid a difficult adequacy proof. But this convenient style of exposition is impossible for a language that stores thunks, because, as Landin showed, recursion can be encoded in terms of thunk storage.

We will therefore define the *data type* generated by $\mathcal{N}$ to be the following class

$$D ::= \textstyle\sum_{i \in I} D_i \mid 1 \mid D \times D \mid \texttt{ref}_C$$

and we divide our exposition into 2 stages.

1 For most of the chapter, we will model cell generation with the "no thunk storage" constraint : $\phi(C)$ is required to be a data type for each $C \in \mathcal{N}$. With this constraint, we can show

Proposition 59 For any world-store w, s and closed w-computation M, there is a unique w', s', T such that $w, s, M \Downarrow w', s', T$. □

2 In Sect. 6.10 we will explain how to adapt our approach to the general "thunk storage" situation, where $\phi(C)$ can be any type.

Even the "no thunk storage" constraint allows storage of cells, so it is more liberal than languages allowing storage of ground values only.

6.6 Introduction

6.6.1 What Is A Possible World Semantics?

We now turn to the real subject of the chapter, denotational semantics using possible worlds.

How can we provide a denotational semantics for a language with cell generation?? One plausible approach is to use the global store semantics of Sect. 5.3, replacing the set of stores by the set of world-stores. However, by contrast with the store in Sect. 5.3, which can change in any way at all, the world-store is constrained in the way it can change by the fact that the world can only increase. Our goal is to provide a model that, unlike the global store model, embodies this important constraint.

The key idea which leads to such a model is that of *possible worlds*: a value type, instead of denoting just one set (or cpo), has a different denotation in each world. For example, $\mathtt{ref}_C$ would denote the 3-element set $\$3$ in a world w where there are 3 C-cells, but the 5-element set $\$5$ in a world w' where there are 5 C-cells. We write $[\![A]\!]w$ for the denotation of A in the world w.

The various sets denoted by a type A are related. In the above example, $[\![C]\!]w$ is clearly a subset of $[\![C]\!]w'$. In general, whenever $w \leqslant w'$, there will be a function $[\![A]\!]^{w}_{w'}$ from $[\![A]\!]w$ to $[\![A]\!]w'$. These functions satisfy the equations

$$\begin{array}{rcll} [\![A]\!]^{w}_{w}a & = & a & \\ [\![A]\!]^{w}_{w''}a & = & [\![A]\!]^{w'}_{w''}([\![A]\!]^{w}_{w'}a) & \text{for } w \leqslant w' \leqslant w'' \end{array}$$

In summary, every value type A denotes a covariant functor from $\mathcal{W}$ to **Set**. For sum, product and cell types, this is all straightforward. Denotations at w are given by

$$\begin{array}{rcl} [\![\sum_{i\in I} A_i]\!]w & = & \sum_{i\in I}[\![A_i]\!]w \\ [\![A \times A']\!]w & = & [\![A]\!]w \times [\![A']\!]w \\ [\![\mathtt{ref}_C]\!]w & = & \$w_C \end{array}$$

while denotations at ${}^{w}_{w'}$ are given in the obvious way.

6.6.2 Contribution Of Chapter

In this chapter, we present a possible world semantics for general storage. It is the second denotational model for general storage, the first being the pointer game semantics of [Abramsky et al., 1998]. A number of possible world models have been described in the literature [Ghica, 1997; Moggi, 90; Oles, 1982; Pitts and Stark, 1993; Reynolds, 1981; Stark, 1994], but these (with the exception of [Ghica, 1997], which models pointers in a CBN setting) interpret storage of ground values only.

The model we will present differs from these in several respects. For example, in these models a type denotes a functor not from $\mathcal{W}$ but from the larger category of *world injections* or the even larger category of *store shapes* [Oles, 1982]. So this chapter is not the usual story of "traditional CBN and CBV semantics decompose into CBPV semantics", at least not in a straightforward way—the relationships between the various models are rather subtle. For that reason, it is probably wise for readers familiar with older models to regard the model in this chapter as essentially different; we discuss this in Sect. 6.9.

6.6.3 Stores

We write Sw for the set

$$\bigtimes_{(C,l) \in \mathsf{cells}\ w} [\![\phi(C)]\!]w$$

We use the $\times$ symbol to emphasize that the product is finite. An element of Sw is called a *semantic* w*-store*, and a syntactic w-store (map from cells to values) *denotes* a semantic w-store. But in the absence of thunk storage, this distinction is not so important, because

Proposition 60 For every data type D and world w, the function $[\![-]\!]$ from the set of closed w-values to to $[\![D]\!]$w is a bijection. Hence the function $[\![-]\!]$ from the set of syntactic w-stores to Sw is a bijection. □

Often, we will speak informally of a "w-store", without specifying "semantic" or "syntactic". It is clear that the 3 operations defined on syntactic world-stores in Sect. 6.3.2 (reading, assignment and cell generation) have evident analogues for semantic world-stores.

Notice that if $w \leqslant w'$, there is no canonical function $Sw \longrightarrow Sw'$ or $Sw' \longrightarrow Sw$, so S is neither covariant nor contravariant. By contrast, when we allow storage of ground values only, there is a canonical function $Sw' \longrightarrow Sw$ given by restriction, so S is contravariant.

That is why our model is so different from the ground storage models mentioned in Sect. 6.6.2.

6.6.4 Possible Worlds Via CBPV Decomposition

One of our heuristics for designing a CBPV semantics is to find a CBV semantics and look for the CBPV decomposition of $\to_{\mathbf{CBV}}$. Such a CBV semantics for cell generation in [Levy, 2002] and was independently considered by O'Hearn. It uses the following interpretation:

$$[\![A \to_{\mathbf{CBV}} B]\!]w = \prod_{w' \geqslant w}(Sw' \to [\![A]\!]w' \to \sum_{w'' \geqslant w'}(Sw'' \times [\![B]\!]w'')) \tag{6.4}$$

The intuition behind (6.4) is as follows. Suppose V is a CBV function in world w; it should denote an element of the LHS of (6.4). Now V can be applied in any future world-store (w', s'), where $w' \geqslant w$ and s' is a w'-store, to an operand which is a w'-value of type A. It will then change the world-store to (w'', s''), where $w'' \geqslant w'$ and s'' is a w''-store, and finally return a w''-value of type B. The RHS of (6.4) precisely describes this narrative.

The CBPV decomposition of $A \to_{\mathbf{CBV}} B$ as $U(A \to FB)$ is immediately apparent in (6.4):

$$\begin{array}{ccccc} \prod_{w' \geqslant w}(Sw' \to & [\![A]\!]w' & \to & \sum_{w'' \geqslant w'}(Sw'' \times & [\![B]\!]w'')) \\ U & (A & \to & F & B) \end{array}$$

This suggests that a computation type will, like a value type, denote a different set in each world, and that these sets are given by

$$\begin{aligned} [\![FA]\!]w &= \textstyle\sum_{w' \geqslant w}(Sw' \times [\![A]\!]w') && (6.5) \\ [\![A \to \underline{B}]\!]w &= [\![A]\!]w \to [\![\underline{B}]\!]w && (6.6) \\ [\![U\underline{B}]\!]w &= \textstyle\prod_{w' \geqslant w}(Sw' \to [\![\underline{B}]\!]w') && (6.7) \end{aligned}$$

How are we to understand these sets intuitively? In the storage model of Sect. 5.3.2, we said that $[\![\underline{C}]\!]$ is the set of *behaviours* for a computation of type $\underline{C}$ in a given store and environment. So it is reasonable to expect $[\![\underline{C}]\!]w$ to be the set of behaviours for a w-computation M of type $\underline{C}$, when executed in a given w-store s and w-environment ρ. Let us now read the above equations in this light.

- If M has type FA, then it terminates in state (w', s') for some $w' \geqslant w$, returning a w'-value V.
- If M has type $A \to \underline{B}$ then it is (equivalent to) a λ-abstraction, which pops a w-value V and then, depending on V, behaves as a w-computation of type $\underline{B}$.

- Similarly, if M has type $\prod_{i \in I} \underline{B}_i$, it is (equivalent to) a λ-abstraction, which pops an $i \in I$ and then, depending on i, behaves as a w-computation of type $\underline{B}_i$. So we will have the equation

$$[\![\textstyle\prod_{i \in I} \underline{B}_i]\!] w = \textstyle\prod_{i \in I} [\![\underline{B}_i]\!] w$$

- If V is a w-value of type $U\underline{B}$ then it can be forced in any state (w', s'), where $w' \geqslant w$, and will then behave as a w'-computation.

As an example, suppose $\phi(C) = \mathtt{nat}$, so a value of type $\mathtt{ref}_C$ is a cell storing a natural number. Let w contain a single C-cell l, represented as $\mathtt{cell}\,0$, and consider the following w-value of type $U(\mathtt{ref}_C \to F\mathtt{ref}_C)$:

```
thunk (  λx.
         read cell 0 as y.
         read x as z.
         cell 0 := z.
         x := y.
         new w := y + (3 × z).
         return z   )
```

This can be forced in any world-store (w', s'), where $w' \geqslant w$. It first pops a cell l' in w'. Then it changes the world-store to (w'', s'') where

- w'' is w' extended with a C-cell l'';

- writing y and z for the contents of cells l and l' in s' respectively, s'' is the w''-store in which

  l stores z
  l' stores y
  l'' stores $y + 3 \times z$
  every other cell stores what it stored in in s', regarded as a w''-value

and returns z.

6.7 Denotational Semantics

6.7.1 Semantics of Values

Suppose we have a value $w|\Gamma \vdash^{\mathsf{v}} V : A$. For each world $w' \geqslant w$ and each environment ρ of closed w'-values, we obtain, by substitution, a closed w'-value of type A. Thus V will denote, for each world $w' \geqslant w$, a function $[\![V]\!]w'$ from $[\![\Gamma]\!]w'$ to $[\![A]\!]w'$. These functions are related: if

$w \leqslant w' \leqslant w''$ then

$$\begin{array}{ccc} [\![\Gamma]\!]w' & \xrightarrow{[\![V]\!]w'} & [\![A]\!]w' \\ {\scriptstyle [\![\Gamma]\!]^{w'}_{w''}} \downarrow & & \downarrow {\scriptstyle [\![A]\!]^{w'}_{w''}} \\ [\![\Gamma]\!]w'' & \xrightarrow[{[\![V]\!]w''}]{} & [\![A]\!]w'' \end{array} \tag{6.8}$$

must commute. In the case that $w = 0$, we can summarize this by saying that V denotes a natural transformation from $[\![\Gamma]\!]$ to $[\![A]\!]$.

Informally, (6.8) says that if we have an environment ρ of closed w'-values, substitute into V and then regard the result as a closed w''-value, we obtain the same as if we regard ρ as an environment of closed w''-values and substitute it into V.

In particular, a *closed* value $w| \vdash V : A$ denotes a family $\{a_{w'}\}_{w' \geqslant w}$, where $a_{w'} \in [\![A]\!]w'$ for each $w' \geqslant w$ and $a_{w''} = [\![A]\!]^{w'}_{w''} a_{w'}$ for each $w \leqslant w' \leqslant w''$. Such a family corresponds to an element of $[\![A]\!]w$. So our earlier intuition that $[\![A]\!]w$ should be thought of as the set of denotations of closed w-values agrees with our interpretation of $\vdash^{\mathsf{v}}$.

6.7.2 Semantics of Computations

Suppose we have a w-computation $w|\Gamma \vdash^{\mathsf{c}} M : \underline{B}$. Now M can be executed in any w-store and w-environment, so we might think that M should denote a function from $Sw \times [\![\Gamma]\!]w$ to $[\![\underline{B}]\!]w$ (the latter set, we recall, being the set of behaviours of a w-computation). But this is not sufficient, because M can also be extended to a w'-computation, for any $w' \geqslant w$, and then executed in any w'-store and w'-environment. So for each $w' \geqslant w$, we want a function $[\![M]\!]w'$ from $Sw' \times [\![\Gamma]\!]w'$ to $[\![\underline{B}]\!]w'$.

What is perhaps surprising is that, unlike the semantics of values, we do not impose any naturality constraint relating $[\![\Gamma]\!]w'$ across different w'. Such a naturality constraint cannot be formulated, because Sw is not functorial in w.

6.7.3 A Computation Type Denotes A Contravariant Functor

If we now attempt to write semantic equations for the various term constructors, all are straightforward except for two: `new` and sequencing. We look at the former. For simplicity, we will suppose that Γ is empty.

Suppose that $\vdash^{\mathsf{v}} V : \phi(C)$ and $\mathtt{x} : \mathtt{ref}_C \vdash^{\mathsf{c}} M : \underline{B}$. We wish to describe the denotation of $\mathtt{new}\ \mathtt{x} := V.\ M$ in the world-store (w, s)—this denotation should be an element of $[\![\underline{B}]\!]w$. First, we extend (w, s) with a C-cell l initialized to $[\![V]\!]w$, giving a world-store (w', s'). Then, we look at the denotation of M (with `x` bound to the new cell l) in the world-store (w', s')—this denotation is an element of $[\![\underline{B}]\!]w'$. How can we obtain an element of $[\![\underline{B}]\!]w$, as required?

The answer is that $[\![\underline{B}]\!]$ must provide extra information. If $w \leqslant w'$, then $[\![\underline{B}]\!]$ must provide a function $[\![\underline{B}]\!]^{w}_{w'}$ from $[\![\underline{B}]\!]w'$ to $[\![\underline{B}]\!]w$. These functions should respect identity and compositions:

$$\begin{aligned} [\![\underline{B}]\!]^{w}_{w} b &= b \\ [\![\underline{B}]\!]^{w}_{w''} b &= [\![\underline{B}]\!]^{w}_{w'}([\![\underline{B}]\!]^{w'}_{w''} b) \end{aligned}$$

In summary, a computation type $\underline{B}$ denotes a *contravariant* functor from $\mathcal{W}$ to **Set**.

As we have done with the rest of the semantics, we would like to provide some intuitive understanding of these functions. A rough idea is that, for $w \leqslant w'$, a w'-behaviour can be turned into a w-behaviour that first generates additional cells. The contravariance is reminiscent of the structure $*$ in the semantics of printing. There, the role of $*$ in $[\![\underline{B}]\!] = (X, *)$ was to "absorb" printing into computations of type $\underline{B}$. Here, the role of $[\![\underline{B}]\!]^{w}_{w'}$ is to "absorb" cell generation into computations of type $\underline{B}$.

The contravariant functors denoted by computation types are given as follows.

- We have already said that $[\![FA]\!]w$ is $\sum_{w' \geqslant w}(Sw' \times [\![A]\!]w')$. Consequently, if $w \leqslant x$ then $[\![FA]\!]x \subseteq [\![FA]\!]w$.
- The denotation of $\prod_{i \in I} \underline{B}_i$ is given pointwise. Since each $[\![\underline{B}_i]\!]w$ is contravariant in w it is clear that $\prod_{i \in I} [\![\underline{B}_i]\!]w$ is contravariant in w.
- The denotation of $A \to \underline{B}$ is given pointwise. Since $[\![A]\!]w$ is covariant in w and $[\![\underline{B}]\!]w$ is contravariant in w, it is clear that $[\![A]\!]w \to [\![\underline{B}]\!]w$ is contravariant in w.

6.7.4 Semantics of Stacks

As in all of our denotational models, we with to interpret stacks $w|\Gamma|\underline{B} \vdash^{\mathsf{k}} K : \underline{C}$ as well as values and computations. Suppose we have a stack $w|\Gamma|\underline{B} \vdash^{\mathsf{k}} K : \underline{C}$. Then for any $w' \geqslant w$ and any w'-environment $\rho \in [\![\Gamma]\!]w'$, it transforms a behaviour of a w'-computation of type $\underline{B}$ into a behaviour of a w'-computation of type $\underline{C}$. As we argued in Sect. 5.3.2, this transformation is not dependent on the initial store, so K will denote, for each $w' \geqslant w$, a function $[\![K]\!]w'$ from $[\![\Gamma]\!]w' \times [\![\underline{B}]\!]w$ to $[\![\underline{C}]\!]w$.

These functions are related: if $w \leqslant w' \leqslant w''$ then

$$\begin{array}{ccccc}
 & & \llbracket \Gamma \rrbracket w' \times \llbracket \underline{B} \rrbracket w' & \xrightarrow{\llbracket K \rrbracket w'} & \llbracket \underline{C} \rrbracket w' \\
 & \nearrow^{\llbracket \Gamma \rrbracket w' \times \llbracket \underline{B} \rrbracket^{w'}_{w''}} & & & \uparrow {\scriptstyle \llbracket \underline{C} \rrbracket^{w'}_{w''}} \\
\llbracket \Gamma \rrbracket w' \times \llbracket \underline{B} \rrbracket w'' & & & & \\
 & \searrow_{\llbracket \Gamma \rrbracket^{w'}_{w''} \times \llbracket \underline{B} \rrbracket w''} & & & \\
 & & \llbracket \Gamma \rrbracket w'' \times \llbracket \underline{B} \rrbracket w'' & \xrightarrow{\llbracket K \rrbracket w''} & \llbracket \underline{C} \rrbracket w''
\end{array} \tag{6.9}$$

must commute. In the case that $w = 0$, we can summarize this by saying that K denotes a dinatural transformation from $\llbracket \Gamma \rrbracket \times \llbracket \underline{B} \rrbracket$ to $\llbracket \underline{C} \rrbracket$.

Informally, (6.9) says that running against K, in an environment $\rho \in \llbracket \Gamma \rrbracket w'$, a w'-computation that first generates cells up to w'' and then behaves as $b \in \llbracket \underline{B} \rrbracket w''$ is equivalent to first generating the cells and then running the remaining computation against K. This is similar to the requirement in the printing model that $\llbracket K \rrbracket$ preserve the printing structure because running `print` $c.\ M$ against K will first print c and then run M against K.

6.7.5 Summary

We now summarize the denotational semantics of CBPV with cell generation, with no thunk storage. The semantics is organized as follows.

- A value type (and likewise a data type and a context) denotes a covariant functor from $\mathcal{W}$ to **Set**.

- A value $w|\Gamma \vdash^{\mathsf{v}} V : A$ denotes, for each $w' \geqslant w''$, a function $\llbracket V \rrbracket w'$ from $\llbracket \Gamma \rrbracket w'$ to $\llbracket A \rrbracket w'$ such that diagram (6.8) commutes for $w \leqslant w' \leqslant w''$.

- A computation type denotes a contravariant functor from $\mathcal{W}$ to **Set**.

- A computation $\Gamma \vdash^{\mathsf{c}} M : \underline{B}$ denotes, for each w, a function from $Sw \times \llbracket \Gamma \rrbracket w$ to $\llbracket \underline{B} \rrbracket w$.

- A computation $w|\Gamma \vdash^{\mathsf{c}} M : \underline{B}$ denotes, for each $w' \geqslant w$ a function from $Sw' \times \llbracket \Gamma \rrbracket w'$ to $\llbracket \underline{B} \rrbracket w'$.

- A stack $w|\Gamma|\underline{B} \vdash^{\mathsf{k}} K : \underline{C}$ denotes, for each $w' \geqslant w$, a function $\llbracket K \rrbracket w'$ from $\llbracket \Gamma \rrbracket w' \times \llbracket \underline{B} \rrbracket w'$ to $\llbracket \underline{C} \rrbracket w'$ such that diagram (6.9) commutes for $w \leqslant w' \leqslant w''$.

The semantics of types is given by

$$\begin{array}{rcl}
Sw & = & \displaystyle\mathop{\times}_{(D,l)\,\in\,\mathsf{cells}\;w} [\![D]\!]w \\
[\![U\underline{B}]\!]w & = & \prod_{w'\geqslant w}(Sw' \to [\![\underline{B}]\!]w') \\
[\![\sum_{i\in I}A_i]\!]w & = & \sum_{i\in I}[\![A_i]\!]w \\
[\![A \times A']\!]w & = & [\![A]\!]w \times [\![A']\!]w \\
[\![\mathtt{ref}_C]\!]w & = & \$w_C \\
[\![FA]\!]w & = & \sum_{w'\geqslant w}(Sw' \times [\![A]\!]w) \\
[\![\prod_{i\in I}\underline{B}_i]\!]w & = & \prod_{i\in I}[\![\underline{B}_i]\!]w \\
[\![A \to \underline{B}]\!]w & = & [\![A]\!]w \to [\![\underline{B}]\!]w
\end{array}$$

Some examples of semantics of terms:

$$\begin{array}{rcl}
[\![\mathtt{return}\ V]\!]ws\rho & = & (w, s, [\![V]\!]w\rho) \\
[\![M\ \mathtt{to}\ \mathtt{x}.\ N]\!]ws\rho & = & \\
\multicolumn{3}{l}{\mathsf{pm}\ [\![M]\!]ws\rho\ \mathsf{as}\ (w', s', a).\ [\![\underline{B}]\!]^{w}_{w'}([\![N]\!]w's'([\![\Gamma]\!]^{w}_{w'}\rho, \mathtt{x} \mapsto a))} \\
[\![\mathtt{thunk}\ M]\!]w\rho & = & \lambda w'.\lambda s'.([\![M]\!]w's'[\![\Gamma]\!]^{w}_{w'}\rho) \\
[\![\mathtt{force}\ V]\!]ws\rho & = & s`w`\ [\![V]\!]w\rho \\
[\![\lambda\mathtt{x}M]\!]ws\rho & = & \lambda a.([\![M]\!]ws(\rho, \mathtt{x} \mapsto a)) \\
[\![V`M]\!]ws\rho & = & ([\![V]\!]w\rho)`([\![M]\!]ws\rho) \\
[\![V := W.\ M]\!]ws\rho & = & [\![M]\!]ws'\rho \\
\multicolumn{3}{c}{\text{where } s' \text{ is } s \text{ with } A\text{-storing cell } [\![V]\!]w\rho \text{ assigned } [\![W]\!]w\rho} \\
[\![\mathtt{read}\ V\ \mathtt{as}\ \mathtt{x}.\ M]\!]ws\rho & = & [\![M]\!]ws(\rho, \mathtt{x} \mapsto a) \\
\multicolumn{3}{c}{\text{where } a \text{ is the contents of } A\text{-storing cell } [\![V]\!]w\rho \text{ in } s} \\
[\![\mathtt{newrec}\ \mathtt{x} := V.\ M]\!]ws\rho & = & [\![\underline{B}]\!]^{w}_{w'}[\![M]\!]w's'([\![\Gamma]\!]^{w}_{w'}\rho, \mathtt{x} \mapsto l) \\
\multicolumn{3}{l}{\text{where } (w', s') \text{ is } (w, s) \text{ extended with cell } l \text{ storing } [\![V]\!]w'([\![\Gamma]\!]^{w}_{w'}\rho, \mathtt{x} \mapsto l)}
\end{array}$$

Proposition 61 (Soundness) Suppose $w, s, M \Downarrow w', s', T$, where M and T have type $\underline{B}$. Then $[\![M]\!]ws = ([\![\underline{B}]\!]^{w}_{w'})([\![T]\!]w's')$. □

The contravariance of $[\![\underline{B}]\!]$ is essential in formulating this statement, just as, for the printing semantics, the structure map of $[\![\underline{B}]\!]$ is essential in formulating Prop. 18.

Corollary 62 (by Prop. 59) If M is a closed ground w-returner then $w, s, M \Downarrow w', s', \mathtt{return}\ n$ iff $[\![M]\!]ws = (w', s', n)$. Hence terms with the same denotation are observationally equivalent. □

6.8 Combining Cell Generation With Other Effects

6.8.1 Introduction

The model for cell generation in Sect. 6.7 generalizes. If we have any CBPV model, we can obtain from it a model for cell generation. We show two examples of this construction: starting with the $\mathcal{A}$-set model for printing, and starting with the Scott model for divergence.

6.8.2 Cell Generation + Printing

When we add both cell generation and printing to CBPV, the big-step semantics will have the form $w, s, M \Downarrow m, w', s', T$. The denotational semantics is organized as follows, writing $\mathcal{A}\mathbf{Set}$ for the category of $\mathcal{A}$-sets and homomorphisms.

- A value type (and likewise a data type and a context) denotes a covariant functor from $\mathcal{W}$ to **Set**.

- A value $w|\Gamma \vdash^{\mathsf{v}} V : A$ denotes, for each $w' \geqslant w''$, a function $[\![V]\!]w'$ from $[\![\Gamma]\!]w'$ to $[\![A]\!]w'$ such that diagram (6.8) commutes for $w \leqslant w' \leqslant w''$.

- A computation type denotes a contravariant functor from $\mathcal{W}$ to $\mathcal{A}\mathbf{Set}$.

- A computation $\Gamma \vdash^{\mathsf{c}} M : \underline{B}$ denotes, for each w, a function from $Sw \times [\![\Gamma]\!]w$ to $[\![\underline{B}]\!]w$.

- A computation $w|\Gamma \vdash^{\mathsf{c}} M : \underline{B}$ denotes, for each $w' \geqslant w$ a function from $Sw' \times [\![\Gamma]\!]w'$ to $[\![\underline{B}]\!]w'$.

- A stack $w|\Gamma|\underline{B} \vdash^{\mathsf{k}} K : \underline{C}$ denotes, for each $w' \geqslant w$, a function $[\![K]\!]w'$ from $[\![\Gamma]\!]w' \times [\![\underline{B}]\!]w'$ to $[\![\underline{C}]\!]w'$ such that $[\![K]\!]w'(\rho, c * x) = c * ([\![K]\!]w(\rho, x))$—i.e. $[\![K]\!]w$ is homomorphic in its second argument—and such that diagram (6.9) commutes for $w \leqslant w' \leqslant w''$.

We use infinitely wide CBPV as a metalanguage describing the printing model, as explained in Sect. 2.8. For example,we write FA for the free $\mathcal{A}$-set on the set A, and we write $U\underline{B}$ for the carrier of the $\mathcal{A}$-set $\underline{B}$. With this notation, the semantics of types is given by

$$\begin{array}{rcl} [\![U\underline{B}]\!]w & = & U\prod_{w' \geqslant w}(Sw' \to [\![\underline{B}]\!]w') \\ [\![FA]\!]w & = & F\sum_{w' \geqslant w}(Sw' \times [\![A]\!]w') \\ [\![\prod_{i \in I}\underline{B}_i]\!]w & = & \prod_{i \in I}[\![\underline{B}_i]\!]w \\ [\![A \to \underline{B}]\!]w & = & [\![A]\!]w \to [\![\underline{B}]\!]w \end{array}$$

Some examples of semantics of terms:

$$
\begin{array}{rcl}
[\![\texttt{return } V]\!]ws\rho & = & \mathsf{return}\ (w, s, [\![V]\!]w\rho) \\
[\![M \texttt{ to x. } N]\!]ws\rho & = & \\
\multicolumn{3}{l}{\quad [\![M]\!]ws\rho \ \mathsf{to}\ (w', s', a).\ [\![\underline{B}]\!]^{w}_{w'}([\![N]\!]w's'([\![\Gamma]\!]^{w}_{w'}\rho, \texttt{x} \mapsto a))} \\
[\![\texttt{thunk } M]\!]w\rho & = & \mathsf{thunk}\ \lambda w'.\lambda s'.([\![M]\!]w's'[\![\Gamma]\!]^{w}_{w'}\rho) \\
[\![\texttt{force } V]\!]ws\rho & = & s`w`\ \mathsf{force}\ [\![V]\!]w\rho \\
[\![\lambda \texttt{x} M]\!]ws\rho & = & \lambda a.([\![M]\!]ws(\rho, \texttt{x} \mapsto a)) \\
[\![V`M]\!]ws\rho & = & ([\![V]\!]w\rho)`([\![M]\!]ws\rho) \\
[\![V := W.\ M]\!]ws\rho & = & [\![M]\!]ws'\rho \\
\multicolumn{3}{l}{\quad \text{where } s' \text{ is } s \text{ with } A\text{-storing cell } [\![V]\!]w\rho \text{ assigned } [\![W]\!]w\rho} \\
[\![\texttt{read } V \texttt{ as x. } M]\!]ws\rho & = & [\![M]\!]ws(\rho, \texttt{x} \mapsto a) \\
\multicolumn{3}{l}{\quad \text{where } a \text{ is the contents of } A\text{-storing cell } [\![V]\!]w\rho \text{ in } s} \\
[\![\texttt{newrec x} := V.\ M]\!]ws\rho & = & [\![\underline{B}]\!]^{w}_{w'}[\![M]\!]w's'([\![\Gamma]\!]^{w}_{w'}\rho, \texttt{x} \mapsto l) \\
\multicolumn{3}{l}{\text{where } (w', s') \text{ is } (w, s) \text{ extended with cell } l \text{ storing } [\![V]\!]w'([\![\Gamma]\!]^{w}_{w'}\rho, \texttt{x} \mapsto l)} \\
[\![\texttt{print } c.\ M]\!]ws\rho & = & c * ([\![M]\!]ws\rho)
\end{array}
$$

Proposition 63 (Soundness) Suppose $w, s, M \Downarrow m, w', s', T$, where M and T have type $\underline{B}$. Then $[\![M]\!]ws = m * ([\![\underline{B}]\!]^{w}_{w'})([\![T]\!]w's')$. □

Corollary 64 (by the analogue of Prop. 59 for printing) If M is a closed ground w-returner then $w, s, M \Downarrow m, w', s', \texttt{return}\ n$ iff $[\![M]\!]ws = (m, w', s', n)$. Hence terms with the same denotation are observationally equivalent. □

6.8.3 Cell Generation + Divergence

We add divergence and recursion to CBPV with cell generation (with no thunk storage). The semantics is organized as follows:

- A data type denotes a covariant functor from $\mathcal{W}$ to **Set**, so Sw is a set for each world w.
- A value type (and likewise a context) denotes a covariant functor from $\mathcal{W}$ to **Cpo**.
- A value $w|\Gamma \vdash^{\mathsf{v}} V : A$ denotes, for each $w' \geqslant w''$, a continuous function $[\![V]\!]w'$ from $[\![\Gamma]\!]w'$ to $[\![A]\!]w'$ such that diagram (6.8) commutes for $w \leqslant w' \leqslant w''$.
- A computation type denotes a contravariant functor from $\mathcal{W}$ to $\mathbf{Cpo}^{\perp}$ (the category of pointed cpos and strict continuous functions).
- A computation $\Gamma \vdash^{\mathsf{c}} M : \underline{B}$ denotes, for each w, a continuous function from $Sw \times [\![\Gamma]\!]w$ to $[\![\underline{B}]\!]w$.

- A computation $w|\Gamma \vdash^{\mathsf{c}} M : \underline{B}$ denotes, for each $w' \geqslant w$ a function from $Sw' \times [\![\Gamma]\!]w'$ to $[\![\underline{B}]\!]w'$.

- A stack $w|\Gamma|\underline{B} \vdash^{\mathsf{k}} K : \underline{C}$ denotes, for each $w' \geqslant w$, a continuous function $[\![K]\!]w'$ from $[\![\Gamma]\!]w' \times [\![\underline{B}]\!]w'$ to $[\![\underline{C}]\!]w'$ such that $[\![K]\!]w'(\rho, \bot) = \bot$—i.e. $[\![K]\!]w$ is strict in its second argument—and such that diagram (6.9) commutes for $w \leqslant w' \leqslant w''$.

The equations giving the semantics of types and terms are exactly the same as in Sect. 6.8.2, except that we now interpret the metalinguistic CBPV constructs as referring to the Scott model rather than the printing model.

Proposition 65 (Soundness/Adequacy) 1 Suppose M has type $\underline{B}$ and $w, s, M \Downarrow w', s', T$. Then $[\![M]\!]ws = ([\![\underline{B}]\!]^{w}_{w'})([\![T]\!]w's')$.

2 Suppose w, s, M diverges. Then $[\![M]\!]ws = \bot$.

□

Proof (in the style of [Tait, 1967]) (1) is straightforward. For (2), define, by mutual induction over types, three families of relations:

- for each A and w, a relation $\leqslant^{\mathsf{v}}_{Aw}$ between $[\![A]\!]$ and closed w-values, such that $\{a \in [\![A]\!] : a \leqslant^{\mathsf{v}}_{Aw} V\}$ is admissible and lift-reflecting, and $a \leqslant^{\mathsf{v}}_{Aw} V$ and $w \leqslant x$ implies $([\![A]\!]^{w}_{x})a \leqslant^{\mathsf{v}}_{Ax} V$

- for each $\underline{B}$ and w, a relation $\leqslant^{\mathsf{t}}_{\underline{B}w}$ between $[\![\underline{B}]\!]w$ and triples w', s, T where $w' \geqslant w$, $s \in Sw'$ and T is a terminal w'-computation, such that $\{b \in [\![\underline{B}]\!]w : b \leqslant^{\mathsf{t}}_{\underline{B}w} x, s, T\}$ is admissible and contains $\bot$, and $b \leqslant^{\mathsf{t}}_{\underline{B}w'} x, s, T$ and $w \leqslant w'$ implies $([\![\underline{B}]\!]^{w}_{w'})b \leqslant^{\mathsf{t}}_{\underline{B}w} x, s, T$

- for each $\underline{B}$ and w, a relation $\leqslant^{\mathsf{c}}_{\underline{B}w}$ between $[\![\underline{B}]\!]w$ and triples w', s, M where $w' \geqslant w$, $s \in Sw'$ and M is a closed w'-computation, such that $\{b \in [\![\underline{B}]\!]w : b \leqslant^{\mathsf{c}}_{\underline{B}w} x, s, M\}$ is admissible and contains $\bot$, and $b \leqslant^{\mathsf{c}}_{\underline{B}w'} x, s, M$ and $w \leqslant w'$ implies $([\![\underline{B}]\!]^{w}_{w'})b \leqslant^{\mathsf{c}}_{\underline{B}w} x, s, M$.

The definition of these relations proceeds as follows:

$a \leqslant^{\mathsf{v}}_{U\underline{B}w}$ thunk M	iff for all $x \geqslant w$ and $s \in Sx$, $s`x`\mathsf{force}\ a \leqslant^{\mathsf{c}}_{\underline{B}x} x, s, M$
$a \leqslant^{\mathsf{v}}_{\sum_{i\in I} A_i w} (\hat{\imath}, V)$	iff $a = (\hat{\imath}, b)$ for some $\hat{\imath} \in I$ and $b \leqslant^{\mathsf{v}}_{A_{\hat{\imath}} w} V$
$a \leqslant^{\mathsf{v}}_{A \times A' w} (V, V')$	iff $a = (b, b')$ for some $b \leqslant^{\mathsf{v}}_{Aw} V$ and $b' \leqslant^{\mathsf{v}}_{A'w} V'$
$b \leqslant^{\mathsf{t}}_{FAw} x, s, \mathtt{return}\ V$	iff $b = \bot$ or $b = \mathsf{return}\ (x, s, a)$ for some $a \leqslant^{\mathsf{v}}_{Ax} V$
$f \leqslant^{\mathsf{t}}_{\prod_{i\in I} \underline{B}_i w} x, s, \lambda\{\ldots, i.M_i, \ldots\}$	iff, for each $\hat{\imath} \in I$, $\hat{\imath}`f \leqslant^{\mathsf{c}}_{\underline{B}_{\hat{\imath}} w} x, s, M_{\hat{\imath}}$
$f \leqslant^{\mathsf{t}}_{A \to \underline{B}w} x, s, \lambda\mathbf{x}.M$	iff, for all $a \in [\![A]\!]w$, $([\![A]\!]^{w}_{x})a \leqslant^{\mathsf{v}}_{Ax} V$ implies $a`f \leqslant^{\mathsf{c}}_{\underline{B}w} x, s, M[V/\mathbf{x}]$
$b \leqslant^{\mathsf{c}}_{\underline{B}w} x, s, M$	iff $b = \bot$ or $x, s, M \Downarrow x', s', T$ and $b \leqslant^{\mathsf{t}}_{\underline{B}w} x', s', T$

Note that for terminal T, $b \leqslant^{\mathsf{c}}_{\underline{B}w} x, s, T$ iff $b \leqslant^{\mathsf{t}}_{\underline{B}w} x, s, T$. We show that, for any datatype D, $a \vdash^{\mathsf{v}}_{Dw} V$ iff $a = [\![V]\!]w$, by induction on D. Finally, we show, by mutual induction on M and V, that

- for any computation $w|A_0, \ldots, A_{n-1} \vdash^{\mathsf{c}} M : \underline{B}$, if $w \leqslant x$, $s \in Sx$ and $a_i \leqslant^{\mathsf{v}}_{A_i x} W_i$ for $i = 0, \ldots, n-1$ then $[\![M]\!]xs\overrightarrow{a_i} \leqslant^{\mathsf{c}}_{\underline{B}x} x, s, M[\overrightarrow{W_i/\mathbf{x}_i}]$
- for any value $w|A_0, \ldots, A_{n-1} \vdash^{\mathsf{v}} V : A$, if $w \leqslant x$ and $a_i \leqslant^{\mathsf{v}}_{A_i x} W_i$ for $i = 0, \ldots, n-1$ then $[\![V]\!]x\overrightarrow{a_i} \leqslant^{\mathsf{v}}_{Ax} V[\overrightarrow{W_i/\mathbf{x}_i}]$.

□

Corollary 66 Let M be a closed ground returner in world w. Then $w, s, M \Downarrow w', s', \mathtt{return}\ n$ iff $[\![M]\!]ws = \mathsf{return}\ (w', s', n)$, and w, s, M diverges iff $[\![M]\!]ws = \bot$. Hence terms with the same denotation are observationally equivalent. □

It is also easy to show that provably equal terms have the same denotation i.e. the model validates the CBPV equations.

We note that this semantics takes place entirely in predomains and domains, in the sense of Def. 4.8, because we required in Sect. 6.4 that the set of cell types $\mathcal{N}$ be countable. If $\mathcal{N}$ were uncountable, then the countable base condition would be violated.

6.9 Related Models and Parametricity

In the models of this chapter, the observational equivalences (6.2) and (6.3) are not validated (although (6.3) is validated when V and W have distinct types). By contrast, the older models do validate these equivalences, so it is immediately clear that there are substantial differences.

For example, the Idealized Algol model of [Oles, 1982] interprets comm—corresponding to[1] the CBPV type $U\underline{\texttt{comm}}$ of thunks of commands—at w by $Sw \to Sw$. Contrast this with our model which interprets $U\underline{\texttt{comm}}$ at w by $\prod_{w' \geqslant w}(Sw' \to \sum_{w'' \geqslant w} Sw'')$. There are two differences between these semantics, both a consequence of the fact that in Idealized Algol only ground values can be stored. (Indeed the cell storage model of [Ghica, 1997] provides a semantics of comm similar to ours.) To understand these differences, suppose that V is a closed w-value of type $U\underline{\texttt{comm}}$.

- Our model specifies the behaviour of V when forced in any future world-store (hence the $\prod$), and we have imposed no relationship between the behaviour in different worlds. In the Idealized Algol model, by contrast, if V is forced at (w', s'), its behaviour is determined by the restriction of s' to a w-store. The extra cells will be unaffected, because they cannot be stored in the cells of w. Thus the contravariance of S (mentioned in Sect. 6.6.3) is essential to this model.
- Our model says that, when V is forced, new cells can be generated (hence the $\sum$). In the Idealized Algol model, any new cells can be garbage-collected when the command is completed, because they will not be stored anywhere. This feature is known as the *stack discipline* of Idealized Algol.

It is because of these differences that, in the Idealized Algol model, the naturality condition is required across all *world-injections*—this is explained in [O'Hearn and Tennent, 1995].

Moggi's model [Moggi, 90] for CBV with `ref 1` uses contravariance of S in a similar way to the Idealized Algol models. But as the language (called ν-calculus in [Stark, 1994]) includes returners other than ground returners (unlike Idealized Algol, where the only returners are commands), there is no stack discipline. So the interpretation of $T1$ (again corresponding to $U\underline{\texttt{comm}}$ in CBPV) at w is $Sw \to \sum_{w' \geqslant w} Sw'$. However, this summation is quotiented by an equivalence relation, and this is how the observational equivalences of Sect. 6.3.4 are validated. Such quotienting is easy when working with sets (as Moggi does), but problematic when working with cpos. For although it is possible to quotient a cpo by an equivalence relation [Jung, 1990], we do not have the simple characterization of elements that we have in the case of sets.

[1] Although Idealized Algol is a CBN language, the possible world model of [Oles, 1982] is essentially a model of thunks—this is why this model does not allow direct interpretation of conditionals at all types. Hence a type of this language is interpreted as a U type, and denotes a covariant functor.

We can recover all of these models of storage by taking the CBV model for storage given in Sect. 6.6.4 and imposing a weak form of relational parametricity [O'Hearn and Tennent, 1995] called "parametricity in initializations". We will not describe this restricted model here; it does not exhibit a simple CBPV decomposition in the way that the basic model does. (This situation is analogous to that of the CBV model for finite nondeterminism, discussed in Sect. 5.5.3.) Furthermore, as with Moggi's model, it is not simple (although possible) to generalize from sets to cpos, because of the quotienting required.

A rather different possible world model, whose relationship to ours we have not investigated, is Odersky's model for `ref` 1 in a CBN language [Odersky, 1994].

6.10 Modelling Thunk Storage And Infinitely Deep Types

We now wish to adapt the model for cell generation + divergence (Sect. 6.8.3) to treat thunk storage, so that $\phi(C)$ can be any value type rather than only a data type. The difficult part of our task is the semantics of types, which should be organized as follows.

- Sw is a cpo for each world w.
- $[\![A]\!]$ is a covariant functor from $\mathcal{W}$ to **Cpo**, for each value type A.
- $[\![\underline{B}]\!]$ is a contravariant functor from $\mathcal{W}$ to $\mathbf{Cpo}^{\perp}$, for each computation type $\underline{B}$.

Collectively, then, the semantics of types will be given by an object in the category

$$\mathcal{B} = \mathbf{Cpo}^{\mathsf{ob}\ \mathcal{W}} \times [\mathcal{W}, \mathbf{Cpo}]^{\mathsf{valtypes}} \times [\mathcal{W}^{\mathsf{op}}, \mathbf{Cpo}^{\perp}]^{\mathsf{comptypes}}$$

These functors should satisfy the isomorphisms

$$\begin{array}{rcl}
S & \cong & \displaystyle\bigtimes_{(C,l)\,\in\,\mathsf{cells}} [\![\phi(C)]\!] \\
[\![U\underline{B}]\!] & \cong & U_S[\![\underline{B}]\!] \\
[\![\sum_{i\in I} A_i]\!] & \cong & \sum_{i\in I}[\![A_i]\!] \\
[\![A \times A']\!] & \cong & [\![A]\!] \times [\![A']\!] \\
[\![\mathtt{ref}_C]\!] & \cong & \$_C \\
[\![FA]\!] & \cong & F_S[\![A]\!] \\
[\![\prod_{i\in I}\underline{B}_i]\!] & \cong & \prod_{i\in I}[\![\underline{B}_i]\!] \\
[\![A \to \underline{B}]\!] & \cong & [\![A]\!] \to [\![\underline{B}]\!]
\end{array}$$

where we use the following terminology.

Definition 6.5 1 For $\underline{X}$ an object of $[\mathcal{W}^{\mathrm{op}}, \mathbf{Cpo}^{\perp}]$ and S an object of $\mathbf{Cpo}^{\mathrm{ob}\ \mathcal{W}}$, we write $U_S\underline{X}$ for the object of $[\mathcal{W}, \mathbf{Cpo}]$ given at w by $U\prod_{w' \geqslant w}(Sw' \to \underline{X}w')$ and at $\frac{w}{x}$ by the restriction.

2 For X an object of $[\mathcal{W}, \mathbf{Cpo}]$ and S an object of $\mathbf{Cpo}^{\mathrm{ob}\ \mathcal{W}}$, we write $F_S X$ for the object of $[\mathcal{W}, \mathbf{Cpo}^{\perp}]$ given at w by $F\sum_{w' \geqslant w}(Sw' \times Xw')$ and at $\frac{w}{x}$ by the inclusion.

3 For a cell type $C \in \mathcal{N}$, we write $\$_C$ for the object of $[\mathcal{W}, \mathbf{Cpo}]$ given by $\$w_C$ at w and by inclusions at $\frac{w}{w'}$.

4 For $\{X_A\}_{A \in \mathcal{N}}$ a family of objects in $[\mathcal{W}, \mathbf{Cpo}]$ we write $\bigtimes_{(C,l) \in \mathsf{cells}} X_A$ for the object of $\mathbf{Cpo}^{\mathrm{ob}\ \mathcal{W}}$ given at w by $\bigtimes_{(C,l) \in \mathsf{cells}\ w} X_C w$.

5 We write $\sum, \times, \prod, \to$ for the evident pointwise operations on covariant and contravariant functors.

□

These isomorphisms collectively describe an isomorphism in $\mathcal{B}$. How can we construct this isomorphism? We apply Prop. 30 and Prop.32, which tell us that

- $\mathbf{Cpo}^{\mathrm{ob}\ \mathcal{W}}$ is sub-bilimit-compact in $\mathbf{pCpo}^{\mathrm{ob}\ \mathcal{W}}$—we call this category $\mathcal{S}$
- $[\mathcal{W}, \mathbf{Cpo}]$ is sub-bilimit-compact in $[\mathcal{W}, \mathbf{Cpo}] \bullet\!\!\to [\mathcal{W}, \mathbf{pCpo}]$—we call this category $\mathcal{A}$
- $[\mathcal{W}^{\mathrm{op}}, \mathbf{Cpo}^{\perp}]$ is bilimit-compact—we call this category $\mathcal{B}$

Consequently $\mathcal{B}$ is sub-bilimit-compact in

$$\mathcal{C} = \mathcal{S} \times \mathcal{A}^{\mathsf{valtypes}} \times \mathcal{B}^{\mathsf{comptypes}}$$

We must therefore specify a locally continuous functor from $\mathcal{C}^{\mathrm{op}} \times \mathcal{C}$ to $\mathcal{C}$; the semantics of types will then be given as the minimal invariant of this functor. The functor is constructed straightforwardly using the following analogue of Prop. 37.

Proposition 67 In the following, we write

- $\hat{X}$ for a morphism in $\mathcal{A}$
- $\hat{\underline{Y}}$ for a morphism in $\mathcal{B}$

- $\hat{S}$ for a morphism in $\mathcal{S}$

and we use the operations on strict continuous and partial continuous functions defined in Prop. 37.

- $\{X_A\}_{A \in \mathsf{valtypes}} \mapsto \bigtimes_{(A,l) \in \mathsf{cells}} X_A$ extends to a locally continuous functor from $\mathcal{A}^{\mathsf{valtypes}}$ to $\mathcal{S}$, when we set

$$(\bigtimes_{(A,l) \in \mathsf{cells}} \hat{X}_A)w = \bigtimes_{(A,l) \in \mathsf{cells}\ w} (\hat{X}_A w)$$

- $(S, \underline{X}) \mapsto U_S \underline{X}$ extends to a locally continuous functor U from $\mathcal{S}^{\mathsf{op}} \times \mathcal{B}$ to $\mathcal{A}$, when we set

$$(U_{\hat{S}} \hat{\underline{X}})w = U \prod_{w' \geqslant w} (\hat{S}w' \to \hat{\underline{X}}w')$$

- $\sum_{i \in I}$ extends to a locally continuous functor from $\mathcal{A}^I$ to $\mathcal{A}$, when we set

$$(\sum_{i \in I} \hat{X}_i)w = \sum_{i \in I} (\hat{X}_i w)$$

- $\times$ extends to a locally continuous functor from $\mathcal{A} \times \mathcal{A}$ to $\mathcal{A}$, whose operation on morphisms is given by

$$(\hat{X} \times \hat{X}')w = \hat{X}w \times \hat{X}'w$$

- $(S, X) \mapsto F_S X$ extends to a locally continuous functor F from $\mathcal{S} \times \mathcal{A}$ to $\mathcal{B}$, when we set

$$(F_{\hat{S}} \hat{X})w = F \sum_{w' \geqslant w} (\hat{S}_{w'} \times \hat{X}_{w'})$$

- $\prod_{i \in I}$ extends to a locally continuous functor from $\mathcal{B}^I$ to $\mathcal{B}$, when we set

$$(\prod_{i \in I} \hat{\underline{Y}}_i)w = \prod_{i \in I} (\hat{\underline{Y}}_i w)$$

- $\to$ extends to a locally continuous functor from $\mathcal{A}^{\mathsf{op}} \times \mathcal{B}$ to $\mathcal{B}$, when we set

$$(\hat{A} \to \hat{\underline{Y}})w = (\hat{A}w) \to (\hat{\underline{Y}}w)$$

□

It is clear from Prop. 67 that the isomorphisms above indeed define a functor from $\mathcal{C}^{\mathsf{op}} \times \mathcal{C}$ to $\mathcal{C}$, and we thus obtain our semantics of types. Having constructed functors satisfying these isomorphisms, we define operations on semantic stores and then define semantics of terms just as in Sect. 6.8.3 (where these isomorphisms are all identities).

Proposition 68 (Soundness/Adequacy) 1 Suppose M has type $\underline{B}$ and $w, s, M \Downarrow w', s', T$. Then $[\![M]\!]w[\![s]\!] = ([\![\underline{B}]\!]^{w}_{w'})([\![T]\!]w'[\![s']\!])$.

2 Suppose w, s, M diverges. Then $[\![M]\!]w[\![s]\!] = \bot$.

□

(1) is straightforward. To prove (2), we adapt the proof of Prop. 65(2). To show the existence of the required logical relation, we apply Pitts' methods [Pitts, 1996] (which work for any bilimit-compact category) to the category $\mathcal{C}$.

Corollary 69 For a closed ground w-returner M, there exists s' such that $w, s, M \Downarrow w', s', \texttt{return}\ n$ iff there exists s' such that $[\![M]\!]w[\![s]\!] = \mathsf{return}\ (w', s', n)$, and w, s, M diverges iff $[\![M]\!]w[\![s]\!] = \bot$. Hence terms with the same denotation are observationally equivalent. □

Finally, we observe that everything we have done in this section works if, in addition to general storage, we allow infinitely deep types.

Chapter 7

JUMP-WITH-ARGUMENT

7.1 Introduction

In Sect. 5.4.6, we presented a continuation semantics for CBPV+control, together with printing. (We use printing as our leading example of a non-control effect in this discussion, but others would work as well.) We used CBPV as the semantic metalanguage, and we gave not just one model, but a whole family, parametrized by an $\mathcal{A}$-set $\underline{\mathsf{Ans}}$. What we described there was, in effect, a translation

$$\begin{array}{c}\text{CBPV + control}\\ +\texttt{print}\end{array} \longrightarrow \begin{array}{c}\text{CBPV} + \underline{\texttt{Ans}}\\ +\texttt{print}\end{array}$$

where $\underline{\texttt{Ans}}$ is a free computation type identifier. Any $\mathcal{A}$-set $\underline{\mathsf{Ans}}$ induces a denotational semantics for CBPV + $\underline{\texttt{Ans}}$ + `print` (where we interpret $\underline{\texttt{Ans}}$ by $\underline{\mathsf{Ans}}$) and so, via this translation, a denotational semantics for CBPV + control + `print`. The translation is called the *stack-passing style* (StkPS) transform, because every computation is regarded as taking its stack as a parameter. (In the CBV setting, it is called the *continuation-passing style* (CPS) transform because all stacks are continuations.)

But a continuation semantics is far more than just a denotational model. As Steele explained in the CBV setting [Steele, 1978], it provides a *jumping implementation.* This is because the range of the StkPS transform is not the whole of CBPV+$\underline{\texttt{Ans}}$, but a very special fragment which we call *Jump-With-Argument* (JWA). This fragment, which is very similar to Steele's intermediate language and various like calculi [Appel, 1991; Appel and Jim, 1989; Danvy, 1992; Sabry and Felleisen, 1993; Thielecke, 1997a], is worth separating out from CBPV+$\underline{\texttt{Ans}}$ and studying independently, because it can be regarded as a language of jump instructions. In Steele's words:

> Continuation-passing style, while apparently applicative in nature, admits a peculiarly imperative interpretation[...] As a result, it is easily converted to an imperative machine language. [Steele, 1978]

To summarize, the StkPS transform goes from CBPV + control + `print` to a jumping language JWA + `print`, which is a fragment of CBPV + Ans + `print`.

$$\begin{array}{cccc} \text{CBPV + control} & \xrightarrow[\text{transform}]{\text{StkPS}} & \text{JWA} & \hookrightarrow & \text{CBPV} + \underline{\texttt{Ans}} \\ +\texttt{print} & & +\texttt{print} & & +\texttt{print} \end{array}$$

The StkPS transform translates **force** V and **return** V into jump commands. It therefore makes apparent the intuitive point that we made in Sect. I.5.3: that **force** and **return** are the only two instructions that cause execution to move to elsewhere in the program. This shows once again the advantage of working with CBPV rather than CBV and CBN, which do not make the flow of control so explicit.

In addition to the jumping operational semantics for JWA, we also describe a more conventional operational semantics, the *C-machine*, and explain their agreement. We define observational equivalence and an equational theory for JWA—as usual, this requires complex values.

We digress in Sect. 7.7 to explain how JWA can be regarded as a type theory for classical logic. This is based on the classic treatments of [Friedman, 1978; Griffin, 1990; Lafont et al., 1993]. But we urge the reader, except during Sect. 7.7, to ignore logical issues—in particular, ignore the fact that the symbol $\neg$ for continuation types is the same as the traditional symbol for logical negation. Furthermore, it would be hard to argue that the logical reading by itself provides a motivation for JWA.

Finally, we see in Sect. 7.8 that CBPV+control is equivalent to JWA, along the StkPS transform. (This is very similar to the "equational correspondence" result of [Sabry and Felleisen, 1993].) Consequently models for CBPV+control are essentially the same as models for JWA—a useful fact, because JWA is so much simpler.

7.2 The Language

JWA is an extension of Thielecke's "CPS calculus" [Thielecke, 1997a]. Unlike Thielecke's language, but like the untyped languages of [Danvy, 1992; Sabry and Felleisen, 1993; Steele, 1978], it includes values and abstraction.

The types of JWA are given by

$$A ::= \neg A \mid \textstyle\sum_{i \in I} A_i \mid 1 \mid A \times A \qquad (7.1)$$

where each set I is finite (or countable, for *infinitely wide* JWA). A value of type $\neg A$ is an "A-accepting jump-point"—intuitively, a point that we jump to taking an argument of type A. For example, a BASIC line number would have type $\neg 1$, because the argument is trivial. Although we have followed the established usage of $\neg$ for jump-point types, we consider it to be a confusing usage because $\neg$ is not exactly logical negation, as we explain in Sect. 7.7. There are two judgements

$$\Gamma \vdash^{\mathsf{v}} V : A \qquad\qquad \Gamma \vdash^{\mathsf{n}} M$$

called respectively *values* and *nonreturning commands* (in the sense of Sect. 2.9.2—we will sometimes omit the word "nonreturning"). The terms of JWA are given in Fig. 7.1.

The embedding of JWA in CBPV+$\underline{\texttt{Ans}}$ (with printing, if desired) is given in Fig. 7.1. We thus obtain, for each $\mathcal{A}$-set $\underline{\mathsf{Ans}}$, denotational semantics for JWA+`print`. By using the empty $\mathcal{A}$-set for $\underline{\mathsf{Ans}}$, we can prove

Proposition 70 There is no closed nonreturning command in JWA. □

Because of this situation, we always consider operational semantics for *non-closed* nonreturning commands; execution terminates when we attempt to jump to or pattern-match a free identifier.

In the absence of effects, JWA can be seen as a fragment of the $\times\sum\prod\to$-calculus (presented in Fig. 3.8), and this has been the usual viewpoint in the literature. However, it is not valid in the presence of effects.

The following will be of great importance to our informal discussion of pointer games in Chap. 8.

Proposition 71 Any closed value in JWA consists of several tags and several jump-points. □

7.3 Jumping

7.3.1 Intuitive Reading of Jump-With-Argument

We said in Sect. 7.2 that an A-accepting jump-point (a value of type $\neg A$) is a point that we jump to taking an argument of type A. We explain the constructs for $\neg$ in a similar way.

- $V \nearrow W$ is the command "jump to the point W taking argument V".
- $\gamma\mathtt{x}.M$ is a point. When we jump to it taking argument V, we bind $\mathtt{x}$ to V and then obey the command M.

The easiest way to grasp this will be to see the execution of an example program. Our explanation of jumping is informal; its aim is to convey the jumping intuitions of JWA.

$$\frac{}{\Gamma, \mathtt{x} : A, \Gamma' \vdash^{\mathsf{v}} \mathtt{x} : A} \qquad \frac{\Gamma \vdash^{\mathsf{v}} V : A \quad \Gamma, \mathtt{x} : A \vdash^{\mathsf{n}} M}{\Gamma \vdash^{\mathsf{n}} \mathtt{let}\ V\ \mathtt{be}\ \mathtt{x}.\ M}$$

$$\frac{\Gamma \vdash^{\mathsf{v}} V : A_{\hat{\imath}}}{\Gamma \vdash^{\mathsf{v}} (\hat{\imath}, V) : \sum_{i \in I} A_i} \qquad \frac{\Gamma \vdash^{\mathsf{v}} V : \sum_{i \in I} A_i \quad \cdots \ \Gamma, \mathtt{x} : A_i \vdash^{\mathsf{n}} M_i \ \cdots_{i \in I}}{\Gamma \vdash^{\mathsf{n}} \mathtt{pm}\ V\ \mathtt{as}\ \{\ldots, (i, \mathtt{x}).M_i, \ldots\}}$$

$$\frac{\Gamma \vdash^{\mathsf{v}} V : A \quad \Gamma \vdash^{\mathsf{v}} V' : A'}{\Gamma \vdash^{\mathsf{v}} (V, V') : A \times A'} \qquad \frac{\Gamma \vdash^{\mathsf{v}} V : A \times A' \quad \Gamma, \mathtt{x} : A, \mathtt{y} : A' \vdash^{\mathsf{n}} M}{\Gamma \vdash^{\mathsf{n}} \mathtt{pm}\ V\ \mathtt{as}\ (\mathtt{x}, \mathtt{y}).M}$$

$$\frac{\Gamma, \mathtt{x} : A \vdash^{\mathsf{n}} M}{\Gamma \vdash^{\mathsf{v}} \gamma\mathtt{x}.M : \neg A} \qquad \frac{\Gamma \vdash^{\mathsf{v}} V : A \quad \Gamma \vdash^{\mathsf{v}} W : \neg A}{\Gamma \vdash^{\mathsf{n}} V \nearrow W}$$

For printing, we add the construct $\frac{\Gamma \vdash^{\mathsf{n}} M}{\Gamma \vdash^{\mathsf{n}} \mathtt{print}\ c.\ M}$

To see Jump-With-Argument as a fragment of CBPV+$\underline{\mathtt{Ans}}$, regard

$$\begin{array}{rcl} \neg A & \text{as} & U(A \to \underline{\mathtt{Ans}}) \\ \Gamma \vdash^{\mathsf{n}} M & \text{as} & \Gamma \vdash^{\mathsf{c}} M : \underline{\mathtt{Ans}} \\ \gamma\mathtt{x}.M & \text{as} & \mathtt{thunk}\ \lambda\mathtt{x}.M \\ V \nearrow W & \text{as} & V\text{`}\mathtt{force}\ W \end{array}$$

Figure 7.1. Terms of Jump-With-Argument, and its embedding into CBPV+$\underline{\mathtt{Ans}}$

7.3.2 Graphical Syntax For JWA

The term syntax (more accurately, the abstract syntax) for JWA is rather inconvenient for jumping around. We therefore use a graphical flowchart-like syntax, illustrated in Fig. 7.2, where

- since γ represents a point, it is written •
- each instruction is enclosed in a rectangle;
- binding occurrences of identifiers and patterns are placed on edges (enclosed in a rounded rectangle).

We give the name *jumpabout* to this kind of tree of rectangles, edges and points. (We will not give a precise definition, as our explanation of jumping is informal.)

7.3.3 Execution

We will illustrate execution using the command `y` : **nat** $\vdash^{\mathsf{n}}$ M where M is the last example in Fig. 7.2. Here is the graphical syntax; we have numbered each point and each rectangle for ease of reference.

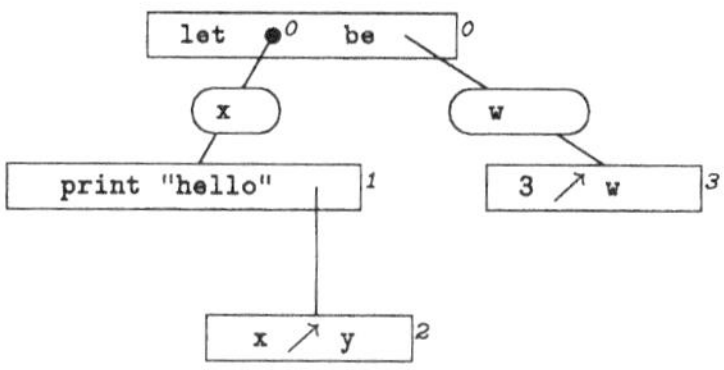

This jumpabout is called the *code*. As we execute it, we form another jumpabout called the *trace*. We call a point in the code jumpabout a *code point*, and a point in the trace a *trace point*; similarly for rectangles. During execution, the code stays fixed but the trace grows. The basic cycle of execution is

- copy the current instruction from the code to the trace
- obey the current instruction in the trace.

We begin at the top; then we follow the sequence of instructions down the code, except when we jump.

cycle 0 copy instruction We copy the instruction `let` • `be` from code rectangle 0, giving

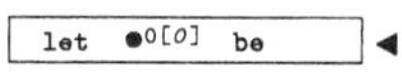

Notice that we number each trace point, and indicate in square brackets the code point that it was copied from—the latter is called the *teacher*. Thus code point 0 is the teacher of trace point 0. To reduce clutter, we omit the numbers and teachers of trace rectangles.

We write ◄ for "where we are now".

obey instruction Now we must obey this instruction, by binding the value `pt0` to `x` giving

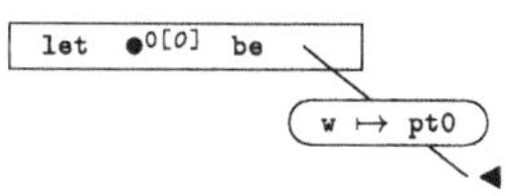

Term Syntax	Graphical Syntax

print "hello". print "goodbye". 3 $\nearrow$ y

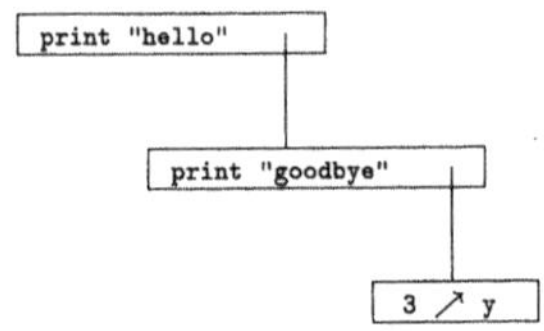

$$\texttt{pm z as} \begin{cases} \texttt{inl y.} & \texttt{print "hello". 3} \nearrow \texttt{y} \\ \texttt{inr y.} & \texttt{diverge} \end{cases}$$

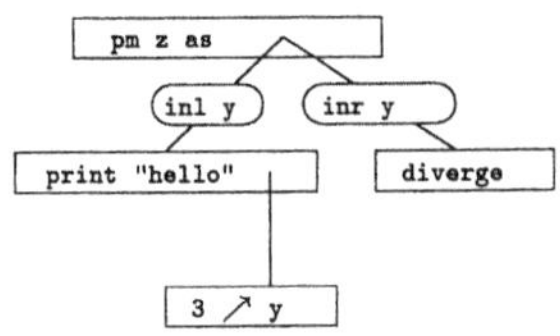

pm z as (x, y). print "hello". 3 $\nearrow$ y

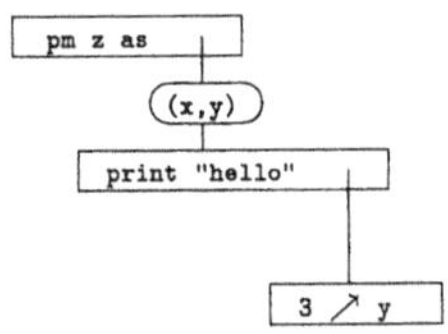

let 3 be x. print "hello". x $\nearrow$ y

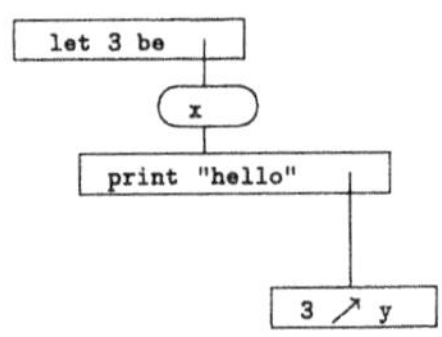

γx.(print "hello". x $\nearrow$ y)

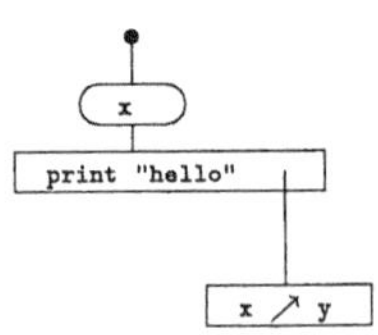

let γx.(print "hello". x $\nearrow$ y) be w. (3 $\nearrow$ w)

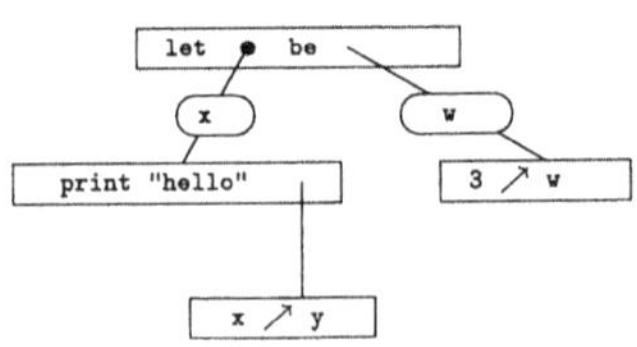

Figure 7.2. Examples of Graphical Syntax for JWA

cycle 1 copy instruction We copy the instruction 3 ↗ `w` from code rectangle 3 to the trace, giving

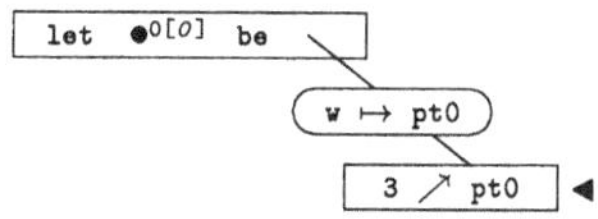

Notice that we have replaced `w` with its binding, point 0. We find this binding by looking up the branch of the trace.

obey instruction The instruction tells us to jump to point 0 (i.e. trace point 0) taking 3 as an argument. We obey it by jumping to trace point 0 and binding `x` to 3, giving

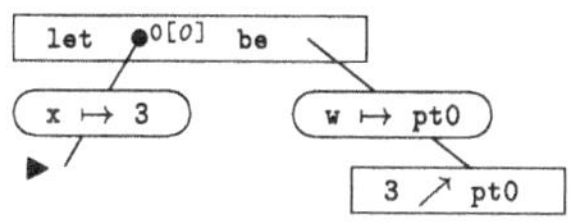

cycle 2 copy instruction We copy the next instruction `print "hello"` from code rectangle 1 to the trace, giving

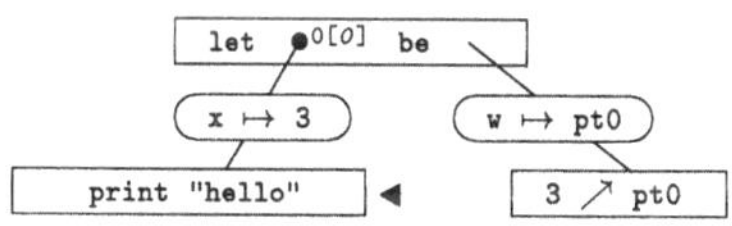

obey instruction We obey this by just printing "`hello`". No binding is made, so we are ready for the next instruction:

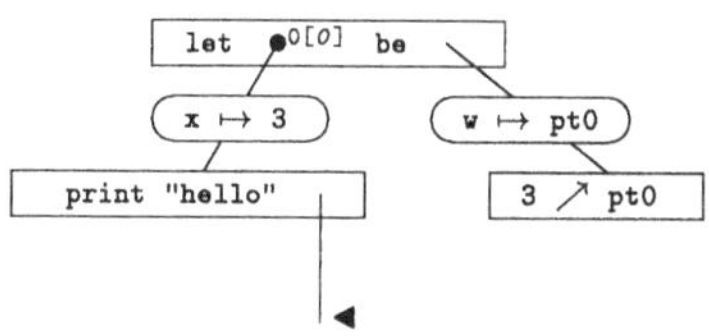

cycle 3 copy instruction We copy the next instruction `x` ↗ `y` from code rectangle 2 to the trace, giving

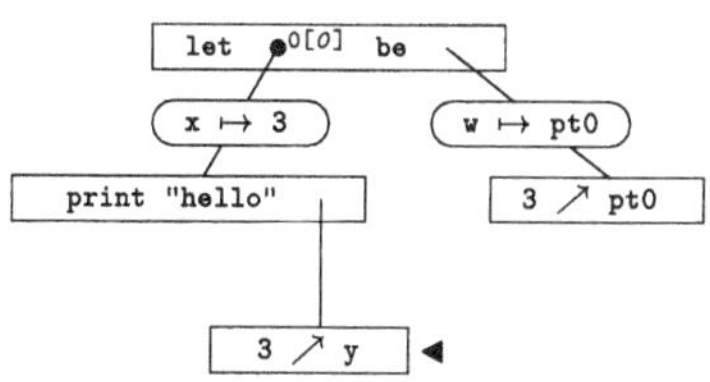

Notice that we have replaced `x` with its binding 3. We find this binding by looking up the branch of the trace.

obey instruction This instruction tells us to jump to y taking 3. But y is free, so execution terminates:

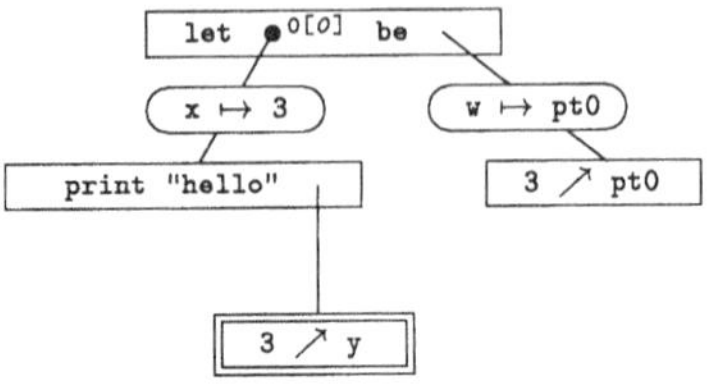

Notice that the function taking trace points and trace rectangles to their teachers describes a jumpabout homomorphism called the *teacher homomorphism* from the trace to the code. As the trace grows during execution, the teacher homomorphism grows with it.

This example program is very simplistic. More interesting situations arise when we jump several times to the same trace point. Each time, a new branch of the trace comes into existence, and so (unlike in our example) there can be many trace points with the same teacher. An example is Fig. 7.3, which the reader is strongly encouraged to execute.

7.4 The StkPS Transform

7.4.1 StkPS Transform As Jumping Implementation

The StkPS transform from CBPV+control to JWA is given in Fig. 7.4. By composing it with the embedding in Fig. 7.1, we recover the continuation semantics given in Sect. 5.4.6.

As we have described (if only informally) a jumping operational semantics for JWA, we obtain from the StkPS transform a jumping implementation of CBPV+control. Notice the following.

- As depicted in Fig. 5.7, there are 2 kinds of values that are translated as jump-points: thunks and continuations. This means that, in CBPV, thunks and continuations are points that we can jump to.
- There are 2 computations that are translated as jumps: `force` V (which is a jump to the thunk V) and `return` V (which is a jump to the current continuation). This makes precise the intuition of Sect. I.5.3.
- The transform of a computation describes its behaviour in the CK-machine. For instance, the computation $\lambda\mathtt{x}.M$ first pattern-matches its stack as $V :: K$, then binds x to V, then performs M with stack K. This is exactly described by the transform of $\lambda\mathtt{x}.M$.

(a) the code—term syntax

```
i : ¬bool ⊢^n let γx.(
                        pm x  as (u, v).
                            let γw.(v ↗ i) be w.
                                pm u as {inl y.(w ↗ y), inr y.(() ↗ y)}
                       ) be f. ((inl γz.((inr z, false) ↗ f), true) ↗ f)
⇝* true ↗ i
```

(b) the code—graphical syntax

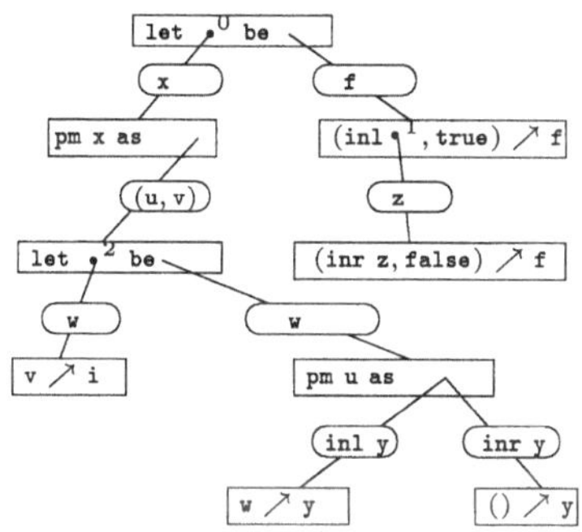

(c) the trace

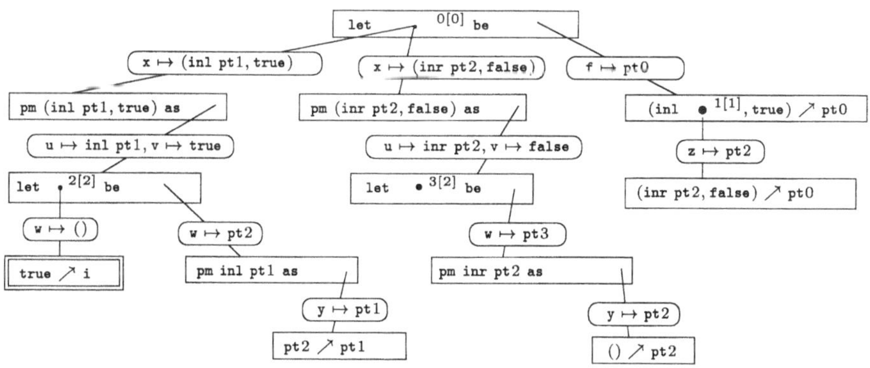

Figure 7.3. Example Program With Repeated Jumps—Please Try This

7.4.2 Related Transforms

The various CPS transforms that appear in the literature, listed and discussed in [Thielecke, 1997a], can all be recovered from the StkPS

A	$\overline{A}$
$U\underline{B}$	$\neg\overline{\underline{B}}$
$\sum_{i\in I} A_i$	$\sum_{i\in I}\overline{A_i}$
$A \times A'$	$\overline{A} \times \overline{A'}$
$\mathtt{stk}\ \underline{B}$	$\overline{\underline{B}}$

$\underline{B}$	$\overline{\underline{B}}$
FA	$\neg\overline{A}$
$\prod_{i\in I} \underline{B}_i$	$\sum_{i\in I}\overline{\underline{B}_i}$
$A \to \underline{B}$	$\overline{A} \times \overline{\underline{B}}$
$\underline{\mathtt{nrcomm}}$	1

$A_0, \ldots, A_{n-1} \vdash^{\mathsf{v}} V : B$	$\overline{A_0}, \ldots, \overline{A_{n-1}} \vdash^{\mathsf{v}} \overline{V} : \overline{B}$
$\mathtt{x}$	$\mathtt{x}$
$(\hat{\imath}, V)$	$(\hat{\imath}, \overline{V})$
(V, V')	$(\overline{V}, \overline{V'})$
$\mathtt{thunk}\ M$	$\gamma\mathtt{k}.\overline{M}$
$[\cdot]\ \mathtt{to}\ \mathtt{x}.\ M :: K$	$\gamma\mathtt{x}.(\overline{M}[\overline{K}/\mathtt{k}])$
$\hat{\imath} :: K$	$(\hat{\imath}, \overline{K})$
$V :: K$	$(\overline{V}, \overline{K})$

$A_0, \ldots, A_{n-1} \vdash^{\mathsf{c}} M : \underline{B}$	$\overline{A_0}, \ldots, \overline{A_{n-1}}, \mathtt{k} : \overline{\underline{B}} \vdash^{\mathsf{n}} \overline{M}$
$\mathtt{let}\ V\ \mathtt{be}\ \mathtt{x}.\ M$	$\mathtt{let}\ \overline{V}\ \mathtt{be}\ \mathtt{x}.\ \overline{M}$
$\mathtt{pm}\ V\ \mathtt{as}\ \{\ldots, (i, \mathtt{x}).M_i, \ldots\}$	$\mathtt{pm}\ \overline{V}\ \mathtt{as}\ \{\ldots, (i, \mathtt{x}).\overline{M_i}, \ldots\}$
$\mathtt{pm}\ V\ \mathtt{as}\ (\mathtt{x}, \mathtt{y}).M$	$\mathtt{pm}\ \overline{V}\ \mathtt{as}\ (\mathtt{x}, \mathtt{y}).\overline{M}$
$\mathtt{return}\ V$	$\overline{V} \nearrow \mathtt{k}$
$M\ \mathtt{to}\ \mathtt{x}.\ N$	$\mathtt{let}\ \gamma\mathtt{x}.\overline{N}\ \mathtt{be}\ \mathtt{k}.\ \overline{M}$
$\lambda\{\ldots, i.M_i, \ldots\}$	$\mathtt{pm}\ \mathtt{k}\ \mathtt{as}\ \{\ldots, (i, \mathtt{k}).\overline{M_i}, \ldots\}$
$\hat{\imath}\text{`}M$	$\mathtt{let}\ (\hat{\imath}, \mathtt{k})\ \mathtt{be}\ \mathtt{k}.\ \overline{M}$
$\lambda\mathtt{x}.M$	$\mathtt{pm}\ \mathtt{k}\ \mathtt{as}\ (\mathtt{x}, \mathtt{k}).\overline{M}$
$V\text{`}M$	$\mathtt{let}\ (\overline{V}, \mathtt{k})\ \mathtt{be}\ \mathtt{k}.\ \overline{M}$
$\mathtt{force}\ V$	$\mathtt{k} \nearrow \overline{V}$
$\mathtt{letstk}\ \mathtt{x}.\ M$	$\mathtt{let}\ \mathtt{k}\ \mathtt{be}\ \mathtt{x}.\ \overline{M}$
$\mathtt{changestk}\ K.\ M$	$\mathtt{let}\ \overline{K}\ \mathtt{be}\ \mathtt{k}.\ \overline{M}$
$\mathtt{print}\ c.\ M$	$\mathtt{print}\ c.\ \overline{M}$

Figure 7.4. The StkPS transform from CBPV+control to JWA, with printing

transform, because in each case the source language is a fragment of CBPV+control.

- Primary among these transforms is the CBV CPS transform [Duba et al., 1991; Plotkin, 1976]. This is recovered from the StkPS transform along the embedding of CBV into CBPV. The CBV control operators (as in ML) are translated into CBPV as follows:

`cont` A	as	`stk` FA
`letcc x.` M	as	`letstk x.` M
`throw` M N	as	M `to x.` N `to y.` (`changestk x. return y`)

- The "CBN CPS transform" [Plotkin, 1976] is, in our terminology of Sect. 1.7.3, not CBN but lazy. Thus, as explained in [Hatcliff and Danvy, 1997], it is recovered from the CBV CPS transform via the thunking transform which we present in Sect. A.5. Many of the other CPS transforms listed in [Thielecke, 1997a] are likewise fragments of CBV.
- The "CBN CPS transform" of [Hofmann and Streicher, 1997] is in our terminology StkPS rather than CPS, but it is genuinely CBN and is recovered from our StkPS transform via the embedding of CBN in CBPV.

This provides yet another illustration of the unifying power of working with CBPV—it provides the *source language* for the StkPS transform from which all the others are obtained.

Furthermore, the *target language* JWA of the StkPS transform is also CBPV, in the sense that it is a fragment of CBPV+$\underline{\texttt{Ans}}$ (as we explained in Fig. 7.1). Admittedly, the transitions of Fig. 7.5 can be seen as CBN reductions or as CBV reductions—this is essentially the *indifference* result of [Plotkin, 1976]. But for denotational/equational purposes, we cannot regard JWA as embedded in CBN, because this embedding does not preserve the η-law for sum types. Regarding JWA as embedded in CBV is also somewhat problematic: $\neg A$ is then regarded as

$$A \to_{\mathbf{CBV}} \texttt{Ans} = U(A \to F\texttt{Ans})$$

Thus $\underline{\texttt{Ans}}$ in Fig. 7.1 has been replaced by $F\texttt{Ans}$, and we have lost generality. In the case of printing, for example, the embedding in Fig. 7.1 provides us with a JWA semantics for *any* $\mathcal{A}$-set, whereas the embedding in CBV allows only a free $\mathcal{A}$-set.

In summary: we have benefited from ensuring that both the source language and the target language of the StkPS transform are CBPV.

7.5 C-Machine

7.5.1 C-Machine for Effect-Free JWA

Whilst the jumping operational semantics for JWA is intuitive, it is useful to have also a more conventional operational semantics based on

Transitions

	Γ	$\texttt{let } V \texttt{ be x}.M$
$\rightsquigarrow$	Γ	$M[V/\texttt{x}]$
	Γ	$\texttt{pm } (\hat{\imath}, V) \texttt{ as } \{\ldots, (i, \texttt{x}).M_i, \ldots\}$
$\rightsquigarrow$	Γ	$M_{\hat{\imath}}[V/\texttt{x}]$
	Γ	$\texttt{pm } (V, V') \texttt{ as } (\texttt{x}, \texttt{y}).M$
$\rightsquigarrow$	Γ	$M[V/\texttt{x}, V'/\texttt{y}]$
	Γ	$V \nearrow \gamma\texttt{x}.M$
$\rightsquigarrow$	Γ	$M[V/\texttt{x}]$

Terminal Configurations

$\Gamma, \texttt{z} : \neg A, \Gamma'$	$V \nearrow \texttt{z}$
$\Gamma, \texttt{z} : \sum_{i \in I} A_i, \Gamma'$	$\texttt{pm z as } \{\ldots, (i, \texttt{x}).M_i, \ldots\}$
$\Gamma, \texttt{z} : A \times A', \Gamma'$	$\texttt{pm z as } (\texttt{x}, \texttt{y}).M$

Figure 7.5. C-Machine for Jump-With-Argument

abstract syntax. The *C-machine* is given in Fig. 7.5. It is similar to the CK-machine, except that there is no need for a stack. Recalling from Sect. 7.2 that we work with non-closed nonreturning commands, we fix a context Γ, and define a Γ-*configuration* to be a nonreturning command $\Gamma \vdash^{\mathsf{n}} M$.

Proposition 72 (determinism) (cf. Prop. 46) For every Γ-configuration M, precisely one of the following holds.

1 M is not terminal, and $M \rightsquigarrow N$ for a unique Γ-configuration N.

2 M is terminal, and there does not exist N such that $M \rightsquigarrow N$.

□

Definition 7.1 (cf. Prop. 2.4) We define the relation $\rightsquigarrow^*$ on configurations inductively:

$$\frac{}{M \rightsquigarrow^* M} \qquad\qquad \frac{M' \rightsquigarrow^* N}{M \rightsquigarrow^* N}\,(M \rightsquigarrow M')$$

□

Proposition 73 (cf. 47) For every Γ configuration M there is a unique terminal Γ-configuration N such that $M \leadsto^* N$, and there is no infinite sequence of transitions from M. □

Proof (in the style of [Tait, 1967]) We fix Γ. We say that a Γ-configuration M with the properties described in Prop. 47 is *reducible*. Thus our aim is to prove that every Γ-configuration is reducible.

For each type A we define a set $\mathsf{red}^{\mathsf{v}}_A$ of *reducible* values $\Gamma \vdash^{\mathsf{v}} V : A$ by induction on types.

- $\Gamma \vdash^{\mathsf{v}} V : \neg A$ is reducible iff V is either a free identifier or $\gamma\mathtt{x}.M$ where for all $V \in \mathsf{red}^{\mathsf{v}}_A$, $M[V/\mathtt{x}]$ is reducible.
- $\Gamma \vdash^{\mathsf{v}} V : \sum_{i \in I} A_i$ is reducible iff V is either a free identifier or $(\hat{\imath}, W)$ where $W \in \mathsf{red}^{\mathsf{v}}_{A_{\hat{\imath}}}$.
- $\Gamma \vdash^{\mathsf{v}} V : A \times A'$ is reducible iff V is either a free identifier or (W, W') where $W \in \mathsf{red}^{\mathsf{v}}_A$ and $W' \in \mathsf{red}^{\mathsf{v}}_{A'}$.

By mutual induction on M and V, we prove the following.

- For any nonreturning command $\Gamma, A_0, \ldots, A_{n-1} \vdash^{\mathsf{n}} M$, and any $W_0 \in \mathsf{red}^{\mathsf{v}}_{A_0}, \ldots, W_{n-1} \in \mathsf{red}^{\mathsf{v}}_{A_{n-1}}$, the Γ-configuration $M[\overrightarrow{W_i/\mathtt{x}_i}]$ is reducible.
- For any value $\Gamma, A_0, \ldots, A_{n-1} \vdash^{\mathsf{v}} V : B$, and $W_0 \in \mathsf{red}^{\mathsf{v}}_{A_0}, \ldots, W_{n-1} \in \mathsf{red}^{\mathsf{v}}_{A_{n-1}}$, the value $V[\overrightarrow{W_i/\mathtt{x}_i}]$ is reducible.

□

7.5.2 Adapting the C-Machine for `print`

In Fig. 7.5 a transition has the form

$$M \qquad \leadsto \qquad M' \quad (7.2)$$

ιen we add `print` to the language, we want a transition to have the form

$$M \qquad \leadsto \qquad m \qquad M'$$

some $m \in \mathcal{A}^*$. We therefore replace each transition (7.2) in Fig. 7.5 by

$$M \qquad \leadsto \qquad \texttt{""} \qquad M'$$

l we add the transition

$$\texttt{print}\ c.\ M \qquad \leadsto \qquad c \qquad M$$

We replace Def. 7.1 by the following.

Definition 7.2 We define the relation $\leadsto^*$, whose form is $M \leadsto^* m, M'$ inductively:

$$\frac{}{M \leadsto^* 1, M} \qquad\qquad \frac{M' \leadsto^* n, N}{M \leadsto^* m * n, N}\ (M \leadsto m, M')$$

□

It is easy to prove analogues of Prop. 72–73.

Proposition 74 (soundness) If $M \leadsto^* m, T$ then, for any $\rho \in [\![\Gamma]\!]$,

$$[\![M]\!]\rho = m * ([\![T]\!]\rho)$$

□

Corollary 75 Suppose $\underline{\mathsf{Ans}}$ has two elements a, b with the property that

$$\begin{array}{rcll} m * a & \neq & m' * a & \text{for } m \neq m' \in \mathcal{A}^* \\ m * a & \neq & m' * b & \text{for } m, m' \in \mathcal{A}^* \end{array}$$

(This property is satisfied by the free $\mathcal{A}$-set on a set of size $\geqslant 2$.) For a countable set N and element n, write $I_{N,n}$ for the function from N to $\underline{\mathsf{Ans}}$ that takes n to a and everything else to b.

Then for any nonreturning command $\mathtt{k} : \neg\sum_{i \in N} 1 \vdash^{\mathsf{n}} M$, we have $M \leadsto^* m, n \nearrow \mathtt{k}$ iff $[\![M]\!](\mathtt{k} \mapsto I_{N,n}) = m * a$. Hence terms with the same denotation are observationally equivalent. □

7.5.3 Relating Operational Semantics

We now have 3 operational semantics, each of which can be extended for `print`:

1 the CK-machine for CBPV+control,

2 the C-machine for JWA,

3 the jumping machine for JWA (which we have described only informally).

The StkPS transform exactly preserves machine progress, from (1) to (2):

Proposition 76 Let M, K be a Γ-configuration of the CK-machine for CBPV+control. Then $\overline{M}[\overline{K}/\mathtt{k}]$ is a $\overline{\Gamma}$-configuration of the C-machine. Furthermore:

- M, K is terminal iff $\overline{M}[\overline{K}/\mathtt{k}]$ is terminal.
- If $M, K \leadsto N, L$, then $\overline{M}[\overline{K}/\mathtt{k}] \leadsto \overline{N}[\overline{L}/\mathtt{k}]$.

□

This enables us to deduce the termination of the CK-machine (Prop. 47 and the weaker Prop. 12) from the termination of the C-machine for JWA (Prop. 73).

Likewise, the relationship between the (2) and (3) is exact: one transition of the C-machine corresponds to one cycle of the jumping machine. (We do not prove this here.) For example, the computation in Sect. 7.3.3 rewrites in 3 transitions to $3 \nearrow \mathtt{y}$, printing `hello` in the last transition. These transitions correspond to cycle 0, cycle 1 and cycle 2 in the jumping execution.

7.5.4 Observational Equivalence

Definition 7.3 A *ground context* is a nonreturning command

$$\mathtt{k} : \neg\textstyle\sum_{i \in I} 1 \vdash^{\mathsf{n}} \mathcal{C}[\cdot]$$

with one or more occurrences of a hole which may be a nonreturning command or a value. □

Definition 7.4 Let $\Gamma \vdash^{\mathsf{n}} M$ and $\Gamma \vdash^{\mathsf{n}} N$ be nonreturning commands. We say that $M \simeq N$ when for any ground context $\mathcal{C}$

$$\mathcal{C}[M] \leadsto^* n \nearrow \mathtt{k} \text{ iff } \mathcal{C}[N] \leadsto^* n \nearrow \mathtt{k}$$

Similarly for values. □

We can easily adapt Def. 7.4 to different computational effects, e.g. printing.

7.6 Equations

7.6.1 Equational Theory

We presented the syntax for JWA in Fig. 7.1. But before we formulate the equational theory, we need, as usual, to add complex values to JWA. This is done just as for CBPV, in Fig. 7.6. As in Sect. 3.2, we exclude these values when considering operational semantics, but include them otherwise. We define the equational theory in Fig. 7.7. As with CBPV, we have

Proposition 77 1 Any computation is provably equal to a computation without complex value constructs.

$$\frac{\Gamma \vdash^{\mathsf{v}} V : A \quad \Gamma, \mathtt{x} : A \vdash^{\mathsf{v}} W : B}{\Gamma \vdash^{\mathsf{v}} \mathtt{let}\ V\ \mathtt{be}\ \mathtt{x}.\ W : B}$$

$$\frac{\Gamma \vdash^{\mathsf{v}} V : \sum_{i \in I} A_i \quad \cdots \ \Gamma, \mathtt{x} : A_i \vdash^{\mathsf{v}} W_i : B \ \cdots_{i \in I}}{\Gamma \vdash^{\mathsf{v}} \mathtt{pm}\ V\ \mathtt{as}\ \{\ldots, (i, \mathtt{x}).W_i, \ldots\} : B}$$

$$\frac{\Gamma \vdash^{\mathsf{v}} V : A \times A' \quad \Gamma, \mathtt{x} : A, \mathtt{y} : A' \vdash^{\mathsf{v}} W : B}{\Gamma \vdash^{\mathsf{v}} \mathtt{pm}\ V\ \mathtt{as}\ (\mathtt{x}, \mathtt{y}).W : B}$$

Figure 7.6. Complex Values For JWA

2 Any closed value is provably equal to a closed value without complex value constructs.

□

The reversible derivations of JWA are presented in Fig. 7.8.

β-laws

$$\begin{array}{rcl} \mathtt{let}\ V\ \mathtt{be}\ \mathtt{x}.\ Q & = & Q[V/\mathtt{x}] \\ \mathtt{pm}\ (\hat{\imath}, V)\ \mathtt{as}\ \{\ldots, (i, \mathtt{x}).Q_i, \ldots\} & = & Q_{\hat{\imath}}[V/\mathtt{x}] \\ \mathtt{pm}\ (V, V')\ \mathtt{as}\ (\mathtt{x}, \mathtt{y}).Q & = & Q[V/\mathtt{x}, V'/\mathtt{y}] \\ V \nearrow \gamma\mathtt{x}.M & = & M[V/\mathtt{x}] \end{array}$$

η-laws

$$\begin{array}{rcl} Q[V/\mathtt{z}] & = & \mathtt{pm}\ V\ \mathtt{as}\ \{\ldots, (i, \mathtt{x}).\ {}^{\mathtt{x}}Q[(i, \mathtt{x})/\mathtt{z}], \ldots\} \\ Q[V/\mathtt{z}] & = & \mathtt{pm}\ V\ \mathtt{as}\ (\mathtt{x}, \mathtt{y}).\ {}^{\mathtt{xy}}Q[(\mathtt{x}, \mathtt{y})/\mathtt{z}] \\ V & = & \gamma\mathtt{x}.(\mathtt{x} \nearrow {}^{\mathtt{x}}V) \end{array}$$

Figure 7.7. JWA equations, using conventions of Sect. I.4.2

7.6.2 Example Isomorphisms

In JWA we can coalesce several continuations into one, using

$$\neg A \times \neg A' \cong \neg(A + A') \tag{7.3}$$

To make this precise, we need to define syntactic isomorphism.

Definition 7.5 An *isomorphism* from A to B is a pair of values $\mathtt{x} : A \vdash^{\mathsf{v}} V : B$ and $\mathtt{y} : B \vdash W : A$ such that $V[W/\mathtt{x}] = \mathtt{y}$ and $W[V/\mathtt{y}] = \mathtt{x}$ are provable. □

Rules for $\sum$

$$\frac{\cdots\ \Gamma, A_i \vdash^{\mathsf{v}} B\ \cdots_{\ i\in I}}{\Gamma, \sum_{i\in I} A_i \vdash^{\mathsf{v}} B}$$

$$\frac{\cdots\ \Gamma, A_i \vdash^{\mathsf{n}}\ \cdots_{\ i\in I}}{\Gamma, \sum_{i\in I} A_i \vdash^{\mathsf{n}}}$$

Rules for $\times$

$$\frac{\Gamma \vdash^{\mathsf{v}} A \quad \Gamma \vdash^{\mathsf{v}} A'}{\Gamma \vdash^{\mathsf{v}} A \times A'}$$

$$\frac{\Gamma, A, A' \vdash^{\mathsf{v}} B}{\Gamma, A \times A' \vdash^{\mathsf{v}} B}$$

$$\frac{\Gamma, A, A' \vdash^{\mathsf{n}}}{\Gamma, A \times A' \vdash^{\mathsf{n}}}$$

Rule for $\neg$

$$\frac{\Gamma, A \vdash^{\mathsf{n}}}{\Gamma \vdash^{\mathsf{v}} \neg A}$$

Figure 7.8. Reversible Derivations For JWA

The isomorphism 7.3 is then easily constructed.

Proposition 78 Every JWA type is equivalent to one in *type canonical form*, meaning a type in the following class:

$$A ::= \quad \textstyle\sum_{i\in I} \neg A_i$$

□

Explicitly, a type canonical form can be written

$$\textstyle\sum_{i\in I} \neg \sum_{j\in J_i} \neg \sum_{k\in K_{ij}} \neg \sum_{l\in L_{ijk}} \neg \cdots \tag{7.4}$$

7.7 JWA As A Type Theory For Classical Logic

We said in Sect. 7.2 that the type constructor $\neg$ in JWA is not exactly the same as logical negation. This is because (for example) there does not exist a value $\neg\neg 0 \vdash^{\mathsf{v}} V : 0$, even though $\neg\neg 0 \vdash 0$ is provable in both intuitionistic and classical logic.

But although the JWA *value* judgement $\vdash^{\mathsf{v}}$ is not related to logic, the JWA *nonreturning command* judgement $\vdash^{\mathsf{n}}$ is related to logic. This is an example of the phenomenon described in Sect. 2.10: many results apply only to computations, and the role of values is merely auxiliary. We now explore this relationship.

Definition 7.6 ■ By *intuitionistic propositional logic* we mean the type theory $\times\sum\prod\to$-calculus defined in Fig. 3.8, omitting the terms. Thus a sequent has the form $\Gamma \vdash A$.

■ By *classical propositional logic* we mean Gentzen's system LK (see e.g. [Girard et al., 1988]), where we write $\sum$ for disjunction and $\times$ for conjunction. Thus a sequent has the form $\Gamma \vdash \Delta$, where both Γ and Δ are "contexts" i.e. finite sequences of propositions. We say that $\Gamma \vdash \Delta$ and $\Gamma' \vdash \Delta'$ are *provably equivalent* when

$$\vdash (\bigwedge \Gamma \Rightarrow \bigvee \Delta) \Leftrightarrow (\bigwedge \Gamma' \Rightarrow \bigvee \Delta')$$

is provable.

□

Some remarks:

- Any other formulation of classical logic, e.g. Hilbert's, or that of the Sheffer stroke, would be just as acceptable for our purposes.
- According to Def. 7.6, $\neg$ is included among the primitive connectives of classical logic, but not among those of intuitionistic logic.
- We have provided an equality relation between intuitionistic proofs (the equational theory of Sect. 3.8), and this equates any two proofs of $\Gamma \vdash 0$ (easy exercise, whose categorical analogue is Prop. 4(3)).
- By contrast, we do not provide any equality relation between classical proofs.

We can now state the key result, central to much work relating control effects and continuations to logic [Friedman, 1978; Griffin, 1990; Hofmann and Streicher, 1997; Lafont et al., 1993; Murthy, 1990].

Proposition 79 1 Any sequent of classical logic is provably equivalent to a sequent of the form $\Gamma \vdash$, where Γ is a JWA context.

2 If Γ is a JWA context, then $\Gamma \vdash$ is provable in classical logic iff there is a nonreturning command $\Gamma \vdash^{\mathsf{n}} M$ in effect-free JWA (with or without complex values, it makes no difference by Prop. 77).

□

We stress that this is a result about provability, not about proofs, and we do not regard as canonical any *particular* translation from LK into JWA. After all, we not would regard as canonical any particular translation from LK to Hilbert's system. All Prop. 79 tells us is that such translations exist, which is all that is needed for JWA to be a system of classical logic.

Proof 1 is straightforward. 2 is a consequence of the following stronger result. Let Γ be a context in Jump-With-Argument. Then the following are equivalent.[1].

1 $\Gamma \vdash$ is provable in classical propositional logic.

2 There exists a nonreturning command $\Gamma \vdash^{\mathsf{n}} M$ in effect-free JWA.

3 $\Gamma[(- \to \mathtt{Ans})/\neg] \vdash \mathtt{Ans}$ is provable in intuitionistic propositional logic extended with a proposition identifier $\mathtt{Ans}$.

4 $\Gamma[(- \to 0)/\neg] \vdash 0$ is provable in intuitionistic propositional logic.

This is proved as follows.

(2)⇒(3) This is given by the composite transform

$$\text{JWA} \hookrightarrow \text{CBPV+}\underline{\mathtt{Ans}} \xrightarrow{\text{trivialization}} \times\sum\prod \to\text{-calculus+}\mathtt{Ans}$$

which takes $\neg$ to $- \to \mathtt{Ans}$ and takes a nonreturning command to a term of type $\mathtt{Ans}$.

(3)⇒(4) We substitute 0 for $\mathtt{Ans}$.

(4)⇒(1) This is because

- intuitionistic provability implies classical provability;
- $\neg A$ and $A \to 0$ are equivalent in classical logic.

(1)⇒(2) We define a transform (again, this is just one choice and it is not canonical) from propositions to JWA types as follows:

A	$\overline{A}$
$\neg A$	$\neg\overline{A}$
$\sum_{i\in I} A_i$	$\sum_{i\in I} \overline{A_i}$
$A \times A'$	$\overline{A} \times \overline{A'}$
$\prod_{i\in I} A_i$	$\neg\sum_{i\in I} \neg\overline{A_i}$
$A \to B$	$\neg(\overline{A} \times \neg\overline{B})$

[1]There is a rather fortuitous result that any (finitely wide) propositional formula of the form $\neg A$ is classically valid iff it is intuitionistically valid. One might think that the equivalence between (1) and (4) is no more than a special case of this fact. But that is not so, because our proof extends to the infinitely wide setting, and can easily be extended to predicate logic.

We then show that if $A_0, \ldots, A_{m-1} \vdash B_0, \ldots, B_{n-1}$ is provable, then there is a nonreturning command $\overline{A_0}, \ldots, \overline{A_{m-1}}, \neg\overline{B_0}, \ldots, \neg\overline{B_{n-1}} \vdash^{\mathsf{n}} M$ in effect-free JWA. The required result is an immediate consequence, because the transform $\overline{-}$ leaves JWA types unchanged.

□

We argue that, in light of Prop. 79, JWA can be viewed as a type theory for classical logic, provided we regard values (which are not logically significant) as merely auxiliary. Whereas some other proposed type theories such as $\lambda\mu$-calculus [Ong, 1996; Parigot, 1992] have to be arbitrarily designated call-by-value or call-by-name, JWA does not have this problem.

7.8 The StkPS Transform Is An Equivalence

7.8.1 The Main Result

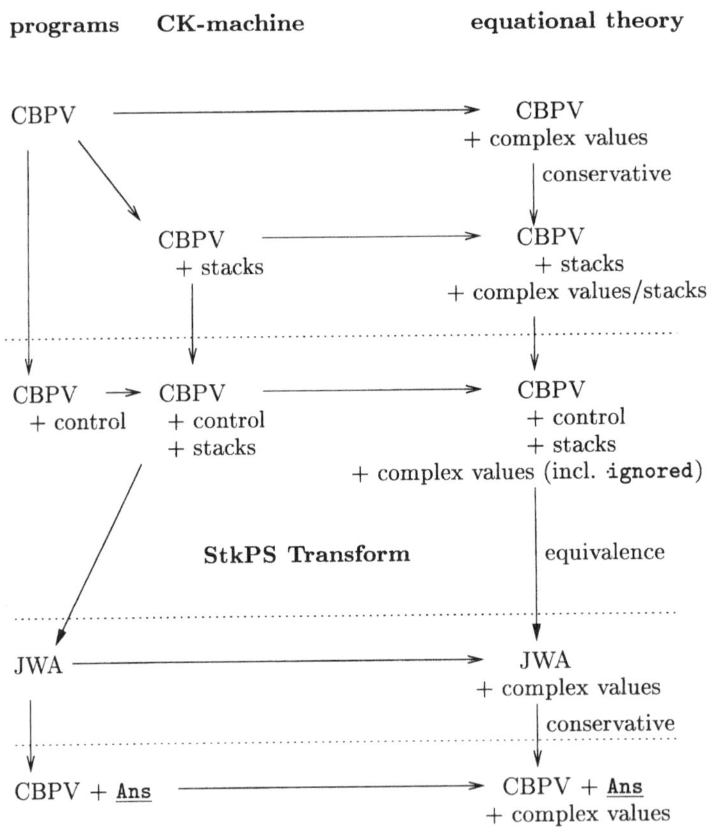

Figure 7.9. The StkPS transform between various languages

In Fig. 7.4 we presented the StkPS transform from CBPV+control to JWA. In this section, we show that it is an equivalence, so models of CBPV+control and models of JWA are essentially the same thing. However, JWA is much simpler and more elegant. We mention again that this material, like all of our material involving equational theory and complex values, is not presently applicable to infinitely deep CBPV/JWA, although it is applicable to infinitely wide CBPV/JWA.

The sense in which the StkPS transform is an equivalence is the following—a typed variant of the "equational correspondence" approach of [Sabry and Felleisen, 1993]:

Proposition 80 1 Every JWA type A is isomorphic to $\overline{B}$ for some value type B.

2 Every JWA type A is isomorphic to $\overline{\underline{B}}$ for some computation type $\underline{B}$.

3 For every context Γ and value type A in CBPV+control, the StkPS transform defines a bijection from the provable-equality classes of values $\Gamma \vdash^{\mathsf{v}} V : A$ to the provable-equality classes of values $\overline{\Gamma} \vdash^{\mathsf{v}} W : \overline{A}$.

4 For every context Γ and computation type $\underline{B}$ in CBPV+control, the StkPS transform defines a bijection from the provable-equality classes of computations $\Gamma \vdash^{\mathsf{c}} M : \underline{B}$ to the provable-equality classes of nonreturning commands $\overline{\Gamma}, \mathtt{k} : \overline{\underline{B}} \vdash^{\mathsf{n}} N$.

(3)–(4) can be extended to values and computations with holes i.e. contexts. □

For Prop. 80 to make sense, we will have to give an equational theory for CBPV+control. We do this in Sect. 7.8.3. We then prove Prop. 80 in Sect. 7.8.6.

Using Prop. 76, we deduce

Corollary 81 The StkPS transform is fully abstract. □

7.8.2 Complex Values

As depicted in Fig. 7.9, we want to contrive an equational theory for CBPV+control (including the stack constructs of Fig. 5.4) that will make Prop. 80 true. As usual, we have to add complex values to get an equational theory with the desired properties, and the complex value constructs are shown in Fig. 7.10. Except for the last one ($\cdot$`ignored`, which we explain presently) these constructs all correspond to those in Fig. 3.5.

In Sect. 2.9.2 we argued that there is essentially one stack from $F0$, called `ignored`. We defined `ignored` there in terms of `nil`. Since `nil`

$$\frac{\Gamma \vdash^{\mathsf{v}} V : A \quad \Gamma, \mathtt{x} : A \vdash^{\mathsf{v}} W : B}{\Gamma \vdash^{\mathsf{v}} \mathtt{let}\ V\ \mathtt{be}\ \mathtt{x}.\ W : B}$$

$$\frac{\Gamma \vdash^{\mathsf{v}} V : \sum_{i \in I} A_i \quad \cdots \ \Gamma, \mathtt{x} : A_i \vdash^{\mathsf{v}} W_i : B \ \cdots_{i \in I}}{\Gamma \vdash^{\mathsf{v}} \mathtt{pm}\ V\ \mathtt{as}\ \{\ldots, (i, \mathtt{x}).W_i, \ldots\} : B}$$

$$\frac{\Gamma \vdash^{\mathsf{v}} V : A \times A' \quad \Gamma, \mathtt{x} : A, \mathtt{y} : A' \vdash^{\mathsf{v}} W : B}{\Gamma \vdash^{\mathsf{v}} \mathtt{pm}\ V\ \mathtt{as}\ (\mathtt{x}, \mathtt{y}).W : B}$$

$$\frac{\Gamma, \mathtt{n} : \mathtt{stk}\ \underline{B} \vdash^{\mathsf{v}} V : C \quad \Gamma \vdash^{\mathsf{v}} L : \mathtt{stk}\ \underline{B}}{\Gamma \vdash^{\mathsf{v}} V\ \ \cdot\mathtt{where}\ \mathtt{n}\ \mathtt{is}\ L : C}$$

$$\frac{\cdots \ \Gamma, \mathtt{n} : \mathtt{stk}\ \underline{B}_i \vdash^{\mathsf{v}} V_i : C \ \cdots_{i \in I} \quad \Gamma \vdash^{\mathsf{v}} L : \mathtt{stk}\ \prod_{i \in I} \underline{B}_i}{\Gamma \vdash^{\mathsf{v}} \{\ldots, V_i\ \ \cdot\mathtt{where}\ i \cdot\!\!:: \mathtt{n}, \ldots\}\ \mathtt{is}\ L : C}$$

$$\frac{\Gamma, \mathtt{x} : A, \mathtt{n} : \mathtt{stk}\ \underline{B} \vdash^{\mathsf{v}} V : C \quad \Gamma \vdash^{\mathsf{v}} L : \mathtt{stk}\ (A \to \underline{B})}{\Gamma \vdash^{\mathsf{v}} V\ \ \cdot\mathtt{where}\ \mathtt{x} \cdot\!\!:: \mathtt{n}\ \mathtt{is}\ L : C}$$

$$\frac{}{\Gamma \vdash^{\mathsf{v}} \cdot\mathtt{ignored} : \mathtt{stk}\ F0}$$

Figure 7.10. Complex Values For CBPV+Control

is a free identifier in the control setting, **ignored** is not closed. But we need a closed value of type $\mathtt{stk}\ F0$ to make Prop. 80(3) true, because $\mathtt{stk}\ F0$ denotes $\neg 0 \cong 1$. We therefore introduce a constant $\cdot\mathtt{ignored}$ as shown in Fig. 7.10.

7.8.3 Equational Theory For CBPV+Control

The equational theory for CBPV+control consists of the usual equations of Fig. 3.3, together with those of Fig. 7.11 (using the conventions of Sect. I.4.2). Notice that these equations are *ad hoc* and inelegant, by contrast with the simplicity of the JWA equational theory, to which they will be shown equivalent. The equations for `print` and `diverge` are of course effect-specific, but there would be a similar equation for other effects.

Looking at Fig. 7.9, we see that there are 2 arrows into this equational theory and 1 arrow out. There are thus 3 tasks we must perform for this theory:

β-laws

$$\begin{array}{rcl}
V \ \cdot\mathtt{where}\ \mathtt{n}\ \mathtt{is}\ L & = & V[L/\mathtt{n}] \\
\{\ldots, V_i \ \cdot\mathtt{where}\ i \cdot\!:: \mathtt{n}, \ldots\}\ \mathtt{is}\ \hat{\imath} \cdot\!:: L & = & V_{\hat{\imath}}[L/\mathtt{n}] \\
V \ \cdot\mathtt{where}\ \mathtt{x} \cdot\!:: \mathtt{n}\ \mathtt{is}\ W \cdot\!:: L & = & V[W/\mathtt{x}, L/\mathtt{n}] \\
\mathtt{changestk}\ ([\cdot]\ \mathtt{to}\ \mathtt{x}.\ N \cdot\!:: K).\ \mathtt{return}\ V & = & \mathtt{changestk}\ K.\ N[V/\mathtt{x}] \\
\mathtt{changestk}\ (\hat{\imath} \cdot\!:: K).\ \lambda\{\ldots, i.M_i, \ldots\} & = & \mathtt{changestk}\ K.\ M_{\hat{\imath}} \\
\mathtt{changestk}\ (V \cdot\!:: K).\ \lambda\mathtt{x}.\ M & = & \mathtt{changestk}\ K.\ M[V/\mathtt{x}] \\
\mathtt{changestk}\ K.\ \mathtt{letstk}\ \mathtt{x}.\ M & = & \mathtt{changestk}\ K.\ M[K/\mathtt{x}] \\
\mathtt{changestk}\ K.\ \mathtt{changestk}\ L.\ M & = & \mathtt{changestk}\ L.\ M
\end{array}$$

η-laws

$$\begin{array}{rcl}
V[L/\mathtt{z}] & = & \{\ldots, {}^{\mathtt{n}}V[i \cdot\!:: \mathtt{n}/\mathtt{z}] \ \cdot\mathtt{where}\ i \cdot\!:: \mathtt{n}, \ldots\}\ \mathtt{is}\ L \\
V[L/\mathtt{z}] & = & {}^{\mathtt{xn}}V[\mathtt{x} \cdot\!:: \mathtt{n}/\mathtt{z}] \ \cdot\mathtt{where}\ \mathtt{x} \cdot\!:: \mathtt{n}\ \mathtt{is}\ L \\
\mathtt{changestk}\ L.\ M[K/\mathtt{z}] & = & \mathtt{changestk}\ K.\ \lambda\{\ldots, i.\mathtt{letstk}\ \mathtt{w}. \\
 & & \quad \mathtt{changestk}\ L.\ {}^{\mathtt{w}}M[i \cdot\!:: \mathtt{w}/\mathtt{z}], \ldots\} \\
\mathtt{changestk}\ L.\ M[K/\mathtt{z}] & = & \mathtt{changestk}\ K.\ \lambda\mathtt{x}.\mathtt{letstk}\ \mathtt{w}. \\
 & & \quad \mathtt{changestk}\ L.\ {}^{\mathtt{xw}}M[\mathtt{x} \cdot\!:: \mathtt{w}/\mathtt{z}] \\
K & = & [\cdot]\ \mathtt{to}\ \mathtt{x}.\ (\mathtt{changestk}\ K.\ \mathtt{return}\ \mathtt{x}) \\
 & & \quad \cdot\!::\ \mathtt{ignored} \\
K & = & \cdot\mathtt{ignored} \\
M & = & \mathtt{letstk}\ \mathtt{x}.\ \mathtt{changestk}\ \mathtt{x}.\ {}^{\mathtt{x}}M \\
M & = & \mathtt{changestk}\ \cdot\mathtt{ignored}.\ M
\end{array}$$

sequencing law

$$\mathtt{changestk}\ K.\ M\ \mathtt{to}\ \mathtt{x}.\ N \quad = \quad M\ \mathtt{to}\ \mathtt{x}.\ \mathtt{changestk}\ {}^{\mathtt{x}}K.\ N$$

`print` law

$$\mathtt{changestk}\ K.\ \mathtt{print}\ c.\ M \quad = \quad \mathtt{print}\ c.\ \mathtt{changestk}\ K.\ M$$

`diverge` law

$$\mathtt{changestk}\ K.\ \mathtt{diverge} \quad = \quad \mathtt{diverge}$$

Figure 7.11. Equational Theory For CBPV+Control

- to represent in it the equational theory for CBPV+stacks (Sect. 7.8.4)
- to show that the embedding in it of CBPV+control is computation-unaffecting (Sect. 7.8.5)
- to describe the StkPS transform from it into the JWA equational theory, and show that this transform is an equivalence (Sect. 7.8.6).

7.8.4 Representing CBPV+Stacks In CBPV+Control

We extend Fig. 7.4 to treat complex stacks in Fig. 7.12.

Proposition 82 The embedding of CBPV+stacks in CBPV+control, shown in Fig. 7.12, has the following properties.

CBPV	CBPV+control
$\Gamma \mid \underline{B} \vdash^{\mathsf{k}} K : \underline{C}$	$\Gamma, \mathtt{nil} : \mathtt{stk}\ \underline{C} \vdash^{\mathsf{v}} K : \mathtt{stk}\ \underline{B}$
$[\cdot]\ \mathtt{to}\ \mathtt{x}.\ N :: K$	$[\cdot]\ \mathtt{to}\ \mathtt{x}.\ {}^{\mathtt{nil}}N \cdot\!:: K$
$\hat{\imath} :: K$	$\hat{\imath} \cdot\!:: K$
$V :: K$	${}^{\mathtt{nil}}V \cdot\!:: K$
$K\ \mathtt{where}\ \mathtt{nil}\ \mathtt{is}\ L$	${}^{\mathtt{nil}}K\ \cdot\mathtt{where}\ \mathtt{nil}\ \mathtt{is}\ L$
$\{\ldots, K_i\ \mathtt{where}\ i :: \mathtt{nil}, \ldots\}\ \mathtt{is}\ L$	$\{\ldots, {}^{\mathtt{nil}}K_i\ \cdot\mathtt{where}\ i \cdot\!:: \mathtt{nil}, \ldots\}\ \mathtt{is}\ L$
$K\ \mathtt{where}\ \mathtt{x} :: \mathtt{nil}\ \mathtt{is}\ L$	${}^{\mathtt{nil}}K\ \cdot\mathtt{where}\ \mathtt{x} \cdot\!:: \mathtt{nil}\ \mathtt{is}\ L$

Figure 7.12. Representing $\vdash^{\mathsf{k}}$

1 It preserves substitution.

2 Dismantling is represented as follows. If $\Gamma \vdash^{\mathsf{c}} M : \underline{B}$ and $\Gamma \mid \underline{B} \vdash^{\mathsf{k}} K : \underline{C}$ then

$$M \bullet K = \mathtt{letstk}\ \mathtt{nil}.\ \mathtt{changestk}\ K.\ M$$

3 Concatenation is represented as follows. If $\Gamma \mid \underline{B} \vdash^{\mathsf{k}} K : \underline{C}$ and $\Gamma \mid \underline{C} \vdash^{\mathsf{k}} L : \underline{D}$ then

$$K +\!\!+ L = K[L/\mathtt{nil}]$$

4 The embedding preserves provable equality.

□

7.8.5 Stacks and Complex Values Are Computation-Unaffecting

We took CBPV with the control operators **letstk/changestk**, and we added first the stacks of Fig. 5.4 (to typecheck the CK-machine) and then the complex values of Fig. 7.10 (to formulate an equational theory). But these additions did not generate any new computations, as we now see. The statement and the proof follow the same lines as Prop. 25(1), but there is no result about closed values corresponding to Prop. 25(2)

Proposition 83 There is an effective procedure that, given a computation $\Gamma \vdash^{\mathsf{c}} M : \underline{B}$, possibly containing stacks and complex values (including $\cdot$**ignored**), returns a computation $\Gamma \vdash^{\mathsf{c}} \tilde{M} : \underline{B}$ without these things, such that $M = \tilde{M}$ is provable. □

Proof We will define one such procedure, and simultaneously, for each value $\Gamma \vdash^{\mathsf{v}} V : A$, possibly containing complex values, and each complex-value-free computation $\Gamma, \mathtt{v} : A \vdash^{\mathsf{c}} N : \underline{B}$, we will define a complex-value-free computation $\Gamma \vdash^{\mathsf{c}} N[V /\!\!/ \mathtt{v}] : \underline{B}$ such that $N[V /\!\!/ \mathtt{v}] = N[V/\mathtt{v}]$ is provable.

By mutual induction on M and V, we define $\tilde{M}$ and $N[V/\!/\mathtt{v}]$ in Fig. 7.13 and we prove the equations

$$\tilde{M} = M$$
$$N[V/\!/\mathtt{v}] \;=\; N[V/\mathtt{v}]$$

□

V	$N[V/\!/\mathtt{v}]$
$\mathtt{x}$	$N[\mathtt{x}/\mathtt{v}]$
$\mathtt{let}\ W\ \mathtt{be}\ \mathtt{x}.\ U$	$(\mathtt{let}\ \mathtt{w}\ \mathtt{be}\ \mathtt{x}.\ N[U/\!/\mathtt{v}])[W/\!/\mathtt{w}]$
$(\hat{\imath}, W)$	$(\mathtt{let}\ (\hat{\imath}, \mathtt{w})\ \mathtt{be}\ \mathtt{v}.\ N)[W/\!/\mathtt{w}]$
$\mathtt{pm}\ W\ \mathtt{as}\ \{\ldots, (i, \mathtt{x}).U_i, \ldots\}$	$(\mathtt{pm}\ \mathtt{w}\ \mathtt{as}\ \{\ldots, (i, \mathtt{x}).N[U_i/\!/\mathtt{v}], \ldots\})[W/\!/\mathtt{w}]$
(W, W')	$(\mathtt{let}\ (\mathtt{w}, \mathtt{x})\ \mathtt{be}\ \mathtt{v}.\ N)[W/\!/\mathtt{w}][W'/\!/\mathtt{x}]$
$\mathtt{pm}\ W\ \mathtt{as}\ (\mathtt{x}, \mathtt{y}).U$	$(\mathtt{pm}\ \mathtt{w}\ \mathtt{as}\ (\mathtt{x}, \mathtt{y}).N[U/\!/\mathtt{v}])[W/\!/\mathtt{w}]$
$\mathtt{thunk}\ M$	$N[\mathtt{thunk}\ \tilde{M}/\mathtt{v}]$
$[\cdot]\ \mathtt{to}\ \mathtt{x}.\ M :: K$	$(\mathtt{letstk}\ \mathtt{q}.\ \mathtt{changestk}\ \mathtt{k}.\ (\mathtt{letstk}\ \mathtt{v}.\ \mathtt{changestk}\ \mathtt{q}.\ N)\ \mathtt{to}\ \mathtt{x}.\ \tilde{M})[K/\!/\mathtt{k}]$
$\hat{\imath} :: K$	$(\mathtt{letstk}\ \mathtt{q}.\mathtt{changestk}\ \mathtt{k}.\ \hat{\imath}`\mathtt{letstk}\ \mathtt{v}.\ \mathtt{changestk}\ \mathtt{q}.N)[K/\!/\mathtt{k}]$
$W :: K$	$(\mathtt{letstk}\ \mathtt{q}.\mathtt{changestk}\ \mathtt{k}.\ \mathtt{w}`\mathtt{letstk}\ \mathtt{v}.\ \mathtt{changestk}\ \mathtt{q}.N)[W/\!/\mathtt{w}, K/\!/\mathtt{k}]$
$W\ \cdot\mathtt{where}\ \mathtt{n}\ \mathtt{is}\ L$	$N[W/\!/\mathtt{v}][L/\!/\mathtt{n}]$
$\{\ldots, W_i\ \cdot\mathtt{where}\ i :: \mathtt{n}, \ldots\}\ \mathtt{is}\ L$	$(\mathtt{letstk}\ \mathtt{q}.\ \mathtt{changestk}\ \mathtt{l}.\ \lambda\{\ldots, i.\mathtt{letstk}\ \mathtt{n}.\ \mathtt{changestk}\ \mathtt{q}.\ (N[W_i/\!/\mathtt{v}]), \ldots\})[L/\!/\mathtt{l}]$
$W\ \cdot\mathtt{where}\ \mathtt{x} :: \mathtt{n}\ \mathtt{is}\ L$	$(\mathtt{letstk}\ \mathtt{q}.\ \mathtt{changestk}\ \mathtt{l}.\ \lambda\mathtt{x}.\ \mathtt{letstk}\ \mathtt{n}.\ \mathtt{changestk}\ \mathtt{q}.\ (N[W/\!/\mathtt{v}]))[L/\!/\mathtt{l}]$

M	$\tilde{M}$
$\mathtt{let}\ W\ \mathtt{be}\ \mathtt{x}.\ N$	$(\mathtt{let}\ \mathtt{w}\ \mathtt{be}\ \mathtt{x}.\ \tilde{N})[W/\!/\mathtt{w}]$
$\mathtt{pm}\ W\ \mathtt{as}\ \{\ldots, (i, \mathtt{x}).N_i, \ldots\}$	$(\mathtt{pm}\ \mathtt{w}\ \mathtt{as}\ \{\ldots, (i, \mathtt{x}).\tilde{N}_i, \ldots\})[W/\!/\mathtt{w}]$
$\mathtt{pm}\ W\ \mathtt{as}\ (\mathtt{x}, \mathtt{y}).N$	$(\mathtt{pm}\ \mathtt{w}\ \mathtt{as}\ (\mathtt{x}, \mathtt{y}).\tilde{N})[W/\!/\mathtt{w}]$
$\lambda\{\ldots, i.N_i, \ldots\}$	$\lambda\{\ldots, i.\tilde{N}_i, \ldots\}$
$\hat{\imath}`N$	$\hat{\imath}`\tilde{N}$
$\lambda\mathtt{x}.N$	$\lambda\mathtt{x}.\tilde{N}$
$V`N$	$(\mathtt{v}`\tilde{N})[V/\!/\mathtt{v}]$
$\mathtt{return}\ V$	$(\mathtt{return}\ \mathtt{v})[V/\!/\mathtt{v}]$
$N\ \mathtt{to}\ \mathtt{x}.\ P$	$\tilde{N}\ \mathtt{to}\ \mathtt{x}.\ \tilde{P}$
$\mathtt{force}\ V$	$(\mathtt{force}\ \mathtt{v})[V/\!/\mathtt{v}]$
$\mathtt{letstk}\ \mathtt{x}.\ N$	$\mathtt{letstk}\ \mathtt{x}.\ \tilde{N}$
$\mathtt{changestk}\ V.\ N$	$\mathtt{changestk}\ \mathtt{v}.\ \tilde{N}[V/\!/\mathtt{v}]$
$\mathtt{print}\ c.\ N$	$\mathtt{print}\ c.\ \tilde{N}$

Figure 7.13. Definitions used in proof of Prop. 83

7.8.6 The StkPS Transform Between Equational Theories

The StkPS transform from CBPV+control (including stacks) to JWA, shown in Fig. 7.4, is extended to complex values in Fig. 7.14.

$A_0, \ldots, A_{n-1} \vdash^{\mathsf{v}} V : B$	$\overline{A_0}, \ldots, \overline{A_{n-1}} \vdash^{\mathsf{v}} \overline{V} : \overline{B}$
$\texttt{let } V \texttt{ be x. } W$	$\texttt{let } \overline{V} \texttt{ be x.} \overline{W}$
$\texttt{pm } V \texttt{ as } \{\ldots, (i, \texttt{x}).W_i, \ldots\}$	$\texttt{pm } \overline{V} \texttt{ as } \{\ldots, (i, \texttt{x}).\overline{W_i}, \ldots\}$
$\texttt{pm } V \texttt{ as } (\texttt{x}, \texttt{y}).W$	$\texttt{pm } \overline{V} \texttt{ as } (\texttt{x}, \texttt{y}).\overline{W}$
$V \;\cdot\texttt{where n is } L$	$\texttt{let n be } \overline{L}.\ \overline{V}$
$\{\ldots, V_i \;\cdot\texttt{where } i :: \texttt{n}, \ldots\} \texttt{ is } L$	$\texttt{pm } \overline{L} \texttt{ as } \{\ldots, (i, \texttt{n}).\overline{V_i}, \ldots\}$
$V \;\cdot\texttt{where x} :: \texttt{n is } L$	$\texttt{pm } \overline{L} \texttt{ as } (\texttt{x}, \texttt{n}).\ \overline{V}$
$\cdot\texttt{ignored}$	$\gamma\{\}$

Figure 7.14. StkPS Transform On Complex Values

Proposition 84 The StkPS transform preserves substitution and provable equality. □

The remainder of this section is a proof of Prop. 80. A natural approach is to try to translate *JWA* back into CBPV+control. The question is how to translate $\neg$. Two possibilities seem natural: either as $U(- \to F0)$ or as $\texttt{stk}\ F-$ (the latter is the $\texttt{cont}$ of NJ-SML). Either of these would work, but we choose the former.

Therefore, our inverse translation from JWA to CBPV + control is given by

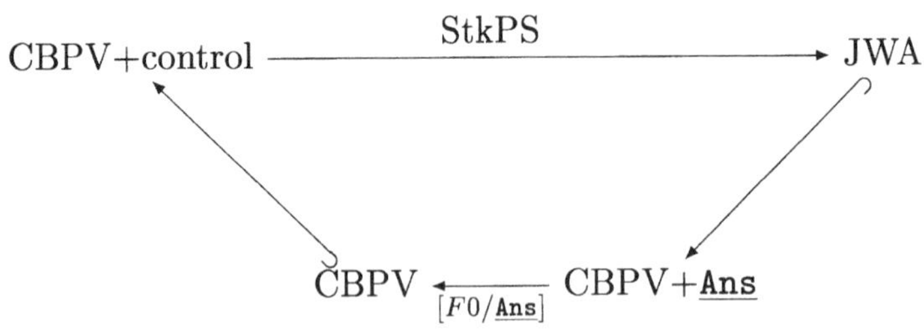

We write $-^{\mathsf{f}}$ for the composite transform from JWA to CBPV+control indicated in this diagram. We will show that this composite transform is inverse to the StkPS transform up to isomorphism. The heart of the proof is these isomorphisms, which, far from being trivial, are full of control effects. In outline, when given a value $\overline{\Gamma} \vdash^{\mathsf{v}} W : \overline{A}$, we first apply

$-^{\mathsf{f}}$, giving a term $\overline{\Gamma}^{\mathsf{f}} \vdash^{\mathsf{v}} W^{\mathsf{f}} : \overline{A}^{\mathsf{f}}$ without control effects, and then apply the isomorphisms to obtain a value $\Gamma \vdash^{\mathsf{v}} \hat{W} : A$ with control effects that transforms back to W (up to provable equality).

We first construct these isomorphisms.

1 For each CBPV value type A, we define a function α_A from CBPV + control values $\Gamma \vdash^{\mathsf{v}} V : A$ to CBPV+control values $\Gamma \vdash^{\mathsf{v}} V : \overline{A}^{\mathsf{f}}$, and a function α_A^{-1} in the opposite direction.

2 For each CBPV computation type $\underline{B}$, we define a function $\alpha_{\underline{B}}$ from CBPV+control values $\Gamma \vdash^{\mathsf{v}} V : \texttt{stk}\ \underline{B}$ to CBPV+control values $\Gamma \vdash^{\mathsf{v}} V : \overline{\texttt{stk}\ \underline{B}}^{\mathsf{f}}$, and a function $\alpha_{\underline{B}}^{-1}$ in the opposite direction.

3 For each JWA type A, we define a function β_A from JWA values $\Gamma \vdash^{\mathsf{v}} V : A$ to JWA values $\Gamma \vdash^{\mathsf{v}} V : \overline{A^{\mathsf{f}}}$, and a function β_A^{-1} in the opposite direction.

The definitions are given in Fig. 7.15. It is straightforward to check that α_A and α_A^{-1} are inverse up to provable equality and commute with substitution in Γ; similarly for $\alpha_{\underline{B}}$ and for β_A.

A	$\alpha_A V$	$\alpha_A^{-1} W$
$\sum_{i\in I} A_i$	$\texttt{pm}\ V\ \texttt{as}\ \{\ldots,(i,\texttt{x}).(i,\alpha_{A_i}\texttt{x}),\ldots\}$	$\texttt{pm}\ W\ \texttt{as}\ \{\ldots,(i,\texttt{x}).(i,\alpha_{A_i}^{-1}\texttt{x}),\ldots\}$
$A \times A'$	$\texttt{pm}\ V\ \texttt{as}\ (\texttt{x},\texttt{y}).(\alpha_A\texttt{x},\alpha_{A'}\texttt{y})$	$\texttt{pm}\ W\ \texttt{as}\ (\texttt{x},\texttt{y}).(\alpha_A^{-1}\texttt{x},\alpha_{A'}^{-1}\texttt{y})$
$U\underline{B}$	$\texttt{thunk}\ \lambda\texttt{k}.(\texttt{changestk}\ \alpha_{\underline{B}}^{-1}\texttt{k}.\texttt{force}\ V)$	$\texttt{thunk letstk k}.(((\alpha_{\underline{B}}\texttt{k})`\texttt{force}\ W)\ \texttt{to x}.\ \{\})$
$\texttt{stk}\ \underline{B}$	$\alpha_{\underline{B}} V$	$\alpha_{\underline{B}}^{-1} W$

$\underline{B}$	$\alpha_{\underline{B}} V$	$\alpha_{\underline{B}}^{-1} W$
FA	$\texttt{thunk}\ \lambda\texttt{x}.(\texttt{changestk}\ V.\ \texttt{return}\ \alpha_A^{-1}\texttt{x})$	$[\cdot]\ \texttt{to x}.\ (\alpha_A\texttt{x}`\texttt{force}\ W) \cdot\!:: \cdot\texttt{ignored}$
$\prod_{i\in I}\underline{B}_i$	$\{\ldots,(i,\alpha_{\underline{B}_i}\texttt{x})\ \cdot\texttt{where}\ i \cdot\!:: \texttt{x},\ldots\}\ \texttt{is}\ V$	$\texttt{pm}\ W\ \texttt{as}\ \{\ldots,(i,\texttt{x}).(i \cdot\!:: \alpha_{\underline{B}_i}^{-1}\texttt{x}),\ldots\}$
$A \to \underline{B}$	$(\alpha_A\texttt{x},\alpha_{\underline{B}}\texttt{y})\ \cdot\texttt{where x} \cdot\!:: \texttt{y is}\ V$	$\texttt{pm}\ W\ \texttt{as}\ (\texttt{x},\texttt{y}).(\alpha_A^{-1}\texttt{x} \cdot\!:: \alpha_{\underline{B}}^{-1}\texttt{y})$

A	$\beta_A V$	$\beta_A^{-1} W$
$\sum_{i\in I} A_i$	$\texttt{pm}\ V\ \texttt{as}\ \{\ldots,(i,\texttt{x}).(i,\beta_{A_i}\texttt{x}),\ldots\}$	$\texttt{pm}\ W\ \texttt{as}\ \{\ldots,(i,\texttt{x}).(i,\beta_{A_i}^{-1}\texttt{x}),\ldots\}$
$A \times A'$	$\texttt{pm}\ V\ \texttt{as}\ (\texttt{x},\texttt{y}).(\beta_A\texttt{x},\beta_{A'}\texttt{y})$	$\texttt{pm}\ W\ \texttt{as}\ (\texttt{x},\texttt{y}).(\beta_A^{-1}\texttt{x},\beta_{A'}^{-1}\texttt{y})$
$\neg A$	$\gamma(\texttt{x},()).(\beta_A^{-1}\texttt{x} \nearrow V)$	$\gamma\texttt{x}.((\beta_A\texttt{x},()) \nearrow W)$

Figure 7.15. Syntactic Isomorphisms α and β used in proof of Prop. 80

We have thus proved Prop. 80(1). It is then straightforward to prove Prop. 80(2) by induction on A.

Lemma 85 1 For any CBPV+control+`print` value $\Gamma \vdash^{\mathsf{v}} V : B$, we can prove

$$V = \alpha_B^{-1}\overline{V}^{\mathsf{f}}[\overrightarrow{\alpha_{A_i}\mathtt{x}_i/\mathtt{x}_i}]$$

2 For any CBPV+control+`print` computation $\Gamma \vdash^{\mathsf{c}} M : \underline{B}$, we can prove

$$M = \mathtt{letstk}\ \mathtt{k}.\mathtt{coerce}\ \overline{M}^{\mathsf{f}}[\overrightarrow{\alpha_{A_i}\mathtt{x}_i/\mathtt{x}_i}, \alpha_{\underline{B}}\mathtt{k}/\mathtt{k}]$$

3 For any JWA+`print` value $\Gamma \vdash^{\mathsf{v}} V : B$, we can prove

$$V = \beta_B^{-1}\overline{V^{\mathsf{f}}}[\overrightarrow{\beta_{A_i}\mathtt{x}_i/\mathtt{x}_i}]$$

4 For any JWA+`print` nonreturning command $\Gamma \vdash^{\mathsf{n}} M$, we can prove

$$M = \overline{M^{\mathsf{f}}}[\overrightarrow{\beta_{A_i}/\mathtt{x}_i}, \gamma\{\}/\mathtt{k}]$$

These results can be extended to computations and values with holes i.e. contexts. □

These are proved by induction over terms.

Lemma 86 1 For any CBPV value type A and value $\Gamma \vdash^{\mathsf{v}} V : A$, we can prove

$$\overline{\alpha_A V} = \beta_{\overline{A}}\overline{V}$$

Similarly for computation types.

□

This is proved by induction over types.

To prove Prop. 80(3), suppose we are given a JWA value $\overline{\Gamma} \vdash^{\mathsf{v}} W : \overline{A}$, where Γ is the context $A_0, \ldots, A_{m-1}$. We set $\Gamma \vdash^{\mathsf{v}} \hat{W} : A$ to be $\alpha_A^{-1}W^{\mathsf{f}}[\overrightarrow{\alpha_{A_i}\mathtt{x}_i/\mathtt{x}_i}]$. Then $\hat{-}$ is inverse to the StkPS transform up to provable equality, and so the latter defines a bijection.

To prove Prop. 80(4), suppose we are given a JWA nonreturning command $\overline{\Gamma}, \mathtt{k} : \overline{\underline{B}} \vdash^{\mathsf{n}} N$, where Γ is the context $A_0, \ldots, A_{m-1}$. We set $\Gamma \vdash^{\mathsf{c}} \hat{N} : \underline{B}$ to be $\mathtt{letstk}\ \mathtt{k}.\mathtt{coerce}\ N^{\mathsf{f}}[\overrightarrow{\alpha_{A_i}\mathtt{x}_i/\mathtt{x}_i}, \alpha_{\underline{B}}\mathtt{k}/\mathtt{k}]$. Then $\hat{-}$ is inverse to the StkPS transform up to provable equality, and so the latter defines a bijection.

These arguments can be adapted to values and computations with holes i.e. contexts.

Chapter 8

POINTER GAMES

8.1 Introduction

8.1.1 Pointer Games And Their Problems

In this chapter we look at the game semantics of Hyland and Ong [Hyland and Ong, 2000], discovered also by Nickau [Nickau, 1996]. It is based on a certain kind of two-player game, where (generally speaking) a player moves by

1 pointing to a previous move of the other player;

2 passing a token.

We call such a game a "pointer game". This is to distinguish it from many other kinds of game, such as the purely token-passing games of [Abramsky et al., 1994], which are quite different and which we shall not be looking at. Pointer games are extremely powerful: in a series of striking results, they have provided universal[1] models for recursion [Hyland and Ong, 2000], type recursion [McCusker, 1997], control effects [Laird, 1997], ground store [Abramsky and McCusker, 1997], general store [Abramsky et al., 1998], erratic choice [Harmer and McCusker, 1999] and more.

However, despite its remarkable successes, the account of pointer game models in the literature suffers from a number of problems. The collective effect of these problems is that the reaction of many readers is negative. They perceive game semantics as complicated and technical,

[1]A model is *universal* [Longley and Plotkin, 1997] when every morphism from $[\![\Gamma]\!]$ to $[\![A]\!]$ is the denotation of some term $\Gamma \vdash M : A$. In the logical setting, this property is called *full completeness* [Abramsky, 1992].

and its elegance is obscured. In this chapter, we aim to rectify *some* of these problems, but not all. We need to describe the various problems with some care, in order to say which are the ones we are aiming to rectify.

1 **lack of intuition** The rules and constraints of play are not clearly motivated by operational intuitions.

2 **difficulty of expression** Even when a strategy is intuitively clear, it is difficult to express it in a clean, rigorous way, as we lack a convenient language for strategies. This is even more the case for strategy combinators (e.g. composition): it is usually obvious how to apply such a combinator in any particular example, but giving a precise description is messy.

3 **difficulty of reasoning** Partly because of 2, it is difficult to reason about game semantics, in particular to prove that strategies are equal.

These problems apply to different parts of the account:

- (1) is a problem for the semantics of *types* and *judgements.*
- (2) is a problem for the semantics of *term constructors.*
- (3) is a problem throughout.

Why do we not consider the semantics of term constructors to be affected by (1)? The reason is that the semantics of term constructors tends to be determined by the semantics of types and judgements—not in a precise technical sense, but rather in the sense that, for any given term, there is only one "reasonable" denotation of the form specified by the semantics of types and judgements. So it is the semantics of types and judgements where the need for intuitive explanation of definitions is most pressing.

8.1.2 Contribution Of This Chapter

The aim of this chapter is to solve problem (1). We aim to give a *motivated* account of pointer game semantics, first for JWA and then for CBPV. We explain the arenas, the moves, the pointers and the question/answer distinction in concrete computational terms.

- An arena gives a type-representation of *several jump-points.*
- A move represents a *jump* in JWA or CBPV.
- A pointer represents the *time of receipt* of the jump-point being jumped to.

- "Asking a question" and "answering" mean *forcing a thunk* and *returning* respectively—the two kinds of jump in CBPV.

We recover, of course, the usual pointer game semantics for CBN and CBV. But the fact that jumps (although not the destination of returns) are explicit it CBPV makes it much more suitable for analyzing pointer games, and the fact that jumps *and* their destinations are explicit in JWA makes it even more suitable.

We make no claim to solve problems (2)–(3). We emphasize that the models *per se* are not new, only the way that we motivate them.

8.1.3 More Is Simpler

It is a remarkable feature of pointer game semantics that the universal model for a language with divergence, general store (excluding equality testing on cells), control effects and infinitary syntax is much simpler than the universal model for a language without all these features (such as PCF). By including these features from the outset, we can dispense with the machinery of views, visibility, innocence and bracketing, which obscure the more important aspects of pointer games. And, as we see in Sect. 8.5.2, we can simplify the definition of "Q/A-labelled arena", because it becomes possible for an answer-token to succeed an answer-token (as a consequence of the `stk` type constructor). The use of infinitary syntax obviates the need for computability restrictions on strategies, and gives us definability theorems for types as well as terms.

A further advantage of this formulation is a theorem which is folklore among games researchers and appears in an indirect formulation in [Laird, 2003]. It states that the pointer game model for control and storage is not just universal but also fully abstract, with no need for the further quotienting which is frequently found in the literature. This means that all the details of a term's denotation can be worked out by applying contexts involving store and control.

8.1.4 Cells As Objects

Our treatment of storage cells exactly follows [Abramsky et al., 1998]; we regard a cell as an "object with a read method and a write method".

Cells As Objects In CBPV

In CBPV, an A-storing cell l can be converted into a thunk that either

- pops the tag write together with a value of type A, assigns this value to l and returns, or
- pops the tag read and returns the contents of l.

This has type

$$U\prod\{\mathsf{write}.(A \to F1), \mathsf{read}.FA\} \tag{8.1}$$

—recall from Sect. 2.9.2 that $F1$ is the type of commands. We give the name `pseudoref` A to (8.1). The thunk associated with the cell l is

$$\texttt{thunk}\ \lambda \begin{cases} \mathsf{write}. & \lambda \texttt{x}.\ l := \texttt{x}.\ \texttt{return} \\ \mathsf{read}. & \texttt{read}\ l\ \texttt{as y. return y} \end{cases}$$

Since we are excluding cell equality testing from the language, this thunk gives all the functionality of l. In the game semantics, we accordingly interpret

- `ref` A the same way as `pseudoref` A
- $V := W.\ M$ the same way as $(W`\mathsf{write}`(\texttt{force}\ V));\ M$
- `read` V `as y.` M the same way as $\mathsf{read}`(\texttt{force}\ V)$ `to y.` M.

Thus, all the difficulty lies in the interpretation of `new`.

Terms Involving Cells

For this kind of semantics, it suffices to interpret terms in the empty world, because we interpret a term in world w and context Γ the same way as a term in context

$$\texttt{x}_0 : \texttt{ref}\ A_0, \ldots, \texttt{x}_{n-1} : \texttt{ref}\ A_{n-1}, \Gamma$$

where w has n cells storing types $A_0, \ldots, A_{n-1}$. (Here, we are supposing an arbitrary ordering of the cells.) In particular, the term `cell` l is interpreted the same way as $\texttt{x}_l$.

"Bad" Cells

As remarked in [Abramsky et al., 1998], there are values of type `pseudoref` A that do not behave like cells. For this reason, definability and full abstraction theorems are valid only at `ref`-free types. To make them valid at all types, we can follow [Abramsky et al., 1998] by introducing a construct

$$\frac{\Gamma \vdash^{\mathsf{v}} V : \texttt{pseudoref}\ A}{\Gamma \vdash^{\mathsf{v}} \texttt{mkpseudo}\ V : \texttt{ref}\ A}$$

adding suitable operational rules. This gives a syntactic isomorphism between `ref` A and `pseudoref` A. Alternatively, we can *define* `ref` A to be `pseudoref` A, again adding suitable operational rules to treat the extra terms that this allows to be formed.

Cells As Objects In JWA

In JWA, an A-storing cell l can be converted into a jump-point that either

- accepts the tag write together with a value of type A and a jump-point, assigns this value to l and then jumps to the jump-point, or
- accepts the tag read together with an A-accepting jump-point, to which it sends the contents of l.

This jump-point has type

$$\neg\sum\{\mathsf{write}.(A \times \neg 1), \mathsf{read}.\neg A\} \tag{8.2}$$

We give the name `pseudoref` A to (8.2). The jump-point associated with the cell l is

$$\gamma\left\{\begin{array}{ll} (\mathsf{write}, \mathtt{x}, \mathtt{k}). & l := \mathtt{x}.\ (() \nearrow \mathtt{k}) \\ (\mathsf{read}, \mathtt{k}). & \mathtt{read}\ l\ \mathtt{as}\ \mathtt{y}.\ (\mathtt{y} \nearrow \mathtt{k}) \end{array}\right.$$

Again, since we are excluding cell equality testing from the language, this thunk gives all the functionality of l. In the game semantics, we accordingly interpret

- `ref` A the same way as `pseudoref` A
- $V := W.\ M$ the same way as $(\mathsf{write}, W, \gamma().M) \nearrow V$
- `read` V `as y.` M the same way as $(\mathsf{read}, \gamma\mathtt{y}.M) \nearrow V$.

Thus, all the difficulty lies in the interpretation of `new`.

8.1.5 Related Work On Abstract Machines

The work of [Danos et al., 1996] is closely related to ours. The language studied there is the simply typed CBN λ-calculus, with a single free type identifier ι and function types but no boolean type. From our viewpoint, this is a fragment of CBPV (the ι type is $F0$), a fragment that does not involve any answer moves.

A jumping abstract machine for this language had been presented in [Danos and Regnier, 1999]. From our viewpoint, it is the CBPV jumping machine obtained via StkPS, restricted to the above-mentioned fragment. (Of course, this is not how it was motivated in [Danos and Regnier, 1999]; it was obtained from a subtle analysis of the Geometry of Interaction machine for the multiplicative-exponential fragment of linear logic.) A difference here is that CBPV, and JWA to an even greater extent, make the control flow explicit (Sect. I.5.3). So the jumping machine for these languages express their operational intuitions—giving a "jumping operational semantics"—in a way that cannot be the case for CBN λ-calculus, where `force` is implicit.

A major result of [Danos et al., 1996] is that the pointer game semantics describes the traces of this machine. This is similar to our use of jumping traces to motivate pointer game semantics in Sect. 8.3.2. But a significant difference is that our account is, at present, only informal.

Also related to our work—although not as closely, since we do not treat innocence—is [Curien, 1998]. The *universal arena* is the arena in which there are countably many roots and each token has countably many successors. An *abstract Böhm tree* is, roughly speaking, a way of representing an innocent function on the universal arena. (There are variations, using other constantly branching arenas.) Because every arena is embeddable in the universal arena, any term of any language that denote an innocent strategy can be represented as an abstract Böhm tree. The way in which abstract Böhm trees are composed involves jumping with arguments, rather like in JWA.

8.1.6 Structure Of Chapter

8.2 Arenas In The Literature

There are some superficial differences between our treatment of arenas and the various treatments that appear in the literature. For the benefit of reader familiar with the other treatments, we discuss them here.

8.2.1 Player/Opponent Labelling

In [Hyland and Ong, 2000] the tokens of an arena are not labelled as Player or Opponent, but such a *polarity* labelling was later introduced in [McCusker, 1996]. This labelling is not actually required for the definition of the game semantics, and we therefore omit it. After a model has been defined, its polarization properties can be studied, and we leave such an analysis—with reference to Laurent's model [Laurent, 2002b]—to future work.

8.2.2 Tokens Of An Arena

In some forms of game semantics, a type denotes a game, which is a set of moves with some additional structure. This is *not* the case for pointer game models, where a type denotes an arena (or family of arenas). It is rather confusing to give the name "move" to an element of an arena, because a player's move consists of a pointer together with an arena element. For this reason, we refer to an element of an arena as a *token*.

8.2.3 Arenas Versus Arena Families

CBV pointer games have been presented in two forms. In [Honda and Yoshida, 1997] a type denotes an arena with every root labelled A, whereas in [Abramsky and McCusker, 1998] a type denotes a family of arenas. It is clear that these are equivalent, since an arena with A-roots corresponds to a family of arenas, by removing the roots. But there are disadvantages in the arena formulation of [Honda and Yoshida, 1997]: for example, the the interpretation of $\to_{\mathbf{CBV}}$ requires changing A-tokens into Q-tokens.

Likewise for CBN: [Hyland and Ong, 2000] and subsequent authors interpret a CBN type as an arena with every root labelled Q, but one could equivalently remove the roots and interpret a CBN type as a family of arenas. In the former variant, a closed term on A denotes an Opponent-first strategy on $[\![A]\!]$ (where a root can be played only in the first move). In the latter formulation, if A denotes $\{R_i\}_{i\in I}$ then a closed CBN term of type A should denote a family of Player-first strategies (where a root can be played at any time), one for each arena R_i.

If we use arena families, then a CBN function type is interpreted as follows. Suppose A denotes $\{R_i\}_{i\in I}$ and B denotes $\{S_j\}_{j\in J}$. Then $A \to_{\mathsf{CBN}} B$ denotes $\{\mathsf{pt}^{\mathsf{Q}}_{i\in I} R_i \uplus S_j\}_{j\in J}$, where we write

- $\uplus$ for disjoint union
- $\mathsf{pt}^{\mathsf{Q}}_{i\in I} R_i$ for the arena with I roots, all labelled Question, and a copy of R_i grafted underneath the ith root.

It is easy to see that this interpretation agrees with that in [Hyland and Ong, 2000], and it makes the CBPV decomposition more apparent.

Of course the difference between the two presentations is superficial, but as with CBV there are some advantages, for our account, in using arena families. In particular, if we interpret as an arena, the initial move where the Opponent plays a root does not correspond to a jump (as moves generally do) but rather to information present in the environment/stack before execution begins. Consequently, it is not really meaningful to label it as Qor A, since that label distinguishes the two kinds of jump.

[Laurent, 2002a] advocates interpreting CBV and CBN types in the same way. This implies we should either interpret both as an arena (as he does, following [Hyland and Ong, 2000]) or interpret both as an arena family. We have argued, for both CBV and CBN, that it is preferable to interpret a type as an arena family, and this is what we shall do throughout.

8.2.4 Forests Versus Graphs

The definition of "arena" varies in the literature. In [Hyland and Ong, 2000] an arena is required to be a forest, but in [McCusker, 1996] this constraint is relaxed and certain more general graphs are allowed. This seems strange, because arenas are isomorphic if their forest-unwindings are isomorphic, so the forest-unwinding seems to be the "true" denotation.

McCusker's reason for this relaxation was to simplify the interpretation of CBN function type, where a CBN type is interpreted as an arena. But, as seen in Sect. 8.2.3, no simplification is necessary when we interpret a CBN type as an arena family. So we will follow [Hyland and Ong, 2000] and insist that an arena be a forest.

8.3 Pointer Game Semantics For Infinitary JWA + Storage

We begin out account of pointer games by formulating a model for infinitary JWA + storage.

8.3.1 Types

Arenas

What information is needed to describe a jump? Let us say that the jump-points we have available are $\mathtt{k}_0, \ldots, \mathtt{k}_{m-1}$. A jump to one of these is described by the following information:

- the *selector* $i \in \$m$, giving the destination $\mathtt{k}_i$
- the argument, which, by Prop. 71, consists of
  - several tags $j_0, \ldots, j_{r-1}$
  - several jump-points $\mathtt{l}_0, \ldots, \mathtt{l}_{s-1}$

The selector together with the tags are called the *data* of the jump and they are immediately observable by the context. The jump-points appear just as points that are available for future jumping.

We depict the jumping possibilities of $\mathtt{k}_0, \ldots, \mathtt{k}_{m-1}$ as a forest—perhaps infinitely deep—in the following recursive manner. Each choice of data $i, j_0, \ldots, j_{r-1}$ is represented as a root (clearly there can be only countably many). Under this root we draw the forest depicting the jump-points $\mathtt{l}_0, \ldots, \mathtt{l}_{s-1}$. In the games literature, this forest is called an "arena".

Definition 8.1 An *arena* R is a structure $(\mathsf{tok}\ R, \mathsf{rt}\ R, \vdash_R)$.

- $\mathsf{tok}\ R$ is a countable set whose elements are called *tokens*.

- rt R is a subset of tok R, whose elements are called *roots* and $\vdash_R$ is a binary relation on tok R. We read $t \vdash_R u$ as "t is the *predecessor* of u" or as "u is a *successor* of t". This must give a "forest", i.e.

  unique predecessor each root has no predecessor and each non-root has a unique predecessor

  well-foundedness there is no infinite chain of predecessors

$$\cdots \vdash_R t_2 \vdash_R t_1 \vdash_R t_0$$

□

Some useful operations are defined as follows.

Definition 8.2 1 We write $\emptyset$ for the empty arena.

2 We write $R \uplus R'$ for the disjoint union of R (writing inl r for the copy of $r \in$ tok R) and R' (writing inr r for the copy of $r \in$ tok R').

3 We write $\mathsf{pt}_{i \in I} R_i$ is the arena with I roots (writing root i for the ith root) and a copy of R_i grafted underneath the ith root (writing under(i, r) for the copy of the token $r \in$ tok R_i under root i).

4 An *isomorphism* from an arena R to an arena S is a bijection from R to S preserving all structure. We write **ArenaIso** for the groupoid of arena isomorphisms.

5 An arena R is a *sub-arena* of an arena S when tok $R \subseteq$ tok S and for all $r \in$ tok R

   - if r is a root in S then r is a root in R
   - if r is a successor of r' in S then $r' \in$ tok R and r is a successor of r' in R

6 Let a be a token in an arena R. Then we write $R \restriction_a$ for the arena consisting of those tokens of R hereditarily preceded by a (excluding a itself), with the successors of a as roots and the $\vdash$ relation inherited from R. (This is *not* a sub-arena of R, in the above sense.)

□

We have explained that an arena represents the jumping possibilities of a finite sequence of jump-points. Since this arena is determined by the type of these jump-points, we shall say that it *type-represents* them. If the arena R type-represents the jump-points $\mathbf{k}_0, \ldots, \mathbf{k}_{m-1}$, and the arena S type-represents the jump-points $\mathbf{k}'_0, \ldots, \mathbf{k}'_{n-1}$, then $R \uplus S$ type-represents the concatenated sequence $\mathbf{k}_0, \ldots, \mathbf{k}_{m-1}, \mathbf{k}'_0, \ldots, \mathbf{k}'_{n-1}$. This is because data for the concatenated sequence must first select which of the two sequences the destination appears in, and then provide data for the selected sequence. Likewise the empty arena $\emptyset$ type-represents the empty sequence of jump-points.

We will give an informal example of an arena representation. Suppose we have just one jump-point **k** accepting $1 + \neg A$ where A is the type

$$\neg(1 + \neg(1+1)) \times \neg(1+1) + 1$$

Then the jumping possibilities of **k** are represented by the arena

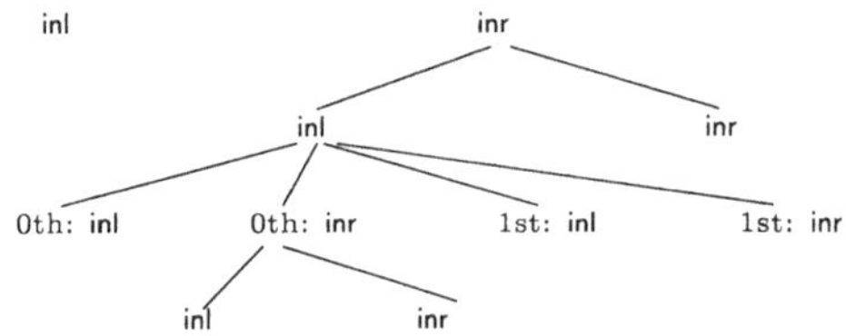

Here we have omitted the selector wherever there is only one jump-point. If we jump to **k** with an inr tag, then there is only one accompanying jump-point **l**, so under the inr root we have not written any selectors. If we jump to **l** with an inl root then there are two accompanying jump-points: we can jump to the zeroth with tags inl or inr, and we can jump to the first with tags inr. This is all represented in the arena.

We emphasize that in a jump whose data is a token a in an arena R, the jump-points passed are type-represented by $R \upharpoonright_a$. Thus, if we subsequently jump to one of these jump-points, the data of this jump will be a successor of a.

Semantics of Types

Now we turn to the semantics of types. A closed value of type A consists of

- several tags $i_0, \ldots, i_{r-1}$
- several jump-points $\mathbf{k}_0, \ldots, \mathbf{k}_{s-1}$

We therefore interpret A as a *countable family of arenas* $\{R_i\}_{i \in I}$; each index $i \in I$ represents a tag sequence $i_0, \ldots, i_{r-1}$, and the arena R_i type-represents the jump-points $\mathbf{k}_0, \ldots, \mathbf{k}_{s-1}$.

Definition 8.3 Let $\{R_i\}_{i \in I}$ and $\{S_j\}_{j \in J}$ be countable families of arenas.

- An *isomorphism* from $\{R_i\}_{i \in I}$ to $\{S_j\}_{j \in J}$ is a bijection $I \xrightarrow{f} J$ together with an isomorphism from R_i to S_{fi} for each $i \in I$.

- $\{R_i\}_{i\in I}$ is a *sub-arena-family* of $\{S_j\}_{j\in J}$ when $I \subseteq J$ and $R_i \subseteq S_i$ for all $i \in I$.

□

We can now give the semantics of types.
We want our semantics of types to have the following properties.

- If A denotes $\{R_i\}_{i\in I}$, then $\neg A$ denotes the singleton family $\{\mathsf{pt}_{i\in I} R_i\}$. This is because a closed value $\gamma\mathbf{x}.M$ of type $\neg A$ consists of no tags and just one jump-point that one jumps to taking tags represented by $i \in I$ and jump-points type-represented by R_i.
- If, for each $i \in I$, the type A_i denotes $\{R_{ij}\}_{j\in J_i}$, then $\sum_{i\in I} A_i$ denotes $\{R_{ij}\}_{(i,j)\in \sum_{i\in I} J_i}$. This is because a closed value (i, V) of type $\sum_{i\in I} A_i$ consists of a tag $i \in I$ and tags represented by j_i, together with jump-points type-represented by R_{ij}.
- If A denotes $\{R_i\}_{i\in I}$ and A' denotes $\{S_j\}_{j\in J}$ then $A \times A'$ denotes $\{R_i \uplus S_j\}_{(i,j)\in I\times J}$. This is because a closed value (V, V') of type $A \times A'$ consists of
  - two sequences of tags, one represented by $i \in I$ and one represented by $j \in J$
  - two sequences of jump-points, one type-represented by R_i and one type-represented by S_j.
- 1 denotes the singleton family containing the empty arena $\emptyset$, because a closed value () of this type consists of no tag and no jump-points.

We use these properties to define the denotation of finitely deep types, and then interpret an infinitely deep type A (using the notation of Sect. 4.4.2) as $\bigcup_{B \lhd_{\mathsf{fin}} A} [\![B]\!]$. The above statements then follow. We can similarly interpret recursive types, up to equality, as in [McCusker, 1996].

Of course, it would be possible to interpret a type as an arena rather than an arena family, but we do not do this because a token in an arena should represent a possible jump, which an index in $[\![A]\!]$ does not.

We can now give a definability result for types.

Proposition 87 Every countable family of arenas is isomorphic to the denotation of a type canonical form $\theta\{R_i\}_{i\in I}$ in *type canonical form*—this is coinductively defined as

$$A ::= \quad \textstyle\sum_{i\in I} \neg A_i$$

□

Proof We define θ coinductively by

$$\theta\{R_i\}_{i\in I} = \textstyle\sum_{i\in I}\neg\theta\{R_i \upharpoonright_a\}_{a\in \mathsf{rt}\ R_i}$$

Next, we construct the isomorphism. Since $\theta\{R_i\}_{i\in I}$ denotes an arena family indexed by $I \times 1$, we set the bijection on indexing sets takes $i \in I$ to $^\circ i = (i, ())$. Then we inductively define a predicate

$$\{R_i\}_{i\in I} : \hat{\imath} : r \mapsto r'$$

where $\hat{\imath} \in I$, $r \in \mathsf{tok}\ R_{\hat{\imath}}$ and $r' \in \mathsf{tok}\ (\theta\{R_i\}_{i\in I})_{^\circ\hat{\imath}}$ as follows:

$$\frac{j \in \mathsf{rt}\ R_{\hat{\imath}}}{\{R_i\}_{i\in I} : \hat{\imath} : j \mapsto_{\mathsf{val}} \mathsf{root}\ j} \qquad \frac{\{R_{\hat{\imath}} \upharpoonright_j\}_{j\in \mathsf{rt}\ R_{\hat{\imath}}} : \hat{\jmath} : s \mapsto s'}{\{R_i\}_{i\in I} : \hat{\imath} : s \mapsto \mathsf{under}(^\circ\hat{\jmath}, s')}$$

We show by induction on n that $\mapsto$ defines an arena isomorphism between the tokens of depth $< n$ in $R_{\hat{\imath}}$ and the tokens of depth $< n$ in $\theta\{R_i\}_{i\in I\,^\circ\hat{\imath}}$. □

It is often helpful to think of an arena family as a representation of a type canonical form in this way.

A context Γ represents a countable family of arenas also, because an environment for Γ consists of several tags and several jump-points. We interpret $\Gamma = A_0, \ldots, A_{n-1}$ the same way as we interpret $A_0 \times \cdots \times A_{n-1}$.

8.3.2 Terms

Semantics of Nonreturning Commands

Our next step is to give the semantics of the judgement $\Gamma \vdash^{\mathsf{n}} M$. It will be recalled from Sect. 2.10 that this is the most important judgement, as the values are merely auxiliary. We will motivate our semantics with the command

$$\mathtt{u} : \neg(1 + \neg A) + 1 \vdash^{\mathsf{n}} M$$

where A is the type

$$\neg(1 + \neg(1 + 1)) \times \neg(1 + 1) + 1$$

and M is shown in graphical form in Fig. 8.1. Notice that it involves a cell `w` storing a boolean, and also a cell `p` storing a jump-point. However, the cell generation, reading and writing is not visible from the outside. We write `new` V =: `x` rather than `new x` := V, to conform to the graphical syntax.

We now wish to describe the externally visible jumping behaviour of the term. We have to treat two cases: where

- `u` is bound to `inr ()`

Program M

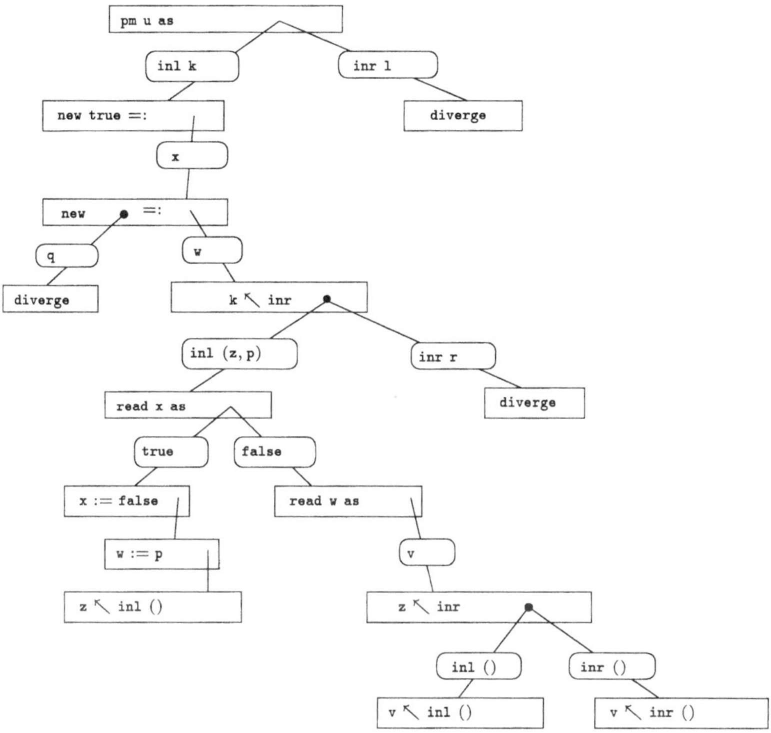

Context $\mathcal{C}[\cdot]$

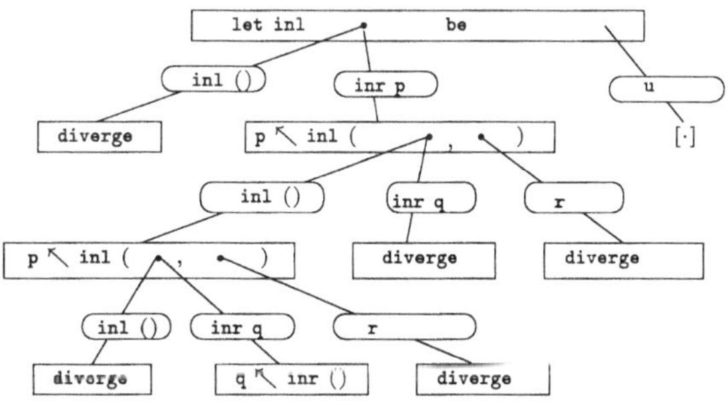

Figure 8.1. Example Program and Context

- u is bound to **inl k**, where **k** is a $1 + \neg A$ accepting jump-point

In the former case, M merely diverges, so we just consider the latter, where the behaviour of M is more interesting. In the following narrative, we shall call a jump performed by the term a "P-move", and a jump performed by the context an "O-move". For the reader's convenience, a context is provided in Fig. 8.1 that behaves in the way we describe.

secretly The term generates a cell `x` storing true and a cell `w` storing a jump-point.

P-move 0 Then it jumps to the sole jump-point present in the environment, viz. `k`, taking an inr tag and an A-accepting jump-point.

O-move 1 Suppose the context jumps to the sole jump-point it received in P-move 0, taking an inl tag and 2 jump-points, which we call the zeroth and the first. The zeroth accepts $1 + \neg(1 + 1)$, while the first accepts $1 + 1$.

secretly Then the term sees that `x` is set to true, sets it to false and stores in `w` the first jump-point received in O-move 1.

P-move 2 The term then jumps to the zeroth jump-point received in O-move 1, taking the tag inl and no jump-points.

O-move 3 Suppose the context jumps again to the sole jump-point it received in P-move 0, taking an inl tag and 2 jump-points, which we call the zeroth and the first. The zeroth accepts $1 + \neg(1 + 1)$, while the first accepts $1 + 1$.

secretly Then term sees that `x` is set to false.

P-move 4 So it jumps to the zeroth jump-point received in O-move 3, taking the tag inr and a $1 + 1$ accepting jump-point.

O-move 5 Suppose the context jumps to the sole jump-point it received in P-move 4, taking an inr tag and no jump-point.

P-move 6 Then the term jumps to the first jump-point it received in O-move 1 (earlier retrieved from `w`), taking a inr tag and no jump-point.

Here is a summary of the story above.

jump-points in environment accepting:			$1 + \neg A$	
move	destination of jump		argument passed consists of	
	received when?	selector	tags	jump-points accepting
P-move 0	environment	sole	inr	A
O-move 1	P-move 0	sole	inl	$1 + \neg(1 + 1)$, $1 + 1$
P-move 2	O-move 1	zeroth	inl	none
O-move 3	P-move 0	sole	inl	$1 + \neg(1 + 1)$, $1 + 1$
P-move 4	O-move 3	zeroth	inr	$1 + 1$
O-move 5	P- move 4	sole	inr	none
P-move 6	O-move 1	first	inr	none

Notice that the information about jump-points passed is redundant, because it is determined by the type of the destination and the tag passed. If, for example, one jumps to a $\neg B \times \neg B' + \neg C$ accepting jump-point and one takes an inl tag, then one has to take also a B-accepting and a B'-accepting jump-point.

We say that a move m *points* to an earlier move n when m jumps to a jump-point received in move n. Notice that an O-move always points to a P-move, whereas a P-move points either to an O-move or—if it jumps to a jump-point in the original environment—to no move at all. This is because we are only concerned with interaction between term and context, so there is no need to record internal jumps within the term or within the context.

Now we turn attention to pointer games. The type of **x** denotes a family of two arenas: the empty arena and the following arena R.

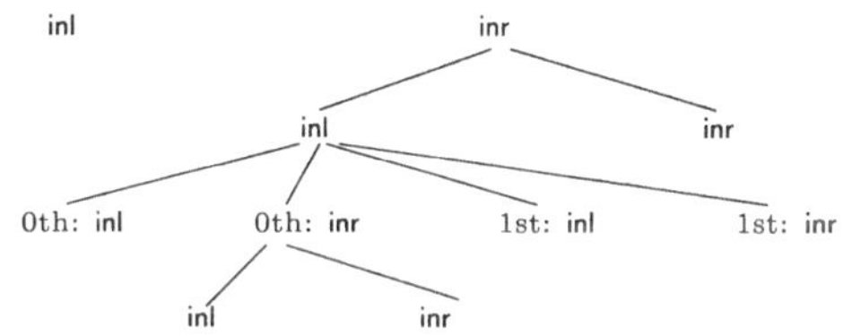

We explained this arena in Sect. 8.3.1.

We are able to describe each move by the pointer to a previous move m (if there is one), a selector and tags. In each move the selector and tags can be represented by a token a, and the accompanying jump-points are type-represented by $R\upharpoonright_a$. The possible tokens in a move are given as follows.

no pointer When we jump to a jump-point in the environment, the token must be a root of R, because R type-represents the jump-points in the environment.

pointer to move m When we jump to a jump-point passed in move m in which token a was passed, we must pass a root of $R\upharpoonright_a$—i.e. a successor of a—because $R\upharpoonright_a$ type-represents the jump-points passed in move m.

Putting all this together, we can see that the sequence described above is a play of a game, which we now describe.

Definition 8.4 (informal) The *nonreturning-command game* on an arena R is described as follows.

- Play alternates between Player and Opponent. Player moves first.

- Player moves by either
  - passing a root r of the source arena R, or
  - pointing to a previous O-move m and passing a successor r of the token passed in move m
- Opponent moves by
  - pointing to a previous P-move m and passing a successor of the token passed in move m.

□

In the next section, we will formally define the notions of *play* and *strategy* for this game.

If a context Γ denotes the arena family $\{R_i\}_{i \in I}$ then a command $\Gamma \vdash^{\mathsf{n}} M$ will denote, for each $i \in I$, a strategy for the nonreturning-command game on R_i. Thus, our example program denotes 2 strategies: one just diverges (represented as the empty set of plays), and the other can behave as in our account above. While we could have made the choice of i into an O-move, we prefer not to do this. For the choice is not part of the execution; rather it is incorporated into the environment long before execution begins.

Formal Definition Of Strategy

We begin with a 1-player game.

Definition 8.5 A *justified sequence* s on an arena R is a triple $(|s|, p, t)$ where

length $|s| \in \mathbb{N} \cup \{\omega\}$—we write $\mathsf{moves}\ s$ for $\$|s|$ and call its elements *moves*

pointers a partial function p from $\mathsf{moves}\ s$ to $\mathsf{moves}\ s$ such that $p(m) < m$ whenever $p(m)$ is defined

tokens a function t from $\mathsf{moves}\ s$ to $\mathsf{tok}\ R$

such that

- if $p(m)$ is undefined then $t(m)$ is a root of R
- if $p(m)$ is defined then $t(m)$ is a successor of $t(p(m))$.

□

We say that a move m plays $n \curvearrowleft r$ when $p(m) = n$ and $t(m) = r$. We say that a move n is *descended oddly* from a move m when there is a odd-length chain of pointers from n to m, and *descended evenly* from m when there is an even-length chain of pointers. In particular, a move is descended evenly from itself.

Now we move to the 2-player game.

Definition 8.6 Let R be an arena. A *nonreturning-command play* (or just "play") on R is a justified sequence $s = (|s|, p, t)$ on R such that for each move m

- if m is even (e.g. 0) then $p(m)$ is undefined or odd
- if m is odd then $p(m)$ is even.

□

We say that a move m in a play is a *Player-move* if m is even (e.g. 0) and an *Opponent-move* if m is odd. We say that a play is *awaiting Player* if its length is even (e.g. 0), *awaiting Opponent* if its length is odd, and *infinite* if its length is infinite (although infinite plays are not really required for our semantics).

Definition 8.7 A *nonreturning-command strategy* (or just "strategy") on R is a set σ of Opponent-awaiting nonreturning-command plays such that

prefix-closed if s' is an Opponent-awaiting prefix of $s \in \sigma$, then $s' \in \sigma$

deterministic if $s, s' \in \sigma$ agree on all but the last move, then they agree on the last move.

We write $\mathsf{nStrat}R$ for the cppo of nonreturning-command strategies on R, ordered under inclusion. Directed joins are given by union, and the least element is the empty set. □

We say that a justified sequence s is *included in* a strategy σ when s is a play and all its Opponent-awaiting prefixes of s are in σ.

Semantics of Values

Let Γ be a context of jump-points $\mathtt{k}_0 : \neg A_0, \ldots, \mathtt{k}_{m-1} : \neg A_{m-1}$ type-represented as the arena R. Suppose we have, defined in context Γ, some jump-points $V_0, \ldots, V_{n-1}$ type-represented as the arena S. The jumping behaviour of these jump-points begins when the context jumps to one of them, with

- a selector $\hat{\imath} \in \$n$ and several tags, represented as a root a of S
- several jump-points, type-represented as $S \restriction_a$.

At any time thereafter, the jump-point $V_{\hat{\imath}}$ can at any time jump to these jump-points, passing a successor of a, or to one of the jump-points in Γ, passing a root of R.

Thus the execution can be described by the following game.

Definition 8.8 (informal) Let R and S be arenas. The *value game* from R (the *source arena*) to S (the *target arena*) is described as follows.

- Play alternates between Player and Opponent. Opponent moves first.
- Each move is classified as
  - a *source move*, where a token in R is passed, or
  - a *target move*, where a token in S is passed.
- Opponent moves initially by passing a root r of S. The initial move must happen—we do not allow an empty play for this game. (Roughly speaking, the play comes into existence only when Opponent makes the initial move.)
- Player moves by either
  - passing a root r of R, or
  - pointing to a previous O-move m, and passing a successor r of the token passed in move m.
- Opponent moves (except in the initial move) by pointing to a previous P-move m, and passing a successor r of the token passed in move m.

We give the name $\mathsf{O}(R, S)$ to the cpo of strategies for this game. □

The reader may wonder why Opponent can pass a root of S is the initial move only. Surely the context can jump to one of these jump-points $V_0, \ldots, V_{n-1}$ many times? Indeed it can, but the behaviour each time will be the same. The jump-points $V_0, \ldots, V_{n-1}$ cannot exhibit a different behaviour each time they are jumped to, because, as values, they have no way of generating a local cell. So by describing the behaviour when the context jumps once to one of these jump-points, we describe the behaviour completely. Alternatively, as in [Abramsky et al., 1998], we could allow the Opponent to pass a root of S in any O-move, but impose a *single-threadedness* condition on strategies to require the behaviour to be the same each time.

We formalize the value game as follows.

Definition 8.9 A *value play* from an arena R to an arena S is a *nonempty* justified sequence $s = (|s|, p, t)$ on $R \uplus S$, such that for each move $m \in \mathsf{moves}\ s$

- if $m = 0$ then $p(m)$ is undefined and m passes a root of S
- if m is even and positive then $p(m)$ is odd
- if m is odd then either $p(m)$ is undefined and m passes a root of R, or $p(m)$ is even.

□

We say that a move m in a value play is a *Player-move* if m is odd and an *Opponent-move* if m is even (e.g. 0). We say that a value play is *awaiting Player* if its length is odd, *awaiting Opponent* if its length is even (but not 0, since the empty sequence is not a play), and *infinite* if its length is infinite (although infinite plays are not really required for our semantics).

Definition 8.10 A *value strategy* from an arena R to an arena S is a set σ of Opponent-awaiting value plays such that

prefix-closed if s' is an Opponent-awaiting (nonempty) prefix of $s \in \sigma$, then $s' \in \sigma$

deterministic if $s, s' \in \sigma$ agree on all but the last move, then they agree on the last move.

We write $\mathsf{vStrat}(R, S)$ for the cpo of value strategies from R to S, ordered under inclusion. Directed joins are given by union. □

Notice that we have the isomorphism

$$\mathsf{vStrat}(R, S) \cong \prod_{a \in \mathsf{rt}\ S} \mathsf{nStrat}(R \uplus S \restriction_a) \tag{8.3}$$

It is convenient to take (8.3) rather than Def. 8.9–8.10 as the definition of $\mathsf{vStrat}(R, S)$, although perhaps this is cheating somewhat, because the initial jump is not represented as a move.

We now consider a more general value $\Gamma \vdash^{\mathsf{v}} V : B$, where Γ denotes $\{R_i\}_{i \in I}$ and B denotes $\{S_j\}_{j \in J}$. An environment for Γ consists of some tags, represented as an $i \in I$, and some jump-points, type-represented as R_i. For a given i, the value V will consist of some tags, represented as $j \in J$, and some jump-points type-represented as S_j. By the above argument, the behaviour of these jump-points will be described by an value strategy from R_i to S_j.

In conclusion, the denotation of V should associate, to each $i \in I$, some $j \in J$ together with a value strategy from R_i to S_j. This is precisely the interpretation of values presented in [Abramsky and McCusker, 1998].

8.3.3 η-Expansion And Copycat Behaviour

Consider now the command M' shown in Fig. 8.2. It is almost the same as the command M in Fig. 8.1—in fact, we can obtain M from M' by η-expanding v. So M' and M must have the same denotation, because our model must validate all the JWA laws, including the η-laws. But M' has a slightly different behaviour from M, because the jump-point that it passes in P-move 4 is *the same as* the first jump-point received in O-move 1. So O-move 5 and P-move 6 are actually the same event.

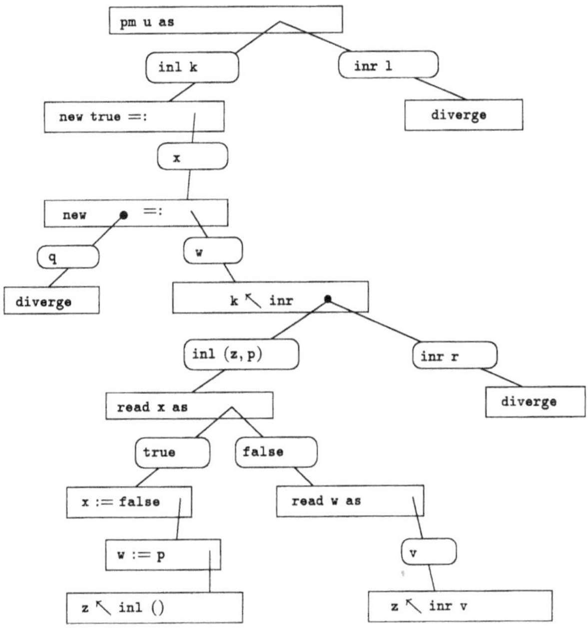

Figure 8.2. Command That η-Expands to Fig. 8.1

The term M can be said to be "η-expanded" in the sense that an argument to a jump always consists of tags and γ-abstractions, never identifiers. It is η-expanded terms whose behaviour is described by pointer game semantics. A term which is not η-expanded has a less straightforward behaviour, because several jump-points can be identified, and hence several consecutive moves can represent a single jump.

It is possible to modify the jumping operational semantics to implement, in effect, the η-expansion of the code, and then the pointer game semantics would describe the behaviour of *all* terms, not just the η-expanded ones. But this modified jumping semantics is less intuitive. It is also less efficient, both in time because one jump is replaced by several jumps, and in space because one point is replaced by several points.

In a similar way, the denotation of an identifier $\mathtt{x} : \neg B \vdash^{\mathsf{v}} \mathtt{x} : \neg B$ is a description of the jumping behaviour of its η-expansion:

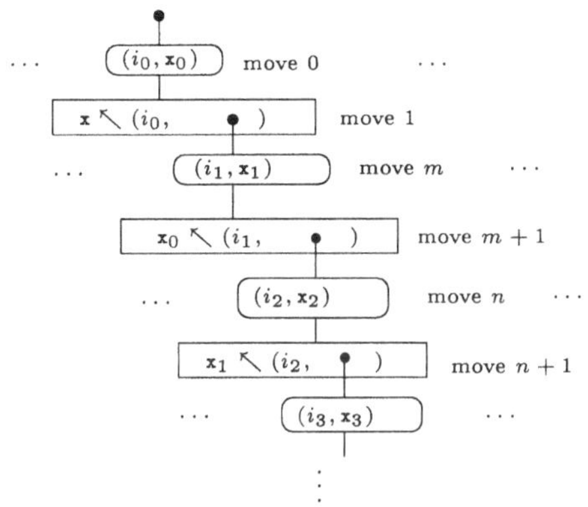

Here, for simplicity, we suppose that B is in type canonical form, so that just one tag and just one jump-point are passed in each move. We know that this term denotes a value strategy from an arena to itself.

It immediately obvious that the η-expansion always mimics the previous move of the context by passing the same tag. We argue by induction that, if the context plays a token in the source arena, then the term plays the corresponding token in the target arena, and vice versa. We see too that the term mimics each pointer specified by the term.

- If, in the initial move, the context passes a tag a to our term (playing a root of the target arena) then the term passes the same tag i to $\mathtt{x}$ (playing a root of the source arena).
- If, in a non-initial move q, the context passes a tag b to the jump-point it received in P-move $p+1$ (i.e. move q points to move $p+1$), then the term in move $q+1$ passes the same tag b to the jump-point it received in O-move p (i.e. move $q+1$ points to move p). Just as moves p and $p+1$ play the same token in opposite arenas, so do moves q and $q+1$.

This behaviour is called a *copycat* strategy.

8.4 Semantics Of Terms

We come to the semantics of term constructors. Whereas all the definitions we have given so far have been carefully motivated by computational ideas, we do not attempt to motivate the strategy combinators in this section. This is because, as we stated in Sect. 8.1.2, there is, for each combinator that we require, only one reasonable definition.

8.4.1 Categorical Structure

As in previous accounts, it is most helpful to use a categorical structure to organize the semantics. We define it briefly, and leave to Part III a more thorough explanation of the various requirements.

Definition 8.11 A *pre-families JWA model* consists of

- a cartesian category $\mathcal{C}$
- a *left $\mathcal{C}$-module* i.e. a functor $\mathcal{N} : \mathcal{C}^{\mathsf{op}} \longrightarrow \mathbf{Set}$
- for each countable family of $\mathcal{C}$-objects $\{R_i\}_{i \in I}$, a *coproduct-jumpwith* i.e. a $\mathcal{C}$-object $\neg_{i \in I} R_i$ together with an isomorphism

$$\prod_{i \in I} \mathcal{N}_{X \times R_i} \cong \mathcal{C}(X, \neg_{i \in I} R_i) \quad \text{natural in } X.$$

□

Such a structure gives us a model of JWA.

- A type (and hence a context) denotes a family of $\mathcal{C}$-objects
- If Γ denotes $\{A_i\}_{i \in I}$ and B denotes $\{B_j\}_{j \in J}$ then a value $\Gamma \vdash^{\mathsf{v}} B$ denotes, for each $i \in I$, some $j \in J$ together with a $\mathcal{C}$-morphism $A_i \longrightarrow B_j$.
- If Γ denotes $\{A_i\}_{i \in I}$, then a nonreturning command $\Gamma \vdash^{\mathsf{n}} M$ denotes, for each $i \in I$, a $\mathcal{N}$-morphism $A_i \longrightarrow$ i.e. an element of $\mathcal{N}_{A_i}$.
- Recursion can be interpreted using enrichment in $\mathbf{Cpo}^{\perp}$. We leave the details to Sect. 9.5.5.

This matches what we have seen of the pointer-game model for JWA: $\mathcal{C}$ is the category of arenas and value strategies, and $\mathcal{N}_R$ is defined to be $\mathsf{nStrat}R$. For convenience, we define $\mathsf{vStrat}(R, S)$ according to (8.3).

We firstly observe that nStrat is a functor from **ArenaIso** to $\mathbf{Cpo}^{\perp}$, and everything we shall describe is natural as arenas range over **ArenaIso**.

To define identity morphisms, we require the following.

Definition 8.12 Let R be an arena and let $a \in \mathsf{rt}\ R$. We define a strategy $\mathsf{copycat}_{R,a}$ on $R \uplus R \upharpoonright_a$ to be the set of Opponent-awaiting plays satisfying the following conditions.

- The initial move 0 plays root $\mathsf{inl}\ a$.
- If move m plays $0 \curvearrowleft \mathsf{inl}\ b$ then move $m+1$ plays root $\mathsf{inr}\ b$. Descendants of m and $m+1$ copy moves back and forth.

back If move q plays $p+1 \curvearrowleft \texttt{inr}\ b$, where $p+1$ is descended evenly from $m+1$, then move $q+1$ plays $p \curvearrowleft \texttt{inl}\ b$.

forth If move q plays $p+1 \curvearrowleft \texttt{inl}\ b$, where $p+1$ is descended oddly from m, then move $q+1$ plays $p \curvearrowleft \texttt{inr}\ b$.

□

The identity on R must be an element of

$$\prod_{a \in \mathsf{rt}\ R} \mathsf{nStrat} R \uplus R \restriction_a$$

and we define it to take $a \in \mathsf{rt}\ R$ to $\mathsf{copycat}_{R,a}$.

The following is used to construct both composition in $\mathcal{C}$ and composition in $\mathcal{N}$.

Definition 8.13 Let R, S, T be arenas. (For readability, we assume all these arenas disjoint; but the formal definition uses tags.) Suppose that $\sigma \in \mathsf{vStrat}(R, S)$, and $\tau \in \mathsf{nStrat} S \uplus T$. We define a strategy $\sigma \backslash \tau \in \mathsf{nStrat} R \uplus T$ as follows.

An *interaction sequence* for these strategies is a justified sequence s on $R \uplus S \uplus T$, together with a "thread-pointer"[2] from each root move in R to an earlier root move in S, such that

- $s \restriction_{R,T}$ is a play on $R \uplus T$
- $s \restriction_{S,T}$ is included in τ
- $s \restriction_m$, is included in σ_a, for each move m playing root $a \in \mathsf{rt}\ S$.

Here, $s \restriction_m$ means all moves hereditarily justified by m, using both pointers and thread-pointers, excluding m itself.

We define $\sigma \backslash \tau$ to be the set of all $s \restriction_{R,T}$, where s is an interaction sequence and $s \restriction_{R,T}$ is Opponent-awaiting. □

To define the composite of $R \xrightarrow{\sigma} S \xrightarrow{\tau} T$ in $\mathcal{C}$, set $(\sigma;\tau)_a$ to be $\sigma \backslash \tau_a$. The composite of $R \xrightarrow{\sigma} S \xrightarrow{\tau}$ in $\mathcal{N}$ is defined to be $\sigma \backslash \tau$. These compositions can be shown to be associative and respect the identities.

[2]These thread-pointers, unlike ordinary pointers, do not indicate the time of receipt of the jump-point being jumped to. They are actually redundant, because, in an interaction sequence, a root move in R belongs to the same thread as the preceding move. However, removing the thread-pointers would complicate our definition. The same applies to the thread-pointers in Def. 8.14.

The binary product is given by $\uplus$, and the isomorphism

$$\mathsf{vStrat}(X, S) \times \mathsf{vStrat}(X, S') \cong \mathsf{vStrat}(X, S \uplus S')$$

is trivial using functoriality in arena isomorphisms. Naturality in X follows from naturality of Def. 8.13 in arena isomorphisms. The terminal object $\emptyset$ is similar.

The coproduct-jumpwith of a family of arenas $\{R_i\}_{i \in J}$ is given by $\mathsf{pt}_{i \in I} R_i$. The isomorphism

$$\prod_{i \in I} \mathsf{nStrat}(X \times R_i) \cong \mathsf{vStrat}(X, \mathsf{pt}_{i \in I} R_i)$$

is trivial using functoriality in arena isomorphisms. Naturality in X follows from naturality of Def. 8.13 in arena isomorphisms.

8.4.2 Cell Generation

We explained in Sect. 8.1.4 that `ref` A is interpreted the same way as (8.2). Thus, if A denotes $\{R_i\}_{i \in I}$ then `ref` A denotes a singleton family, whose sole element we call $\mathsf{ref}_{i \in I} R_i$. It has 6 classes of token

write_i where $i \in I$
$\mathsf{written}_i$ where $i \in I$
$\mathsf{written}_{i,r}$ where $i \in I, r \in \mathsf{tok}\ R_i$
read
read_i where $i \in I$
$\mathsf{read}_{i,r}$ where $i \in I, r \in \mathsf{tok}\ R_i$

and is depicted as

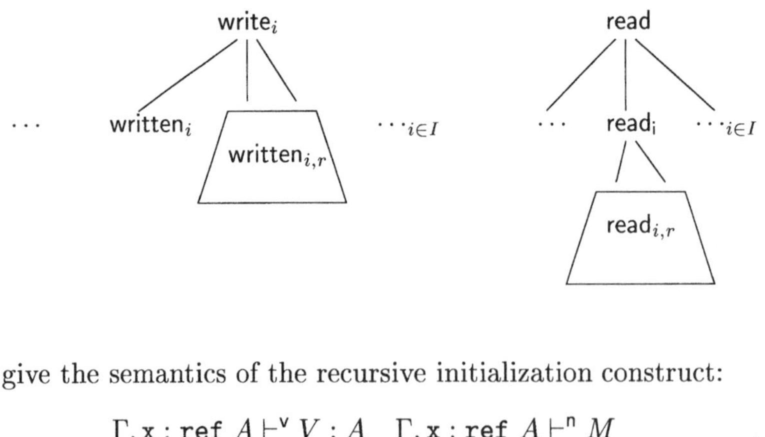

We give the semantics of the recursive initialization construct:

$$\frac{\Gamma, \mathtt{x} : \mathtt{ref}\ A \vdash^{\mathsf{v}} V : A \quad \Gamma, \mathtt{x} : \mathtt{ref}\ A \vdash^{\mathsf{n}} M}{\Gamma \vdash^{\mathsf{n}} \mathtt{newrec}\ \mathtt{x} := V.\ M} \tag{8.4}$$

For this we require the following operation.

Definition 8.14 Let S be an arena and let $\{R_i\}_{i \in I}$ be a family of arenas. (For readability, we assume all these arenas disjoint; but the formal definition uses tags.) Let $\hat{\imath} \in I$, let $\sigma \in \mathsf{vStrat}(S \uplus \mathsf{ref}_{i\in I} R_i, R_{\hat{\imath}})$ and let $\tau \in \mathsf{nStrat}(S \uplus \mathsf{ref}_{i\in I} R_i)$. We define $\mathsf{newrec}_{\hat{\imath},\sigma}\tau \in \mathsf{nStrat}(S)$ as follows.

An *interaction sequence* for these strategies is a justified sequence s on $S \uplus \mathsf{ref}_{i\in I} R_i \uplus R_{\hat{\imath}}$, together a "thread-pointer" from certain root moves in $S \uplus \mathsf{ref}_{i\in I} R_i$ to an earlier root move in $R_{\hat{\imath}}$, such that

- $s \upharpoonright_S$ is a play on S
- $s \upharpoonright_{S,\mathsf{ref}_{i\in I} R_i,\mathrm{nothreadptr}}$ (that means: we exclude moves hereditarily justified by a root move with a thread-pointer) is included in τ
- If move m plays root write_i—we call this a *write move*—then move $m+1$ plays $m \curvearrowleft \mathsf{written}_i$.
- If move m plays root read, and there is no preceding write move, then move $m+1$ plays $m \curvearrowleft \mathsf{read}_{\hat{\imath}}$. If a later move n plays $m+1 \curvearrowleft \mathsf{read}_{\hat{\imath},a}$ then move $n+1$ plays root a, and $s \upharpoonright_{n+1}$ is included in σ_a.
- If move m plays root read, and m' playing write_i is the most recent preceding write move, then move $m+1$ plays $m \curvearrowleft \mathsf{read}_{\mathsf{i}}$. If a later move n plays $m+1 \curvearrowleft \mathsf{read}_{i,a}$ then move $n+1$ plays $m' \curvearrowleft \mathsf{written}_{i,a}$, and descendants of n and $n+1$ copy moves back and forth.

  back If move q plays $p \curvearrowleft \mathsf{written}_{i,b}$, where $p+1$ is descended evenly from $n+1$, then move $q+1$ plays $p \curvearrowleft \mathsf{read}_{i,b}$.

  forth If move q plays $p+1 \curvearrowleft \mathsf{read}_{i,b}$, where $p+1$ is descended oddly from n, then move $q+1$ plays $p \curvearrowleft \mathsf{written}_{i,b}$.

We define $\mathsf{newrec}_{\hat{\imath},\sigma}\tau$ to be the set of all $s \upharpoonright_S$, where s is an interaction sequence and $s \upharpoonright_s$ is Opponent-awaiting. □

To interpret (8.4), suppose Γ denotes $\{S_j\}_{j\in J}$ and A denotes $\{R_i\}_{i\in I}$. Take $\hat{\jmath} \in J$. Write $(\hat{\imath}, \sigma)$ for $[\![V]\!](\hat{\jmath}, ())$ and write τ for $[\![M]\!](\hat{\jmath}, ())$ denotes τ. Then $[\![\texttt{newrec x} := V.\ M]\!]\hat{\jmath}$ is $\mathsf{newrec}_{\hat{\imath},\sigma}\tau$.

Mutually recursive initialization, as defined in Sect. 6.2, is treated similarly—we omit details.

8.4.3 Claims

We have not proved the following results, but conjecture that the same methods used for call-by-value in [Abramsky et al., 1998] work for JWA. In the following, recall from Sect. 8.1.4 how we interpret terms in world w.

Claim 88 (soundness and adequacy) Given a configuration w, s, M, we write $[\![w, s, M]\!]$ for $[\![\texttt{newrec}\ \overrightarrow{\texttt{x}_i := s_i}.M]\!]$. Then

- if $w, s, M \rightsquigarrow^* w', s', M'$ then $[\![w, s, M]\!] = [\![w', s', M']\!]$
- if w, s, M diverges then $[\![w, s, M]\!] = \bot$.

□

Corollary 89 If $\texttt{k} : \neg\sum_{i\in I}1 \vdash^{\mathsf{k}} M$ then $M \rightsquigarrow^* (i, ()) \nearrow \texttt{k}$ iff $[\![M]\!]()$ plays the ith root. Therefore, terms with the same denotation are observationally equivalent. □

Claim 90 (definability) Suppose that Γ denotes $\{R_i\}_{i\in I}$. Then any element of $\prod_{i\in I}\mathsf{nStrat}(R_i)$ is the denotation of some nonreturning command $\Gamma \vdash^{\mathsf{n}} M$. □

We do not have a corresponding definability result for values, because there is no value from $\mu\texttt{X}.(\texttt{bool} \times \texttt{X})$ to 0.

Claim 91 (full abstraction) Observationally equivalent terms have the same denotation. □

8.5 Pointer Game For Infinitary CBPV + Control + Storage

8.5.1 Pointer Game Via StkPS

The StkPS transform immediately gives us a model for infinitary CBPV + control + storage (without cell equality testing). Thus every value type A, every computation type $\underline{B}$, every context Γ and every w, Δ denotes a countable family of arenas, and in particular $[\![\texttt{stk}\ \underline{B}]\!] = [\![\underline{B}]\!]$.

For reference, here is the semantics of judgements.

- Suppose the context Γ denotes $\{R_i\}_{i\in I}$ and the type $\underline{B}$ denotes $\{S_j\}_{j\in J}$. Then a computation $\Gamma \vdash^{\mathsf{c}} M : \underline{B}$ denotes an element of

$$\prod_{i\in I}\prod_{j\in J}\mathsf{nStrat}(R_i \uplus S_j)$$

- Suppose the context Γ denotes $\{R_i\}_{i\in I}$ and the type A denotes $\{S_j\}_{j\in J}$. Then a value $\Gamma \vdash^{\mathsf{v}} V : A$ denotes an element of

$$\prod_{i\in I}\sum_{j\in J}\mathsf{vStrat}(R_i, S_j)$$

- Suppose the context Γ denotes $\{R_i\}_{i\in I}$, the type $\underline{B}$ denotes $\{S_j\}_{j\in J}$ and the type $\underline{C}$ denotes $\{T_h\}_{h\in H}$. Then a stack $\Gamma|\underline{B} \vdash^{\mathsf{k}} K : \underline{C}$ (is by definition a value $\Gamma, \texttt{nil} : \texttt{stk}\ \underline{C} \vdash^{\mathsf{v}} K : \texttt{stk}\ \underline{B}$ and therefore) denotes an element of

$$\prod_{i\in I}\prod_{h\in H}\sum_{j\in J}\mathsf{vStrat}(R_i \uplus T_h, S_j)$$

Soundness and adequacy for the CBPV model follow from that of the JWA model; this appears to be true also for definability and full abstraction, but the absence of complex values makes this derivation more difficult.

8.5.2 Introducing Questions and Answers

In Sect. I.5.3 we saw informally that there are two kinds of instruction in CBPV that cause a jump to another kind of a program, viz. `force` V and `return` V. In Sect. 7.4.1, we made this more precise: these are indeed the computations transformed by StkPS into jump commands in JWA.

Let us now seek to make our pointer game CBPV model more informative by distinguishing the two kinds of jump. We will give the name

- *question-move* to a move that represents forcing
- *answer-move* to a move that represents returning.

The appropriateness of these names, especially the former, is debatable. Perhaps we can defend the "question" terminology by saying that when we force a thunk, we always pass a continuation, in addition to passing operands; but this does not seem to be the most significant aspect of forcing. In any case, we use these names for the sake of consistency with the literature.

We can identify whether a move is a question-move or an answer-move by looking just at the token that is passed in it. Recall that a token a represents the data of a jump, i.e. a selector i together with some tags $j_0, \ldots, j_{r-1}$. If the jump-point selected by i is a thunk, then any jump that passes a is a question-move, so the token should be labelled Q. If the jump-point selected by i is a continuation, then any jump that passes a is an answer-move, so the token should be labelled A.

This motivates the following definition.

Definition 8.15 ▪ A Q/A-*labelled arena* (sometimes just called "arena") is an arena R together with a labelling function $\lambda^{\mathsf{QA}} : \mathsf{tok}\ R \longrightarrow \{\mathsf{Q}, \mathsf{A}\}$.

- $\emptyset$ is the empty Q/A-labelled arena.

- $R \uplus R'$ is the disjoint union of R and R'.
- $\mathsf{pt}^{\mathsf{Q}}_{i \in I} R_i$ (where I is countable) is the arena with I roots all labelled Q, and a copy of R_i grafted underneath the ith root.
- $\mathsf{pt}^{\mathsf{A}}_{i \in I} R_i$ (where I is countable) is the arena with I roots all labelled A, and a copy of R_i grafted underneath the ith root.
- We define Q/A-*respecting isomorphism*, *sub-arena* and $R \upharpoonright_a$ as in Def. 8.2.

□

Now we give the semantics of types, with explanation. Remember from Sect. 5.4.5 that in a continuation semantics like this one, when we read $[\![\underline{B}]\!]$, we should think $[\![\texttt{stk}\ \underline{B}]\!]$.

- $\texttt{stk}\ \underline{B}$ denotes the same as $\underline{B}$.
- If $\underline{B}$ denotes $\{R_i\}_{i \in I}$ then $U\underline{B}$ denotes the singleton family containing $\mathsf{pt}^{\mathsf{Q}}_{i \in I} R_i$. This is because a closed value of type $U\underline{B}$ consists of no tag and just one jump-point, which is a thunk. When we jump to it, we take (implicitly in CBPV, explicitly after StkPS) a stack for $\underline{B}$, and this consists of some tags represented by $i \in I$ and some jump-points type-represented by R_i.
- If A denotes $\{R_i\}_{i \in I}$ then FA denotes the singleton family containing $\mathsf{pt}^{\mathsf{A}}_{i \in I} R_i$. This is because a stack for FA consists of no tag and just one jump-point which is a continuation. When we jump to it, we take a closed value of type A, consisting of some tags represented by $i \in I$ and some jump-points type-represented by R_i.
- If, for each $i \in I$, the type A_i denotes $\{R_{ij}\}_{j \in J_i}$, then $\sum_{i \in I} A_i$ denotes $\{R_{ij}\}_{(i,j) \in \sum_{i \in I} J_i}$. This is because a closed value (i, V) of type $\sum_{i \in I} A_i$ consists of a tag $i \in I$ and tags represented by j_i, together with jump-points type-represented by R_{ij}.
- If A denotes $\{R_i\}_{i \in I}$ and A' denotes $\{S_j\}_{j \in J}$ then $A \times A'$ denotes $\{R_i \uplus S_j\}_{(i,j) \in I \times J}$. This is because a closed value (V, V') of type $A \times A'$ consists of
  - two sequences of tags, one represented by $i \in I$ and one represented by $j \in J$
  - two sequences of jump-points, one type-represented by R_i and one type-represented by S_j.

- 1 denotes the singleton family containing the empty arena $\emptyset$, because a closed value () of this type consists of a trivial tag and no jump-points.

- If, for each $i \in I$, the type $\underline{B}_i$ denotes $\{R_{ij}\}_{j \in J_i}$, then $\prod_{i \in I} \underline{B}_i$ denotes $\{R_{ij}\}_{(i,j) \in \sum_{i \in I} J_i}$. This is because a stack $i :: K$ for $\prod_{i \in I} \underline{B}_i$ consists of a tag $i \in I$ and tags represented by j_i, together with jump-points type-represented by R_{ij}.

- If A denotes $\{R_i\}_{i \in I}$ and $\underline{B}$ denotes $\{S_j\}_{j \in J}$ then $A \to \underline{B}$ denotes $\{R_i \uplus S_j\}_{(i,j) \in I \times J}$. This is because a stack $V :: K$ for $A \to \underline{B}$ consists of

  - two sequences of tags, one represented by $i \in I$ and one represented by $j \in J$
  - two sequences of jump-points, one type-represented by R_i and one type-represented by S_j.

We interpret infinitely deep types as for JWA in Sect. 8.3.1.

Proposition 92 Let $\{R_i\}_{i \in I}$ be a countable family of Q/A-labelled arenas. It is isomorphic to the denotation of a value type, and to the denotation of a computation type. □

Proof We obtain types in the following class

$$\begin{aligned} A &::= \textstyle\sum_{i \in I}(U\underline{B}_i \times \mathtt{stk}\ FA_i) \\ \underline{B} &::= \textstyle\prod_{i \in I}(U\underline{B}_i \to FA_i) \end{aligned}$$

and isomorphisms, in the manner of the proof of Prop. 87. □
Beware, however: distinct types in this class can be isomorphic types in the CBPV+control equational theory. An example is $FU\underline{1}$ and $U(U\underline{1} \times F0) \to F0$. Their denotation differs in the Q/A-labelling.

This makes it clear that the Q/A-labelling is purely informative—to distinguish the two kinds of jump. Technically speaking, the model is the same as the one without Q/A-labelling, because arenas with different Q/A-labelling are isomorphic in the category of value strategies.

The categorical structure that organizes Q/A-labelled pointer games is presented in Sect. 9.8.3.

8.5.3 Answer-Move Pointing To Answer-Move

We give an example of a strategy in which an A-move points to an A-move, because this possibility has not been treated previously in the games literature. Let $\underline{B}$ be the computation type UF`stk` F`bool` $\to$ F`nat`. This denotes the singleton arena family

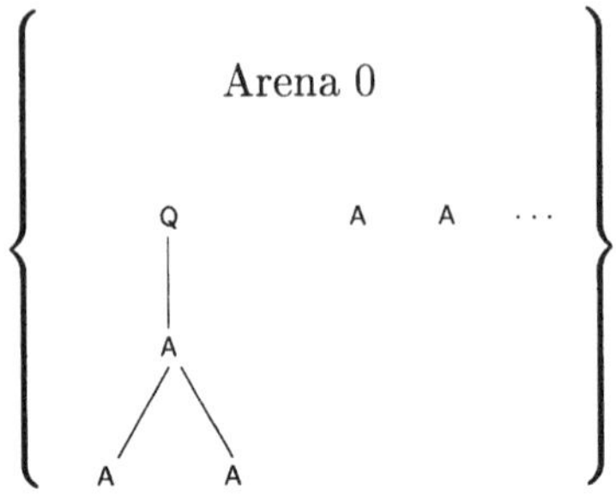

A closed computation of this type is

λ`x.(force x to y. (changestk y; return true))`

This (at 0) denotes a P-first strategy for Arena 0^{P}, as follows.

P-move 0 CBPV terminology The computation pops the thunk `x` and forces it.

game terminology So Player plays the root Q-token.

O-move 1 game terminology Now suppose Opponent points to move 0 and plays the A-token.

CBPV terminology This means that the thunk `x`, forced in move 0, returns a continuation `y` to the continuation that was then (and in fact still is) current.

P-move 2 CBPV terminology Then the computation returns `true` to this continuation `y` returned in move 1.

game terminology So Player points to move 1 and passes the A-token for true.

8.6 Removing Control

The most important part of the chapter, the model for JWA and CBPV, is now complete, and we turn to a more technical topic. Looking at our model for CBPV with control and storage, we can ask some questions.

- What objects are definable without `stk`? Between these,
  - what morphisms are definable without control operators?
- What morphisms are definable without non-ground storage? Among these,

– what morphisms are definable without any storage?
– what morphisms (between finitely deep arenas) are definable without divergence/recursion?

We will not discuss the storage and divergence questions here, since they are in no way specific to CBPV. But the control question is relevant to CBPV, because

- it leads us to a pointer game CBPV model which is not (technically) a JWA-derived model
- it demonstrates the utility of Q/A-labelling.

Firstly, objects. For a countable family of Q/A-labelled arenas $\{R_i\}_{i \in I}$, when is it—up to isomorphism—the denotation of a `stk`-free value type? When is it—up to isomorphism—the denotation of a `stk`-free computation type?

Definition 8.16 An arena is *stackfree-in-value* when no root and no successor of an A-token is an A-token. An arena is *stackfree-in-stack* when no successor of an A-token is an A-token. □

Proposition 93 In the absence of `stk`,

- the denotation of a value type A is a countable family of stackfree-in-value arenas.
- the denotation of a computation type $\underline{B}$ is a countable family of stackfree-in-stack arenas.

Conversely, every countable family of stackfree-in-value arenas is isomorphic to the denotation of a value type A, and every countable family of stackfree-in-stack arenas is isomorphic to the denotation of a computation type $\underline{B}$, of the following *type canonical form*

$$\begin{array}{rcl} A & ::= & \sum_{i \in I} U\underline{B}_i \\ \underline{B} & ::= & \prod_{i \in I}(U\underline{B}_i \to FA_i) \end{array}$$

□

Proof Similar to the proof of Prop. 87. □

We explain these as follows. In the absence of `stk`, a value consists of some tags and some thunks, but no continuations. In the denotation of a value type, R_i type-represents the jump-points accompanying the tags described by i. Since these jump-points are all thunks, it is impossible to jump to one of them as an answer-move, and all the roots of R_i are labelled Q. Similarly, when we return a value (passing an A-token a in an arena R) the jump-points we pass, type-represented by $R\restriction_a$, are all thunks, so again a successor of a cannot be labelled A. On the other hand, the denotation of a computation type $\underline{B}$ describes stacks from $\underline{B}$, and such a stack will contain a continuation, so the roots of the arenas in $[\![\underline{B}]\!]$ might be labelled A.

For morphisms, we require the following.

Definition 8.17 A *bracketable* game is any game G where moves are classified as Q/A, some moves include a pointer to an earlier move, and no A-move points to an A-move. □

Definition 8.18 Let c be a finite play in a bracketable game G. The *pending* Q-*move* of c is defined as follows.

- If c is of the form

  then the pending Q-move is Q_0.

- If c is of the form

  or of the form

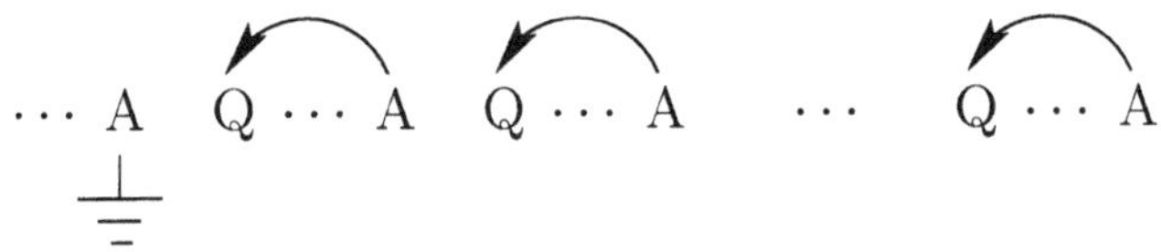

  (the earth symbol indicates the absence of a pointer), then the pending Q-move does not exist.

□

Thus the pending Q-move of c, if it exists, is

- an O-move if c is awaiting P
- a P-move if c is awaiting O.

Definition 8.19 Let c be a play in a bracketable game G.

1 Let m be an A-labelled move in c. We say that m is *bracketed* when it points

- to the pending Q-move of $c \restriction_{\$m}$, if it exists
- to nothing, otherwise.

2 c is P-*bracketed* when every A-labelled P-move is bracketed.

3 c is OP-*bracketed* when every A-labelled move is bracketed.

□

Notice that the bracketing conditions allow us to omit the pointers from A-moves (either for Player, or for both Player and Opponent) when describing a play.

Definition 8.20 ▪ A strategy σ for a bracketable game G is P-*bracketed* when every play in σ is P-bracketed.

- If R is stackfree-in-stack, we write $\mathsf{nStrat}^{\mathsf{Pbrack}}\ R$ for the set of P-bracketed strategies in $\mathsf{nStrat}R$.
- If R and S are stackfree-in-stack, we write $\mathsf{vStrat}^{\mathsf{Pbrack}}\ (R, S)$ for the set of P-bracketed strategies

□

Proposition 94 Suppose Γ is $\mathtt{stk}$-free and denotes $\{R_i\}_{i \in I}$.

- Suppose $\underline{B}$ is $\mathtt{stk}$-free and denotes $\{S_j\}_{j \in J}$. Then the denotation of a control-free computation $\Gamma \vdash^{\mathsf{c}} M : \underline{B}$ lies in

$$\prod_{i \in I} \prod_{j \in J} \mathsf{nStrat}^{\mathsf{Pbrack}}\ R_i \uplus S_j$$

- Suppose $\underline{B}$ and $\underline{C}$ are $\mathtt{stk}$-free and denote $\{S_j\}_{j \in J}$ and $\{T_k\}_{k \in K}$ respectively. Then the denotation of a control-free stack $\Gamma | \underline{B} \vdash^{\mathsf{k}} K : \underline{C}$ lies in

$$\prod_{i \in I} \prod_{k \in K} \sum_{j \in J} \mathsf{vStrat}^{\mathsf{Pbrack}}\ (R_i \uplus T_k, S_j)$$

- Suppose A is $\mathtt{stk}$-free and lies in $\{S_j\}_{j \in J}$. Then the denotation of a control-free value $\Gamma \vdash^{\mathsf{v}} V : A$ lies in

$$\prod_{i \in I} \sum_{j \in J} \mathsf{vStrat}^{\mathsf{Pbrack}}\ (R_i, S_j)$$

□

To prove this, each strategy combinator must be shown to preserve the bracketing condition, and this is accomplished using the techniques of [Hyland and Ong, 2000].

In the case of computations, we conjecture a converse. The detailed proof still needs to be done, but we expect that the call-by-value techniques adapt.

Claim 95 Suppose Γ is `stk`-free and denotes $\{R_i\}_{i \in I}$, and that $\underline{B}$ is `stk`-free and denotes $\{S_j\}_{j \in J}$.. Then any element of

$$\prod_{i \in I} \prod_{j \in J} \mathsf{nStrat}^{\mathsf{Pbrack}}\ R_i \uplus S_j$$

is the denotation of a control-free computation $\Gamma \vdash^{\mathsf{c}} M : \underline{B}$. □

As an illustration of the bracketing condition, consider the following:

```
thunk λx.
 (U'force y) to z.
  (V'force y') to z'
   return W
```

Its StkPS transform (after some simplification) is

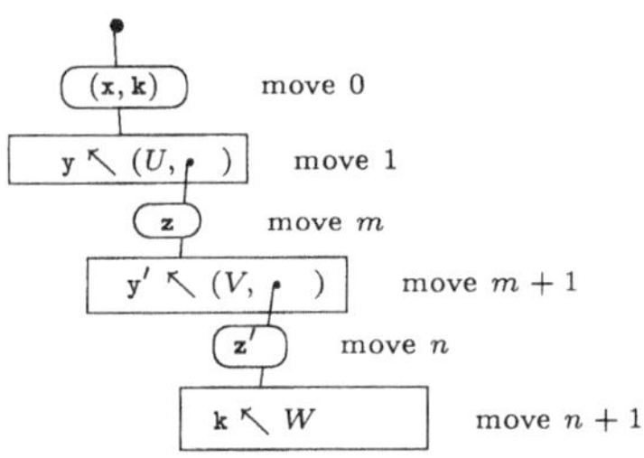

Let us think about the sequence of events that causes `return` W to be executed.

O-move 0, labelled Q The context forces the thunk. According to the StkPS transform, it passes a stack, which consists of an operand `x` and a continuation `k`.

P-move 1, labelled Q The term forces a thunk `y`. Again, it passes an operand U and a continuation viz. `to z. ...`

O-move m, labelled A The context (specifically `y`) returns a value, to the continuation passed in move 1, so this move points to move 1.

P-move $m+1$, labelled Q The term forces a thunk `y'`. Again, it passes an operand V and a continuation viz. `to z'. ...`

O-move n, labelled A The context (specifically `y'`) returns a value, to the continuation passed in move $m+1$, so this move points to move $m+1$

P-move $n+1$ The term returns W to the continuation `k` received in move 0, so this move points to move 0.

Each time the context returns, any `return` instruction would have `k` as its destination, so move 0 is the pending question-move.

III

CATEGORICAL SEMANTICS

Chapter 9

SEMANTICS IN ELEMENT STYLE

Certain parts of this chapter are designated **abstract material**. They are needed for Chap. 11 but not for Chap. 10, and may be omitted if desired.

9.1 Countable vs. Finite

Throughout Part III, we model infinitely wide CBPV and JWA. The reader interested only in finitary syntax (e.g. for studying realizability models) should substitute "finite" for "countable" throughout. The reverse substitution would not be possible, because, even for infinitely wide languages, only finite products are required in a value category.

9.2 Introduction

Many categorical structures, e.g. product, can be defined in two styles:

- *element style* uses a universal property
- *naturality style* uses a natural isomorphism.

The element style is easy to define and work with, so this is what we use in this chapter. We briefly mention the naturality style formulations in each case, but a proper treatment is deferred to Chap. 11.

After presenting in Sect. 9.3 the category theory we require, we review the semantics of effect-free language in Fig. 3.8—in particular, the use of *distributive coproducts* to model sum types.

Then in Sect. 9.5, we then present the basic structure for interpreting JWA judgements and for interpreting CBPV judgements. At this point, the only connectives we can interpret are 1 and $\times$. So our next step, in Sect. 9.6, is to work through the various connectives of CBPV and

JWA giving for each an appropriate categorical structure in element style. When we have the structure for each connective, we have a *CBPV adjunction model* or a *JWA module model.*

Finally in Sect. 9.7 we work through the various examples of denotational semantics for CBPV and JWA presented in Part I–II, and see how they are instances of CBPV adjunction models and JWA module models.

At each stage, we describe how to interpret syntax within these categorical structures, and show that the equational laws are validated. The reverse process—where we show that *any* model of the equational theory is such a categorical structure—is less important. So we defer it to Chap. 10.

9.3 Categorical Preliminaries

9.3.1 Modules

The notion of a module and bimodule for a category appear in many places e.g. [Borceux, 1994; Carboni et al., 1998] (There the relationship with modules in ring theory is explained, but this does not concern us.)

Definition 9.1 Let $\mathcal{C}$ be a category. A *left $\mathcal{C}$-module $\mathcal{N}$* consists of

- for each $A \in \mathsf{ob}\ \mathcal{C}$, a set whose elements we call *$\mathcal{N}$-morphisms from A*—we write $A \xrightarrow{\quad g \quad}$ when g is such a morphism
- for each $\mathcal{C}$-morphism $A \xrightarrow{\quad f \quad} A'$ and $\mathcal{N}$-morphism $A' \xrightarrow{\quad g \quad}$, a *composite* $\mathcal{N}$-morphism $A \xrightarrow{\quad f;g \quad}$

satisfying identity and associativity laws

$$\begin{aligned} \mathsf{id}; g &= g \\ (f; f'); g &= f; (f'; g) \end{aligned}$$

where g is a $\mathcal{N}$-morphism. □

Definition 9.2 Let $\mathcal{D}$ be a category. A *right $\mathcal{D}$-module $\mathcal{O}$* consists of

- for each $B \in \mathsf{ob}\ \mathcal{D}$, a set whose elements we call *$\mathcal{O}$-morphisms to B*—we write $\xrightarrow{\quad g \quad} B$ when g is such a morphism
- for each $\mathcal{O}$-morphism $\xrightarrow{\quad g \quad} B$ and $\mathcal{D}$-morphism $B \xrightarrow{\quad h \quad} B'$, a *composite* $\mathcal{O}$-morphism $\xrightarrow{\quad g;h \quad} B'$

satisfying identity and associativity laws

$$\begin{aligned} g; \mathsf{id} &= g \\ g; (h; h') &= (g; h); h' \end{aligned}$$

where g is a $\mathcal{O}$-morphism. □

We can combine these two structures into a *bimodule*, although we do not use bimodules until Chap. 11.

Definition 9.3 Let $\mathcal{C}$ and $\mathcal{D}$ be categories. A $(\mathcal{C},\mathcal{D})$-bimodule $\mathcal{M}$ consists of

- for each $A \in \mathsf{ob}\ \mathcal{C}$ and each $B \in \mathsf{ob}\ \mathcal{D}$, a set whose elements we call *$\mathcal{M}$-morphisms from A to B*—we write $A \xrightarrow{g} B$ when g is such a morphism
- for each $\mathcal{C}$-morphism $A \xrightarrow{f} A'$ and $\mathcal{M}$-morphism $A' \xrightarrow{g} B$, a *composite* $\mathcal{M}$-morphism $A \xrightarrow{f;g} B$
- for each $\mathcal{M}$-morphism $A \xrightarrow{g} B$ and $\mathcal{D}$-morphism $B \xrightarrow{h} B'$, a *composite* $\mathcal{M}$-morphism $A \xrightarrow{g;h} B'$

satisfying identity and associativity laws

$$\begin{aligned}
\mathsf{id}; g &= g \\
(f; f'); g &= f; (f; g) \\
g; \mathsf{id} &= g \\
g; (h; h') &= (g; h); h' \\
(f; g); h &= f; (g; h)
\end{aligned}$$

where g is a $\mathcal{M}$-morphism. □

Abstract material we can characterize

- a right $\mathcal{D}$-module as a functor from $\mathcal{D}$ to **Set**
- a left $\mathcal{B}$-module as a functor from $\mathcal{B}^{\mathsf{op}}$ to **Set**
- a $(\mathcal{B},\mathcal{D})$-bimodule as a functor from $\mathcal{B}^{\mathsf{op}} \times \mathcal{D}$ to **Set**.

9.3.2 Homset Functors

This section is **abstract material**. It reviews some familiar definitions.

Definition 9.4 For any category $\mathcal{D}$, we write $\mathsf{hom}_{\mathcal{D}}$ for the *homset functor*

$$\mathcal{D}^{\mathsf{op}} \times \mathcal{D} \longrightarrow \mathbf{Set}$$

$$(X, Y) \longmapsto \mathcal{D}(X, Y)$$

$$(f, h) \longmapsto \lambda g.(f; g; h)$$

□

Definition 9.5 1 If A is an object of a category $\mathcal{D}$, we write $\mathsf{obj}A$ for the functor $1 \longrightarrow \mathcal{D}$ taking the sole object () of 1 to A.

2 Given a functor $\mathcal{M} : \mathcal{B}^{\mathrm{op}} \times \mathcal{D} \longrightarrow \mathbf{Set}$ and a $\mathcal{B}$-object B, we write $\mathcal{M}(B, -)$ for the composite

$$\mathcal{D} \xrightarrow{i} 1 \times \mathcal{D} \xrightarrow{(\mathsf{obj}B)\times\mathcal{D}} \mathcal{B}^{\mathrm{op}} \times \mathcal{D} \xrightarrow{\mathcal{M}} \mathbf{Set}$$

3 Given a functor $\mathcal{M} : \mathcal{B}^{\mathrm{op}} \times \mathcal{D} \longrightarrow \mathbf{Set}$ and a $\mathcal{D}$-object D, we write $\mathcal{M}(-, D)$ for the composite

$$\mathcal{B}^{\mathrm{op}} \xrightarrow{i'} \mathcal{B}^{\mathrm{op}} \times 1 \xrightarrow{\mathcal{B}^{\mathrm{op}}\times(\mathsf{obj}D)} \mathcal{B}^{\mathrm{op}} \times \mathcal{D} \xrightarrow{\mathcal{M}} \mathbf{Set}$$

4 In particular, for any object D of a category $\mathcal{D}$, we obtain functors $\mathcal{D}(D, -)$ and $\mathcal{D}(-, D)$ from $\mathsf{hom}_{\mathcal{D}}$.

□

9.3.3 Some Notation

Suppose $\mathcal{C}$ is a cartesian category. We will often write

- k^*f instead of $k; f$ for the composite of $A \xrightarrow{k} B \xrightarrow{f} C$
- $\mathcal{C}_A B$ instead of $\mathcal{C}(A, B)$ for the set of $\mathcal{C}$-morphisms from A to B
- (**abstract material**) $\mathcal{C}_{-}B$ instead of $\mathcal{C}(-, B)$ for the functor from $\mathcal{C}^{\mathrm{op}}$ to **Set** defined in Def. 9.5 (3).

Suppose $\mathcal{C}$ is a cartesian category and $\mathcal{N}$ is a left $\mathcal{C}$-module. We will often write

- k^*g instead of $k; g$ for the composite of $A \xrightarrow{k} B \xrightarrow{g} C$
- $\mathcal{N}_A$ instead of $\mathcal{N}A$ for the set of morphisms from A.

We will use these notations *only* when $\mathcal{C}$ is cartesian. They enable us to unify some definitions in Sect. 9.6 and to avoid confusion between the homset functor for cartesian categories (Sect. 11.3.3) and ordinary categories.

9.3.4 Locally Indexed Categories

The notion of locally indexed category is essential for interpreting stacks in CBPV. But, as we shall see in Sect. 11.3.3, it is important (although not to the same extent) even in the effect-free setting: for example, in order to describe distributive coproducts in naturality style.

Let $\mathcal{C}$ be a category.

- For readers familiar with indexed categories, a *locally $\mathcal{C}$-indexed category* $\mathcal{D}$ is a strict $\mathcal{C}$-indexed category—i.e. a functor $\mathcal{D} : \mathcal{C}^{\mathsf{op}} \longrightarrow$ **Cat**—in which all the fibres $\mathcal{D}_X$ have the same class of objects $\mathsf{ob}\,\mathcal{D}$ and all the reindexing functors $\mathcal{D}_k$ are identity-on-objects.
- For readers familiar with enriched categories, a *locally $\mathcal{C}$-indexed category* is a $[\mathcal{C}^{\mathsf{op}}, \mathbf{Set}]$-enriched category.

But for the general reader, we give an explicit definition.

Definition 9.6 A *locally $\mathcal{C}$-indexed category* consists of

- a class $\mathsf{ob}\,\mathcal{D}$, whose elements we call *$\mathcal{D}$-objects* and we underline (except in cases where $\mathsf{ob}\,\mathcal{D} = \mathsf{ob}\,\mathcal{C}$)
- for each object $X \in \mathsf{ob}\,\mathcal{C}$ and each pair of objects $\underline{Y}, \underline{Z} \in \mathsf{ob}\,\mathcal{D}$, a set $\mathcal{D}_X(\underline{Y}, \underline{Z})$, an element of which we call a *$\mathcal{D}$-morphism* and write $\underline{Y} \xrightarrow[X]{f} \underline{Z}$
- for each object $X \in \mathsf{ob}\,\mathcal{C}$ and each object $\underline{Y} \in \mathsf{ob}\,\mathcal{D}$, an *identity* morphism $\underline{Y} \xrightarrow[X]{\mathsf{id}} \underline{Y}$
- for $\mathcal{D}$-morphisms $\underline{Y} \xrightarrow[X]{f} \underline{Z} \xrightarrow[X]{g} \underline{W}$ a *composite* morphism $\underline{Y} \xrightarrow[X]{f;g} \underline{W}$
- for each $\mathcal{D}$-morphism $\underline{Y} \xrightarrow[X]{f} \underline{Z}$ and each $\mathcal{C}$-morphism $X' \xrightarrow{k} X$, a *reindexed* $\mathcal{D}$-morphism $\underline{Y} \xrightarrow[X']{k^*f} \underline{Z}$

such that

$$\begin{array}{rclrclrcl}
\mathsf{id}; f & = & f & k^*\mathsf{id} & = & \mathsf{id} & \mathsf{id}^*f & = & f \\
f; \mathsf{id} & = & f & k^*(f;g) & = & (k^*f);(k^*g) & (l;k)^*f & = & l^*(k^*f) \\
(f;g);h & = & f;(g;h) & & & & & &
\end{array}$$

For a $\mathcal{C}$-object X, we write $\mathcal{D}_X$ for the category of $\mathcal{D}$-objects and morphisms over X, also called the "fibre over X". If $\mathcal{C}$ has a terminal object 1, morphisms over it are called *global morphisms*. For a $\mathcal{C}$-morphism $Y \xrightarrow{k} X$, we write $\mathcal{D}_X \xrightarrow{\mathcal{D}_k} \mathcal{D}_Y$ for the functor that is identity on objects and k^* on morphisms, also called the "reindexing functor over k". □

As with ordinary categories, we can easily define a notion of "functor".

Definition 9.7 Let $\mathcal{D}$ and $\mathcal{D}'$ be locally $\mathcal{C}$-indexed categories. A (locally $\mathcal{C}$-indexed) *functor* F from $\mathcal{D}$ to $\mathcal{D}'$ associates

- to each object $\underline{Y} \in \mathsf{ob}\ \mathcal{D}$ an object $F\underline{Y} \in \mathsf{ob}\ \mathcal{D}'$;
- to each morphism $X \xrightarrow[X]{f} \underline{Z}$ in $\mathcal{D}$ a morphism $F\underline{Y} \xrightarrow[X]{Ff} F\underline{Z}$ in $\mathcal{D}'$

such that

$$\begin{aligned} F\mathsf{id} &= \mathsf{id} \\ F(f;g) &= (Ff);(Fg) \\ F(k^*f) &= k^*(Ff) \end{aligned}$$

□

Building Locally $\mathcal{C}$-Indexed Categories

We look at some ways of building locally $\mathcal{C}$-indexed categories. The most important is the following construction (sometimes called the *simple fibration*):

Definition 9.8 Let $\mathcal{C}$ be a cartesian category. We form a locally $\mathcal{C}$-indexed category $\mathsf{self}\,\mathcal{C}$ as follows:

- the objects are $\mathsf{ob}\ \mathcal{C}$;
- a morphism $X \xrightarrow[Z]{f} Y$ is a $\mathcal{C}$-morphism $X \times Y \xrightarrow{f} Z$;
- the identity on Y over X is given by $X \times Y \xrightarrow{\pi'} Y$;
- the composite of

$$Y \xrightarrow[X]{f} Z \xrightarrow[X]{g} W$$

is given by

$$X \times Y \xrightarrow{(\pi,f)} X \times Z \xrightarrow{g} W$$

- the reindexing of $X \xrightarrow[Z]{f} Y$ along $X' \xrightarrow{k} X$ is given by

$$X' \times Y \xrightarrow{k \times Y} X \times Y \xrightarrow{f} Z$$

It is easy to verify the identity, associativity and reindexing laws. □

Numerous examples are given as *Eilenberg-Moore* categories:

Definition 9.9 Let (T, η, μ, t) be a strong monad on a cartesian category $\mathcal{C}$. We write $\mathcal{C}^T$ for the locally $\mathcal{C}$-indexed categories in which an object is a T-algebra and a morphism over X from (Y, θ) to (Z, ϕ) is a T-algebra homomorphism in the sense of Def. 2.8. □

As special cases, we have the locally **Cpo**-indexed category of cppos and strict maps, and the locally **Set**-indexed category of $\mathcal{A}$-sets and $\mathcal{A}$-set homomorphisms.

As with ordinary categories, we can form products and opposites of locally $\mathcal{C}$-indexed categories:

Definition 9.10 1 Given two locally $\mathcal{C}$-indexed categories $\mathcal{D}$ and $\mathcal{D}'$, we define the locally $\mathcal{C}$-indexed category $\mathcal{D} \times \mathcal{D}'$ by

$$\begin{aligned} \mathsf{ob}\ (\mathcal{D} \times \mathcal{D}') &= \mathsf{ob}\ \mathcal{D} \times \mathsf{ob}\ \mathcal{D}' \\ (\mathcal{D} \times \mathcal{D}')_X((A, A'), (B, B')) &= \mathcal{D}_X(A, B) \times \mathcal{D}'_X(A', B') \end{aligned}$$

with the evident identities, composition and reindexing.

2 Given a locally $\mathcal{C}$-indexed category $\mathcal{D}$, define the locally $\mathcal{C}$-indexed category $\mathcal{D}^{\mathsf{op}}$ by

$$\begin{aligned} \mathsf{ob}\ \mathcal{D}^{\mathsf{op}} &= \mathsf{ob}\ \mathcal{D} \\ \mathcal{D}^{\mathsf{op}}_X(A, B) &= \mathcal{D}_X(B, A) \end{aligned}$$

with the evident identities, composition and reindexing.

□

9.3.5 OpGrothendieck Construction And Homset Functors

This section is **abstract material.**

When we study ordinary categories, we frequently have to consider functors to **Set**, in the setting of homset functors, representable functors, adjunctions, modules etc.

However, we cannot speak of a functor from a locally indexed category to **Set**, because **Set** is an ordinary category. To remedy this problem, we use the *opGrothendieck construction.*

Definition 9.11 Let $\mathcal{D}$ be a locally $\mathcal{C}$-indexed category, where $\mathcal{C}$ is cartesian. Then $\mathsf{opGr}\ \mathcal{D}$ is the ordinary category defined as follows:

- an object of $\mathsf{opGr}\ \mathcal{D}$ is a pair ${}_X\underline{Y}$ where $X \in \mathsf{ob}\ \mathcal{C}$ and $\underline{Y} \in \mathsf{ob}\ \mathcal{D}$;

- a morphism from ${}_X\underline{Y}$ to ${}_{X'}\underline{Z}$ in $\mathsf{opGr}\ \mathcal{D}$ consists of a pair ${}_k f$ where $X' \xrightarrow{k} X$ in $\mathcal{C}$ and $\underline{Y} \xrightarrow[X']{g} \underline{Z}$ in $\mathcal{D}$;
- the identity on ${}_X\underline{Y}$ is given by ${}_{\mathsf{id}}\mathsf{id}$;
- the composite of

$$ {}_X\underline{Y} \xrightarrow{{}_k f} {}_{X'}\underline{Z} \xrightarrow{{}_l g} {}_{X''}\underline{W} $$

is given by ${}_{l;k}((l^* f); g)$.

It is easy to verify the identity and associativity laws. □

Wherever we see a functor from $\mathcal{D}$ to **Set** in ordinary category theory, we expect to see a functor from $\mathsf{opGr}\ \mathcal{D}$ to **Set** in locally indexed category theory. As an example, here is the homset functor (cf. Def. 9.4).

Definition 9.12 Let $\mathcal{D}$ be a locally $\mathcal{C}$-indexed category. We write $\mathsf{hom}_{\mathcal{D}}$ for the functor

$$ \mathsf{opGr}(\mathcal{D}^{\mathsf{op}} \times \mathcal{D}) \longrightarrow \mathbf{Set} $$

$$ {}_X(\underline{Y}, \underline{Z}) \longmapsto \mathcal{D}_X(\underline{Y}, \underline{Z}) $$

$$ {}_k(f, h) \longmapsto \lambda g.(f; (k^* g); h) $$

□

The opGrothendieck construction can also be applied in the obvious way to a locally $\mathcal{C}$-indexed functor $\mathcal{D} \xrightarrow{F} \mathcal{D}'$, giving a functor $\mathsf{opGr}\ \mathcal{D} \xrightarrow{\mathsf{opGr}\ F} \mathsf{opGr}\ \mathcal{D}'$. Using this, we can adapt Def. 9.5 to the locally $\mathcal{C}$-indexed setting.

Definition 9.13 1 If $\underline{A}$ is an object of a locally $\mathcal{C}$-indexed category $\mathcal{D}$, we write $\mathsf{obj}\underline{A}$ for the functor $1 \longrightarrow \mathcal{D}$ taking the sole object $()$ of 1 to $\underline{A}$.

2 Given a functor $\mathcal{M} : \mathsf{opGr}(\mathcal{B}^{\mathsf{op}} \times \mathcal{D}) \longrightarrow \mathbf{Set}$ and a $\mathcal{B}$-object $\underline{B}$, we write $\mathcal{M}_-(\underline{B}, -)$ for the composite

$$ \mathsf{opGr}\ \mathcal{D} \xrightarrow{\mathsf{opGr}\ i} \mathsf{opGr}(1 \times \mathcal{D}) \xrightarrow{\mathsf{opGr}((\mathsf{obj}\underline{B}) \times \mathcal{D})} \mathsf{opGr}(\mathcal{B}^{\mathsf{op}} \times \mathcal{D}) \xrightarrow{\mathcal{M}} \mathbf{Set} $$

3 Given a functor $\mathcal{M} : \mathsf{opGr}(\mathcal{B}^{\mathsf{op}} \times \mathcal{D}) \longrightarrow \mathbf{Set}$ and a $\mathcal{D}$-object $\underline{D}$, we write $\mathcal{M}_-(\underline{D}, -)$ for the composite

$$ \mathsf{opGr}\ \mathcal{B} \xrightarrow{\mathsf{opGr}\ i'} \mathsf{opGr}(\mathcal{B} \times 1) \xrightarrow{\mathsf{opGr}(\mathcal{B} \times (\mathsf{obj}\underline{D}))} \mathsf{opGr}(\mathcal{B}^{\mathsf{op}} \times \mathcal{D}) \xrightarrow{\mathcal{M}} \mathbf{Set} $$

4 In particular, for any object $\underline{D}$ of a locally $\mathcal{C}$-indexed category $\mathcal{D}$, we obtain functors $\mathcal{D}_{-}(\underline{D}, -)$ and $\mathcal{D}_{\mathcal{D}\,-}(-, \underline{D})$ from $\mathsf{hom}_{\mathcal{D}}$.

5 Given a functor $\mathcal{O} : \mathsf{opGr}\ \mathcal{D} \longrightarrow \mathbf{Set}$, and a $\mathcal{D}$-object $\underline{B}$, we write $\mathcal{O}_{-}\underline{B}$ for the composite

$$\mathcal{C}^{\mathsf{op}} \xrightarrow{\cong} \mathsf{opGr}\ 1 \xrightarrow{\mathsf{opGr\ obj}\underline{B}} \mathsf{opGr}\ \mathcal{D} \xrightarrow{\mathcal{O}} \mathbf{Set}$$

□

9.3.6 Locally Indexed Modules

Just as we define modules for an ordinary category, so we can define them for a locally indexed category.

Definition 9.14 Let $\mathcal{D}$ be a locally $\mathcal{C}$-indexed category. A (locally $\mathcal{C}$-indexed) *right* $\mathcal{D}$*-module* $\mathcal{O}$ consists of

- for each $X \in \mathsf{ob}\ \mathcal{C}$ and $\underline{Y} \in \mathsf{ob}\ \mathcal{B}$, a set whose elements we call $\mathcal{O}$-*morphisms over* X *to* $\underline{Y}$—we write $\xrightarrow[X]{g} \underline{B}$ when g is such a morphism
- for each $\mathcal{O}$-morphism $\xrightarrow[X]{g} \underline{Y}$ and $\mathcal{D}$-morphism $\underline{Y} \xrightarrow[X]{h} \underline{Z}$ a *composite* $\mathcal{O}$-morphism $\xrightarrow[X]{g;h} \underline{Z}$
- for each $\mathcal{C}$-morphism $X' \xrightarrow{k} X$ and $\mathcal{O}$-morphism $\xrightarrow[X]{g} \underline{Y}$, a *reindexed* $\mathcal{O}$-morphism $\xrightarrow[X']{k^*g} \underline{Y}$

satisfying identity, associativity and reindexing laws

$$\begin{aligned} g; \mathsf{id} &= g \\ g; (h; h') &= (g; h); h' \\ \mathsf{id}^* g &= g \\ (k; k')^* g &= k^*(k'^* g) \\ k^*(g; h) &= (k^* g); (k^* h) \end{aligned}$$

where g is a $\mathcal{O}$-morphism. □

Similarly, we can define a left $\mathcal{B}$-module and a $(\mathcal{B}, \mathcal{D})$-bimodule.

Abstract material we can characterize

- a right $\mathcal{D}$-module as a functor from $\mathsf{opGr}\ \mathcal{D}$ to **Set**
- a left $\mathcal{B}$-module as a functor from $\mathsf{opGr}(\mathcal{B}^{\mathsf{op}})$ to **Set**
- a $(\mathcal{B}, \mathcal{D})$-bimodule as a functor from $\mathsf{opGr}(\mathcal{B}^{\mathsf{op}} \times \mathcal{D})$ to **Set**.

9.3.7 Locally Indexed Natural Transformations

This section is **abstract material.**

Just as we form a 2-category **Cat** of categories, functors and natural transformations, so we can do the same with locally $\mathcal{C}$-indexed categories.

Definition 9.15 Suppose $\mathcal{C}$ has a terminal object. Let $\mathcal{D}$ and $\mathcal{D}'$ be locally $\mathcal{C}$-indexed categories and let F and G be functors from $\mathcal{D}$ to $\mathcal{D}'$. A (locally $\mathcal{C}$-indexed) *natural transformation* α from F to G provides, for each object $\underline{Y} \in \mathsf{ob}\,\mathcal{D}$ a global morphism $F\underline{Y} \xrightarrow[1]{\alpha_X \underline{Y}} G\underline{Y}$ such that for each $X \in \mathsf{ob}\,\mathcal{C}$ and each $\underline{Y} \xrightarrow[X]{f} \underline{Z}$ the diagram

$$\begin{array}{ccc} F\underline{Y} & \xrightarrow{()^*\alpha\underline{Y}} & G\underline{Y} \\ {\scriptstyle Ff}\downarrow & & \downarrow{\scriptstyle Gf} \\ F\underline{Z} & \xrightarrow[()^*\alpha\underline{Z}]{} & G\underline{Z} \end{array} \quad \text{commutes in } \mathcal{D}'_X.$$

□

Definition 9.16 We write $\mathcal{C}$**LICat** for the 2-category of locally $\mathcal{C}$-indexed categories, functors and natural transformations, with the obvious vertical and horizontal compositions. We extend the opGrothendieck construction to a 2-functor opGr from $\mathcal{C}$**LICat** to **Cat** in the obvious way. □

We can talk about monads in $\mathcal{C}$**LICat**, because Def. 1.4 make sense in any 2-category. The following result was noted in [Moggi, 1991].

Proposition 96 Let $\mathcal{C}$ be a cartesian category. A strong monad on $\mathcal{C}$ is precisely a monad on $\mathsf{self}\,\mathcal{C}$ in $\mathcal{C}$**LICat**. □

9.4 Modelling Effect-Free Languages

9.4.1 Cartesian Categories and Substitution Lemma

By (simply typed) *effect-free languages*, we mean $\times\sum\prod\rightarrow$-calculus presented in Fig. 3.8, and various fragments of it, which we name by their connectives. Thus "$\times$-calculus" is the fragment whose only connectives are $1, \times$. These are the key connectives for categorical semantics, because they allow us to turn a context $A_0, \ldots, A_{n-1}$ into a type $A_0 \times \cdots \times A_{n-1}$.

For $\times$-calculus, and other languages, we say that a *context morphism* [Lawvere, 1963] q from $\Gamma = A_0, \ldots, A_{m-1}$ to $\Delta = B_0, \ldots, B_{n-1}$ is a sequence of terms $M_0, \ldots, M_{n-1}$ where $\Gamma \vdash^{\mathsf{v}} M_i : B_i$. (In the case of CBPV and JWA, a context morphism will be a sequence of *values*.) If we are given such a morphism q and a term $\Delta \vdash N : C$, we can substitute M_i for $\mathtt{x}_i$ in N, and obtain a term which we write $\Gamma \vdash q^*M : C$. (Technically this operation is defined by two inductions: one for weakening, and then one for general substitution, of which weakening is a special case. See [Altenkirch and Reus, 1999; Fiore et al., 1999] for details.)

The fundamental theorem of (simply typed) categorical semantics is that "models of $\times$-calculus and cartesian categories are equivalent". We make this statement precise in Sect. 10.2, but for the moment we will merely describe the interpretation of $\times$-calculus in a cartesian category $\mathcal{C}$.

Each type denotes a $\mathcal{C}$-object, defined by induction, and the context $A_0, \ldots, A_{n-1}$ denotes the object $[\![A_0]\!] \times \cdots \times [\![A_{n-1}]\!]$. Each term $\Gamma \vdash M : A$ denotes a $\mathcal{C}$-morphism $[\![\Gamma]\!] \xrightarrow{[\![M]\!]} [\![A]\!]$, defined by induction.

$$\begin{array}{rcl}
[\![\mathtt{x}_i]\!] & = & \pi_i \\
[\![\mathtt{let}\ M\ \mathtt{be}\ \mathtt{x}.\ N]\!] & = & (\mathsf{id}, [\![M]\!])^*[\![N]\!] \\
[\![(M, M')]\!] & = & ([\![M]\!], [\![M']\!]) \\
[\![\mathtt{pm}\ M\ \mathtt{as}\ (\mathtt{x}, \mathtt{y}).\ N]\!] & = & ((\mathsf{id}, ([\![M]\!]; \pi)), ([\![M]\!]; \pi'))^*[\![N]\!]
\end{array}$$

It is vital to prove that this soundly interprets the equational theory, i.e. that if $M = N$ is provable then $[\![M]\!] = [\![N]\!]$. But the laws mention substitution, and so, in order to do this, we first need a *substitution lemma* saying what a substitution denotes:

$$[\![q^*M]\!] = [\![q]\!]^*[\![M]\!]$$

where, if $q = (M_0, \ldots, M_{n-1})$, we write $[\![q]\!]$ for $([\![M_0]\!], \ldots, [\![M_{n-1}]\!])$ This is proved straightforwardly. (Again, two inductions are required—see [Fiore et al., 1999] for details.) It is then straightforward to show that the equations of the theory are validated.

9.4.2 Products, Exponentials and Distributive Coproducts

If we now extend the $\times$-calculus by adding countable product types, function types or (finite or countable) sum types to the $\times$-calculus, we have to add appropriate structure to the cartesian category $\mathcal{C}$, viz. countable products, exponentials or distributive coproducts. We have already defined these structures (in Sect. 1.6.4), but we now recapitulate these definitions in a concise form. This is to familiarize the reader with this way of describing universal properties.

Definition 9.17 Let $\mathcal{C}$ be a cartesian category.

1 A *product* for a family of objects $\{A_i\}_{i \in I}$ consists of an object V (the vertex), together with a morphism $V \xrightarrow{\pi_i} A_i$ for each $i \in I$, such that the functions

$$\begin{array}{rcll} \mathcal{C}_X V & \longrightarrow & \prod_{i \in I} \mathcal{C}_X A_i & \text{for all } X \\ f & \longmapsto & \lambda i.(f; \pi_i) & \end{array}$$

are isomorphisms—we write $\mathsf{q}^{\prod}{}_X$ for the inverse. This is exactly Def. 1.3.

2 An *exponential* from an object A to an object B consists of an object V (the vertex), together with a morphism $V \times A \xrightarrow{\mathsf{ev}} B$ such that the functions

$$\begin{array}{rcll} \mathcal{C}_X V & \longrightarrow & \mathcal{C}_{X \times A} B & \text{for all } X \\ f & \longmapsto & (f \times A); \mathsf{ev} & \end{array}$$

are isomorphisms—we write $\mathsf{q}^{\to}{}_X$ for the inverse. This is exactly Def. 1.6.

3 A *distributive coproduct* for a family of objects $\{A_i\}_{i \in I}$ is an object V (the vertex) together with morphisms $A_i \xrightarrow{\mathsf{in}_i} V$ such that the functions

$$\begin{array}{rcll} \mathcal{C}_{X \times V} Y & \longrightarrow & \prod_{i \in I} \mathcal{C}_{X \times A_i} V & \text{for all } X,Y \\ f & \longmapsto & \lambda i.((X \times \mathsf{in}_i)^* f) & \end{array}$$

are isomorphisms—we write $\mathsf{q}^{\sum}{}_X Y$ for the inverse. This is exactly Def. 1.7.

□

It is clear that the isomorphisms in these definitions correspond to the reversible derivations for $\prod$, for $\to$ and for $\sum$

$$\frac{\cdots \quad \Gamma \vdash B_i \quad \cdots_{\, i \in I}}{\Gamma \vdash \prod_{i \in I} B_i}$$

$$\frac{\Gamma, A \vdash B}{\Gamma \vdash A \to B}$$

$$\frac{\cdots \quad \Gamma, A_i \vdash B \quad \cdots_{\, i \in I}}{\Gamma, \sum_{i \in I} A_i \vdash B}$$

This is a common pattern throughout the categorical semantics we shall consider.

The semantics of terms for countable product types using countable products is given as follows.

$$\begin{array}{rcl} [\![\hat{\imath}`M]\!] & = & [\![M]\!];\pi_{\hat{\imath}} \\ [\![\lambda\{\ldots,i.M_i,\ldots\}]\!] & = & \mathsf{q}^{\prod}\lambda i.[\![M_i]\!] \end{array}$$

The semantics of terms for function types using exponentials is given as follows.

$$\begin{array}{rcl} [\![V`M]\!] & = & ([\![M]\!],[\![V]\!]);\mathsf{ev} \\ [\![\lambda \mathtt{x}.M]\!] & = & \mathsf{q}^{\rightarrow}[\![M]\!] \end{array}$$

The semantics of terms for sum types using distributive coproducts is given as follows.

$$\begin{array}{rcl} [\![(\hat{\imath},M)]\!] & = & [\![M]\!];\mathsf{in}_{\hat{\imath}} \\ [\![\mathsf{pm}\ M\ \mathsf{as}\ \{\ldots,(i,\mathtt{x}).\ N_i,\ldots\}]\!] & = & (\mathsf{id},[\![M]\!])^*\mathsf{q}^{\sum}\lambda i.[\![N_i]\!] \end{array}$$

We call the definitions in Def. 9.17 *element style* because we explicitly describe what the isomorphism does to an element of the homset. But we shall show in Sect. 11.3.1 and Sect. 11.3.3 that these structures can also be described in *naturality style*, where, in addition to the vertex V, we provide

- for a product of $\{A_i\}_{i\in I}$, an isomorphism

$$\mathcal{C}_X V \cong \prod_{i\in I}\mathcal{C}_X A_i \quad \text{natural in } X$$

- for an exponential from A to B, an isomorphism

$$\mathcal{C}_X V \cong \mathcal{C}_{X\times A} B \quad \text{natural in } X$$

- for a distributive coproduct of $\{A_i\}_{i\in I}$, an isomorphism

$$\mathcal{C}_{X\times V}\pi^*Y \cong \prod_{i\in I}\mathcal{C}_{X\times A_i}\pi^*Y \quad \text{natural in } X \text{ and } Y$$

This "naturality in Y" uses the following notation. Given a $\mathcal{C}$-morphism $X\times Y \xrightarrow{h} Z$, we write

$$\mathcal{C}_X Y \xrightarrow{\mathcal{C}_X h} \mathcal{C}_X Z$$

for the function taking g to $(\mathsf{id},g)^*h$.

9.5 Categorical Structures For Judgements Of JWA And CBPV

9.5.1 Introduction

In this section we interpret the JWA fragment and the CBPV fragment whose only connectives are $1, \times$. In each case, values are interpreted in a cartesian category $\mathcal{C}$.

9.5.2 Judgement Model of JWA

We first recall some facts about the JWA equational theory. There are 2 judgements, for *values* and *nonreturning commands*, written

$$\Gamma \vdash^{\mathsf{v}} V : A \qquad\qquad \Gamma \vdash^{\mathsf{n}} M$$

Values are to be interpreted in a cartesian category $\mathcal{C}$, but what about nonreturning commands? We recall that we can substitute a context morphism $\Gamma \xrightarrow{q} \Delta$ into a nonreturning command $\Delta \vdash^{\mathsf{n}} M$ to obtain a nonreturning command $\Gamma \vdash^{\mathsf{n}} q^*M$. Substitution satisfies the equations

$$\begin{aligned} \mathsf{id}^*M &= M \\ p^*q^*M &= (p;q)^*M \end{aligned}$$

It is therefore clear that we should interpret nonreturning commands in a left $\mathcal{C}$-module $\mathcal{N}$ (Def. 9.1).

Definition 9.18 A *JWA judgement model* consists of

- a cartesian category $\mathcal{C}$
- a left $\mathcal{C}$-module $\mathcal{N}$.

□

Thus we interpret

- a value $\Gamma \vdash^{\mathsf{v}} V : A$ as a $\mathcal{C}$-morphism from $[\![\Gamma]\!]$ to $[\![A]\!]$
- a nonreturning command $\Gamma \vdash^{\mathsf{n}} M$ as a $\mathcal{N}$-morphism from $[\![\Gamma]\!]$

As usual, we will want a substitution lemma. This will state that for any context morphism $\Gamma \xrightarrow{q} \Delta$, and any term P in context Δ (whether a value or a nonreturning command)

$$[\![q^*P]\!] = [\![q]\!]^*[\![P]\!]$$

Here is the semantics in a JWA judgement model of the JWA fragment whose only connectives are $1, \times$.

$$\begin{array}{rcl} [\![\mathtt{x}_i]\!] & = & \pi_i \\ [\![\mathtt{let}\ V\ \mathtt{be\ x.}\ P]\!] & = & (\mathsf{id}, [\![V]\!])^*[\![P]\!] \\ [\![(V, V')]\!] & = & ([\![V]\!], [\![V']\!]) \\ [\![\mathtt{pm}\ V\ \mathtt{as}\ (\mathtt{x}, \mathtt{y}).\ P]\!] & = & ((\mathsf{id}, ([\![V]\!]; \pi)), ([\![V]\!]; \pi'))^*[\![P]\!] \end{array}$$

9.5.3 Stacks In CBPV

Like in JWA, values in CBPV are interpreted in a cartesian category $\mathcal{C}$. But how do we interpret the stack judgement $\Gamma|\underline{B} \vdash^{\mathsf{k}} K : \underline{C}$? Let us recall some of the operations on stacks.

Firstly we can concatenate a stack $\Gamma|\underline{A} \vdash^{\mathsf{k}} K : \underline{B}$ with a stack $\Gamma|\underline{B} \vdash^{\mathsf{k}} L : \underline{C}$ to give a stack $\Gamma|\underline{A} \vdash^{\mathsf{k}} K +\!\!+ L : \underline{C}$. Concatenation satisfies the equations

$$\begin{array}{rcl} \mathtt{nil} +\!\!+ K & = & K \\ K +\!\!+ \mathtt{nil} & = & K \\ (K +\!\!+ L) +\!\!+ L' & = & K +\!\!+ (L +\!\!+ L') \end{array}$$

Secondly, we can substitute a context morphism $\Gamma \xrightarrow{q} \Delta$ (denoting a $\mathcal{C}$-morphism) into a stack $\Delta|\underline{B} \vdash^{\mathsf{k}} K : \underline{C}$ to give a stack $\Gamma|\underline{B} \vdash^{\mathsf{k}} q^*K : \underline{C}$. Substitution satisfies the equations

$$\begin{array}{rcl} \mathsf{id}^*K & = & K \\ p^*q^*K & = & (p; q)^*K \end{array}$$

Concatenation and substitution are related by the equations

$$\begin{array}{rcl} p^*\mathtt{nil} & = & \mathtt{nil} \\ p^*(K +\!\!+ L) & = & (p^*K) +\!\!+ (p^*L) \end{array}$$

It is therefore clear that we require a *locally $\mathcal{C}$-indexed category* (Def. 9.6). We will interpret a stack $\Gamma|\underline{B} \vdash^{\mathsf{k}} K : \underline{C}$ as a $\mathcal{D}$-morphism over $[\![\Gamma]\!]$ from $[\![\underline{B}]\!]$ to $[\![\underline{C}]\!]$. The empty stack nil denotes an identity morphism, and we will want a concatenation lemma saying

$$[\![K +\!\!+ L]\!] = [\![K]\!]; [\![L]\!]$$

and a reindexing lemma saying

$$[\![q^*K]\!] = [\![q]\!]^*[\![K]\!]$$

9.5.4 Computations In CBPV

Next we want to interpret the computation judgement $\Gamma \vdash^{\mathsf{c}} M : \underline{B}$. We recall the operations that can be performed on computations.

Firstly we can dismantle a stack $\Gamma|\underline{B} \vdash^{\mathsf{k}} K : \underline{C}$ onto a computation $\Gamma \vdash^{\mathsf{c}} M : \underline{B}$ to give a computation $\Gamma \vdash^{\mathsf{c}} M \bullet K : \underline{C}$. Dismantling satisfies the equations

$$\begin{aligned} M \bullet \mathtt{nil} &= M \\ M \bullet (K +\!\!+ K') &= (M \bullet K) \bullet K' \end{aligned}$$

Secondly we can substitute a a context morphism $\Gamma \xrightarrow{q} \Delta$ (denoting a $\mathcal{C}$-morphism) into a computation $\Delta \vdash^{\mathsf{c}} M : \underline{B}$ to give a stack $\Gamma \vdash^{\mathsf{c}} q^* M : \underline{B}$. Substitution satisfies the equations

$$\begin{aligned} \mathsf{id}^* M &= M \\ p^* q^* M &= (p;q)^* M \end{aligned}$$

Dismantling and substitution are related by the equation

$$p^*(M \bullet K) = (p^* M) \bullet (p^* K)$$

It is therefore clear that we should interpret computations in a right $\mathcal{D}$-module (Def. 9.14). We summarize as follows.

Definition 9.19 A *CBPV judgement model* consists of

- a cartesian category $\mathcal{C}$
- a locally $\mathcal{C}$-indexed category $\mathcal{D}$
- a right $\mathcal{D}$-module $\mathcal{O}$.

Objects of $\mathcal{C}$ are called *val-objects*, and objects of $\mathcal{D}$ are called *comp-objects*. □

We will interpret a computation $\Gamma \vdash^{\mathsf{c}} M : \underline{B}$ as an $\mathcal{O}$-morphism from $[\![\Gamma]\!]$ to $[\![\underline{B}]\!]$, and we will want a dismantling lemma

$$[\![M \bullet K]\!] = [\![M]\!] \bullet [\![K]\!]$$

and a substitution lemma

$$[\![q^* M]\!] = [\![q]\!]^* [\![M]\!]$$

Here is the semantics in a CBPV judgement model of the CBPV fragment whose only connectives are $1, \times$.

$$\begin{aligned} [\![\mathtt{x}_i]\!] &= \pi_i \\ [\![\mathtt{let}\ V\ \mathtt{be}\ \mathtt{x}.\ P]\!] &= (\mathsf{id}, [\![V]\!])^* [\![P]\!] \\ [\![(V, V')]\!] &= ([\![V]\!], [\![V']\!]) \\ [\![\mathtt{pm}\ V\ \mathtt{as}\ (\mathtt{x}, \mathtt{y}).\ P]\!] &= ((\mathsf{id}, ([\![V]\!]; \pi)), ([\![V]\!]; \pi'))^* [\![P]\!] \end{aligned}$$

9.5.5 Enrichment

We will not look at enriched models of CBPV and JWA in general, as the theory has not been developed, but just look at 3 examples of enrichment. Our main reason for doing this is that in Sect. 9.7 we will study various constructions of CBPV and JWA models, and we want to see that these constructions enrich (i.e. can be applied to enriched models to give other enriched models).

Suppose we want a CBPV judgement model $(\mathcal{C}, \mathcal{D}, \mathcal{O})$ to interpret CBPV with printing. Then it is not sufficient for each $\mathcal{O}$-homset $\mathcal{O}_A\underline{B}$ to be merely a set; it must be equipped with an $\mathcal{A}$-set structure, so that we can interpret $\mathtt{print}\ c.\ M$ as $c * [\![M]\!]$. Furthermore, in order to validate the laws

$$\begin{array}{rcl} q^*(\mathtt{print}\ c.\ M) & = & \mathtt{print}\ c.\ (q^*M) \\ (\mathtt{print}\ c.\ M) \bullet K & = & \mathtt{print}\ c.\ (M \bullet K) \end{array}$$

we require similar equations in the model.

Likewise if we want a JWA judgement model $(\mathcal{C}, \mathcal{N})$ to interpret JWA with printing, we require each $\mathcal{N}_A$ to be an $\mathcal{A}$-set, and in order to validate

$$q^*(\mathtt{print}\ c.\ M) \quad = \quad \mathtt{print}\ c.\ (q^*M)$$

we require a similar equation in the model. This leads us to the following definition.

Definition 9.20 1 An *$\mathcal{A}$-set enriched CBPV judgement model* is a CBPV judgement model $(\mathcal{C}, \mathcal{D}, \mathcal{O})$ where each $\mathcal{O}_X\underline{Y}$ is equipped with a structure $* : \mathcal{A} \times \mathcal{O}_X\underline{Y} \longrightarrow \mathcal{O}_X\underline{Y}$ such that the equations

$$\begin{array}{rcl} k^*(c * g) & = & c * (k^*g) \\ (c * g); h & = & c * (g; h) \end{array}$$

hold where g is a computation morphism.

2 An *$\mathcal{A}$-set enriched JWA judgement model* is a JWA judgement model $(\mathcal{C}, \mathcal{N})$ where each $\mathcal{N}_X$ is equipped with a structure $* : \mathcal{A} \times \mathcal{N}_X \longrightarrow \mathcal{N}_X$ such that the equation

$$k^*(c * g) \quad = \quad c * (k^*g)$$

holds where g is a nonreturning command morphism.

□

In the same way, if we want to model CBPV+recursion or JWA+recursion in such a way that recursion is interpreted as a least prefixed point in a cpo, then we require the following structure.

Definition 9.21 1 A *cppo-enriched CBPV judgement model* is a CBPV judgement model $(\mathcal{C}, \mathcal{D}, \mathcal{O})$ where

- all the value homsets $\mathcal{C}_X Y$, stack homsets $\mathcal{D}_X(\underline{Y}, \underline{Z})$ and computation homsets $\mathcal{O}_X \underline{Y}$ are equipped with a partial order making them cpos, and all operations (composition and reindexing) are continuous
- all the computation homsets $\mathcal{O}_X \underline{Y}$ are pointed and the equations

$$\begin{array}{rcl} k^*\bot & = & \bot \\ \bot; h & = & \bot \end{array}$$

are satisfied for the least element $\bot$ of the computation homsets.

2 A *cppo-enriched JWA judgement model* is a JWA judgement model $(\mathcal{C}, \mathcal{N})$ where

- all the value homsets $\mathcal{C}_X Y$ and nonreturning command homsets $\mathcal{N}_X$ are equipped with a partial order making them cpos, and all compositions are continuous
- all the nonreturning command homsets $\mathcal{N}_X$ are pointed and the equation

$$k^*\bot \quad = \quad \bot$$

is satisfied for the least element $\bot$ of the nonreturning command homsets.

□

If we want to restrict attention to domains (as in Def. 4.8), then we use the following.

Definition 9.22 1 A *domain-enriched CBPV judgement model* is a cppo-enriched CBPV judgement model $(\mathcal{C}, \mathcal{D}, \mathcal{O})$ where all the computation homsets $\mathcal{O}_X \underline{Y}$ are domains.

2 A *domain-enriched JWA judgement model* is a cppo-enriched JWA judgement model $(\mathcal{C}, \mathcal{N})$ where all the nonreturning command homsets $\mathcal{N}_X$ are domains.

□

Notice that we do *not* require that value homsets $\mathcal{C}_X Y$ be predomains, as this would not be the case e.g. in the predomain/domain model of CBPV, where the value homset from $\mathbb{N}$ to $\mathbb{N}$ is an uncountable flat cpo. The fact that it is only computation homsets (or nonreturning command homsets) that are well-behaved is an instance of the phenomenon discussed in Sect. 2.10: computations are the important judgement.

Abstract material

- An $\mathcal{A}$-set enriched CBPV judgement model $(\mathcal{C}, \mathcal{D}, \mathcal{O})$ is a CBPV judgement model in which $\mathcal{O}$ is a functor from $\mathsf{opGr}\ \mathcal{D}$ to $\mathcal{A}\mathbf{Set}$, rather than just to **Set**.
- An $\mathcal{A}$-set enriched JWA judgement model $(\mathcal{C}, \mathcal{N})$ is a JWA judgement model in which $\mathcal{N}$ is a functor from $\mathcal{C}^{\mathsf{op}}$ to $\mathcal{A}\mathbf{Set}$, rather than just to **Set**.
- A cppo enriched CBPV judgement model $(\mathcal{C}, \mathcal{D}, \mathcal{O})$ is a CBPV judgement model in a **Cpo**-enriched sense, in which $\mathcal{O}$ is a functor from $\mathsf{opGr}\ \mathcal{D}$ to $\mathbf{Cpo}^{\perp}$, rather than just to **Cpo**. It is domain enriched if moreover $\mathcal{O}$ is a functor to **DomStr**.
- A cppo enriched JWA judgement model $(\mathcal{C}, \mathcal{N})$ is a JWA judgement model in a **Cpo**-enriched sense, in which $\mathcal{N}$ is a functor from $\mathcal{C}^{\mathsf{op}}$ to $\mathbf{Cpo}^{\perp}$, rather than just to **Cpo**. It is domain enriched if moreover $\mathcal{N}$ is a functor to **DomStr**.

9.6 Connectives

9.6.1 Right Adjunctives—Interpreting U

In CBPV, we recall that the U connective has the following reversible derivation

$$\frac{\Gamma \vdash^{\mathsf{c}} \underline{B}}{\Gamma \vdash^{\mathsf{v}} U\underline{B}}$$

This suggests the following definition.

Definition 9.23 (element style) A *right adjunctive* for a comp-object $\underline{B}$ in a CBPV judgement model $(\mathcal{C}, \mathcal{D}, \mathcal{O})$ consists of a val-object V (the vertex) together with an $\mathcal{O}$-morphism $\xrightarrow[V]{\mathsf{force}} \underline{B}$, such that the functions

$$\begin{array}{rcl} \mathcal{C}_X V & \longrightarrow & \mathcal{O}_X \underline{B} \\ f & \longmapsto & f^*\mathsf{force} \end{array} \qquad \text{for all } X$$

are isomorphisms—we write $\mathsf{q}^U{}_X$ for the inverse. □

Here is the semantics of U using right adjunctives.

$$\begin{aligned} [\![\texttt{force } V]\!] &= [\![V]\!]^*\mathsf{force} \\ [\![\texttt{thunk } M]\!] &= \mathsf{q}^U[\![M]\!] \end{aligned}$$

We shall see in Sect. 11.3.1 that the naturality style description of a right adjunctive for $\underline{B}$ is the vertex V together with an isomorphism

$$\mathcal{C}_X V \cong \mathcal{O}_X \underline{B} \quad \text{natural in } X.$$

9.6.2 Jumpwiths—Interpreting $\neg$

The interpretation of $\neg$ in JWA is similar to that of U in CBPV. We recall that $\neg$ has the following reversible derivation

$$\frac{\Gamma, A \vdash^{\mathsf{n}}}{\overline{\Gamma \quad \vdash^{\mathsf{v}} \neg A}}$$

This suggests the following definition. (The idea of giving a direct categorical characterization of $\neg$ first appeared in [Thielecke, 1997b], although the characterization given there was different.)

Definition 9.24 (element style) A *jumpwith* for an object A in a JWA judgement model $(\mathcal{C}, \mathcal{N})$ consists of an object V (the vertex) together with a $\mathcal{N}$-morphism $V \times A \xrightarrow{\mathsf{jump}}$ such that the functions

$$\begin{aligned} \mathcal{C}_X V &\longrightarrow \mathcal{N}_{X \times A} \\ f &\longmapsto (f \times A)^*\mathsf{jump} \end{aligned} \qquad \text{for all } X$$

are isomorphisms—we write $\mathsf{q}^{\neg}{}_X$ for the inverse. □

Here is the semantics of $\neg$ using jumpwiths.

$$\begin{aligned} [\![V \nearrow W]\!] &= ([\![W]\!], [\![V]\!])^*\mathsf{jump} \\ [\![\gamma \mathtt{x}.M]\!] &= \mathsf{q}^{\neg}[\![M]\!] \end{aligned}$$

We shall see in Sect. 11.3.1 that the naturality style description of a jumpwith for A is the vertex V together with an isomorphism

$$\mathcal{C}_X V \cong \mathcal{N}_{X \times A} \quad \text{natural in } X$$

9.6.3 Left Adjunctives—Interpreting F

The interpretation of F in CBPV is somewhat more complicated, but not much. We recall that it has the reversible derivation

$$\frac{\Gamma, A \vdash^{\mathsf{c}} \underline{B}}{\overline{\Gamma | FA \vdash^{\mathsf{k}} \underline{B}}}$$

This motivates the following definition.

Definition 9.25 (element style) A *left adjunctive* for a val-object A in a CBPV judgement model $(\mathcal{C}, \mathcal{D}, \mathcal{O})$ consists of a comp-object $\underline{V}$ (the vertex) together with an $\mathcal{O}$-morphism $\xrightarrow[A]{\mathsf{return}} \underline{V}$, such that the functions

$$\begin{array}{rcll} \mathcal{D}_X(\underline{V}, \underline{Y}) & \longrightarrow & \mathcal{O}_{X \times A}\underline{Y} & \text{for all } X, \underline{Y} \\ h & \longmapsto & (\pi'^{*}\mathsf{return}); (\pi^{*} h) & \end{array}$$

are isomorphisms—we write $\mathsf{q}^{F}{}_{X}\underline{Y}$ for the inverse. □

Here is the semantics of F using left adjunctives.

$$\begin{array}{rcl} [\![\mathtt{return}\ V]\!] & = & [\![V]\!]^{*}\mathsf{return} \\ [\![M\ \mathtt{to\ x.}\ N]\!] & = & [\![M]\!]; \mathsf{q}^{F}[\![N]\!] \\ [\![[\cdot]\ \mathtt{to\ x.}\ N :: K]\!] & = & (\mathsf{q}^{F}[\![N]\!]); [\![K]\!] \end{array}$$

We shall see in Sect. 11.3.2 that the naturality style description of a left adjunctive for A is the vertex $\underline{V}$ together with an isomorphism

$$\mathcal{C}_X(\underline{V}, \underline{Y}) \cong \mathcal{O}_{X \times A}\pi^{*}\underline{Y} \quad \text{natural in } X \text{ and } \underline{Y}.$$

9.6.4 Products—Interpreting $\prod$

We first extend the notion of categorical product (Def. 9.17(1)) to a situation where, as well as a category, there are some right modules.

Definition 9.26 Let $\mathcal{D}$ be a category, and let $\{\mathcal{O}^p\}_{p \in P}$ be a family of right $\mathcal{D}$-modules. A *product* in $(\mathcal{D}, \{\mathcal{O}^p\}_{p \in P})$ for a family of $\mathcal{D}$-objects $\{B_i\}_{i \in I}$ consists of a $\mathcal{D}$-object V (the vertex) and a morphism $V \xrightarrow{\pi_i} A_i$ for each $i \in I$, such that the functions

$$\begin{array}{rcll} \mathcal{D}(Y, V) & \longrightarrow & \prod_{i \in I} \mathcal{D}(Y, B_i) & \text{for all } Y \\ \mathcal{O}^p V & \longrightarrow & \prod_{i \in I} \mathcal{O}^p B_i & \text{for all } p \\ h & \longmapsto & \lambda i.(h; \pi_i) & \end{array}$$

are isomorphisms. □

Of course, in the special case that the module family is empty, Def. 9.26 gives the usual definition of categorical product in $\mathcal{D}$. But, in general, it makes a stronger requirement. We now adapt Def. 9.26 to the locally indexed setting.

Definition 9.27 Let $\mathcal{C}$ be a category, let $\mathcal{D}$ be a locally $\mathcal{C}$-indexed category, and let $\{\mathcal{O}^p\}_{p \in P}$ be a family of right $\mathcal{D}$-modules. A *product* in

$(\mathcal{D}, \{\mathcal{O}^p\}_{p\in P})$ for a family of $\mathcal{D}$-objects $\{\underline{B}_i\}_{i\in I}$ consists of a $\mathcal{D}$-object $\underline{V}$ (the vertex) together with a $\mathcal{D}$-morphism $\underline{V} \xrightarrow[1]{\pi_i} \underline{B}_i$, such that the functions

$$\begin{array}{rcll} \mathcal{D}_X(\underline{Y}, \underline{V}) & \longrightarrow & \prod_{i\in I} \mathcal{D}_X(\underline{Y}, \underline{B}_i) & \text{for all } X, \underline{Y} \\ \mathcal{O}^p_X \underline{V} & \longrightarrow & \prod_{i\in I} \mathcal{O}^p_X \underline{B}_i & \text{for all } X, p \\ h & \longmapsto & \lambda i.(h; ()^* \pi_i) & \end{array}$$

are isomorphisms. □

The two cases that are of interest are

- where the module family is empty, giving a definition of product in a locally indexed category
- where the module family is singleton $\{\mathcal{O}\}$, giving a definition of product in a CBPV judgement model $(\mathcal{C}, \mathcal{D}, \mathcal{O})$.

In the latter case, the isomorphism condition states that the functions

$$\begin{array}{lrcll} \textbf{stacks} & \mathcal{D}_X(\underline{Y}, \underline{V}) & \longrightarrow & \prod_{i\in I} \mathcal{D}_X(\underline{Y}, \underline{B}_i) & \text{for all } X, \underline{Y} \\ \textbf{computations} & \mathcal{O}_X \underline{V} & \longrightarrow & \prod_{i\in I} \mathcal{O}_X \underline{B}_i & \text{for all } X \\ & h & \longmapsto & \lambda i.(h; ()^* \pi_i) & \end{array}$$

are isomorphisms—we write $\mathsf{q}\Pi^{\mathsf{k}}_X \underline{Y}$ and $\mathsf{q}\Pi^{\mathsf{c}}_X$ for the inverse. This condition corresponds to the reversible derivations for the $\prod$ connective in CBPV:

$$\frac{\cdots \quad \Gamma | \underline{A} \vdash^{\mathsf{k}} \underline{B}_i \quad \cdots \; i\in I}{\Gamma | \underline{A} \vdash^{\mathsf{k}} \prod_{i\in I} \underline{B}_i}$$

$$\frac{\cdots \quad \Gamma \vdash^{\mathsf{c}} \underline{B}_i \quad \cdots \; i\in I}{\Gamma \vdash^{\mathsf{c}} \prod_{i\in I} \underline{B}_i}$$

In fact, the isomorphism condition for computations is redundant in the presence of left adjunctives:

Proposition 97 Let $(\mathcal{C}, \mathcal{D}, \mathcal{O})$ be a CBPV judgement model with all left adjunctives. Then any product in $\mathcal{D}$ is automatically a product in $(\mathcal{C}, \mathcal{D}, \mathcal{O})$. (The converse is trivial.) □

Proof The isomorphism

$$\mathcal{D}_X(F1, \underline{Y}) \xrightarrow{\cong} \mathcal{O}_{X\times 1} \underline{Y} \xrightarrow{\cong} \mathcal{O}_X \underline{Y}$$

maps h to $(()^*\mathsf{return}\ ); h$, by calculation. So the diagram

$$\begin{array}{ccc} \mathcal{D}_X(F1,\underline{V}) & \longrightarrow & \prod_{i\in I}\mathcal{D}_X(F1,\underline{B}_i) \\ \downarrow\cong & & \downarrow\cong \\ \mathcal{O}_X(F1,\underline{V}) & \longrightarrow & \prod_{i\in I}\mathcal{O}_X\underline{B}_i \end{array}$$

commutes, by calculation. So if the top arrow is an isomorphism, the bottom arrow must be too. □

Here is the semantics of $\prod$, including complex stacks, using products.

$$\begin{array}{rl} [\![\hat{\imath}`M]\!] & [\![M]\!]; ()^*\pi_{\hat{\imath}} \\ [\![\lambda\{\ldots,i.M_i,\ldots\}]\!] & \mathsf{q}^{\prod^{\mathsf{c}}}\lambda i.[\![M_i]\!] \\ [\![\hat{\imath} :: K]\!] & (()^*\pi_{\hat{\imath}}); [\![K]\!] \\ [\![\{\ldots,K_i\ \mathtt{where}\ i :: \mathtt{nil},\ldots\}\ \mathtt{is}\ L]\!] & (\mathsf{q}^{\prod^{\mathsf{k}}}\lambda i.[\![K_i]\!]); [\![L]\!] \end{array}$$

We shall see in Sect. 11.3.4 that the naturality style description of a product for $\{\underline{B}_i\}_{i\in I}$ in $(\mathcal{D},\{\mathcal{O}^p\}_{p\in P})$ is the vertex $\underline{V}$ together with isomorphisms

$$\begin{array}{rcll} \mathcal{D}_X(\underline{Y},\underline{V}) & \cong & \prod_{i\in I}\mathcal{D}_X(\underline{Y},\underline{B}_i) & \text{natural in } X \text{ and } \underline{Y} \\ \mathcal{O}^p_X\underline{V} & \cong & \prod_{i\in I}\mathcal{O}^p_X\underline{B}_i & \text{natural in } X \end{array}$$

mutually natural in $\underline{Y}$. This "mutually natural in $\underline{Y}$" condition uses the following notation. Given an $\mathcal{O}^p$-morphism $\xrightarrow[X]{g} \underline{Y}$, we write

$$\mathcal{D}_X(\underline{Y},\underline{Z}) \xrightarrow{\mathcal{O}^p\mathcal{D}_X(g,\underline{Z})} \mathcal{O}^p_X\underline{Z}$$

for the function taking h to $g; h$.

9.6.5 Exponentials—Interpreting →

We define exponentials in a similar way to Def. 9.26.

Definition 9.28 Let $\mathcal{C}$ be a cartesian category, let $\mathcal{D}$ be a locally $\mathcal{C}$-indexed category, and let $\{\mathcal{O}^p\}_{p\in P}$ be a family of right $\mathcal{D}$-modules. An *exponential* from a $\mathcal{C}$-object A to a $\mathcal{D}$-object $\underline{B}$ consists of a comp-object $\underline{V}$ together with a $\mathcal{D}$-morphism $\underline{V} \xrightarrow[A]{\mathsf{ev}} \underline{B}$, such that the functions

$$\begin{array}{rcll} \mathcal{D}_X(\underline{Y},\underline{V}) & \longrightarrow & \mathcal{D}_{X\times A}(\underline{Y},\underline{B}) & \text{for all } X,\underline{Y} \\ \mathcal{O}^p_X\underline{V} & \longrightarrow & \mathcal{O}^p_{X\times A}\underline{B} & \text{for all } X,p \\ h & \longmapsto & (\pi^*h); (\pi'^*\mathsf{ev}) & \end{array}$$

are isomorphisms. □

Again, the two cases that are of interest are

- where the module family is empty, giving a definition of exponential in a locally $\mathcal{C}$-indexed category, where $\mathcal{C}$ is cartesian

- where the module family is singleton $\{\mathcal{O}\}$, giving a definition of exponential in a CBPV judgement model $(\mathcal{C}, \mathcal{D}, \mathcal{O})$.

In the latter case, the isomorphism condition states that the functions

$$\begin{array}{llcll} \textbf{stacks} & \mathcal{D}_X(\underline{Y}, \underline{V}) & \longrightarrow & \mathcal{D}_{X \times A}(\underline{Y}, \underline{B}) & \text{for all } X, \underline{Y} \\ \textbf{computations} & \mathcal{O}_X \underline{V} & \longrightarrow & \mathcal{O}_{X \times A} \underline{B} & \text{for all } X \\ & h & \longmapsto & (\pi^* h); (\pi'^* \mathsf{ev}) & \end{array}$$

are isomorphisms—we write $\mathsf{q}^{\to\mathsf{k}}_{X}\underline{Y}$ and $\mathsf{q}^{\to\mathsf{c}}_{X}$ for the inverse. This condition corresponds to the reversible derivations for the $\to$ connective in CBPV:

$$\frac{\Gamma, A | \underline{C} \vdash^{\mathsf{k}} \quad \underline{B}}{\Gamma \quad | \underline{C} \vdash^{\mathsf{k}} A \to \underline{B}}$$

$$\frac{\Gamma, A \vdash^{\mathsf{c}} \quad \underline{B}}{\Gamma \quad \vdash^{\mathsf{c}} A \to \underline{B}}$$

As with products, the isomorphism for computations is redundant in the presence of left adjunctives:

Proposition 98 Let $(\mathcal{C}, \mathcal{D}, \mathcal{O})$ be a CBPV judgement model with all left adjunctives. Then any exponential in $\mathcal{D}$ is automatically an exponential in $(\mathcal{C}, \mathcal{D}, \mathcal{O})$. (The converse is trivial.) □

Proof Like in the proof of Prop. 97, the diagram

$$\begin{array}{ccc} \mathcal{D}_X(F1, \underline{V}) & \longrightarrow & \mathcal{D}_{X \times A}(F1, \underline{B}) \\ \downarrow & & \downarrow \\ \mathcal{O}_X \underline{V} & \longrightarrow & \mathcal{O}_{X \times A} \underline{B} \end{array}$$

commutes, by calculation. So if the top arrow is an isomorphism, the bottom arrow must be too. □

Here is the semantics of $\to$, including complex stacks, using exponentials.

$$\begin{array}{rcl} [\![V`M]\!] & = & [\![M]\!];[\![V]\!]^*\mathsf{ev} \\ [\![\lambda\mathtt{x}.M]\!] & = & \mathsf{q}^{\to\mathsf{c}}[\![M]\!] \\ [\![V :: K]\!] & = & ([\![V]\!]^*\mathsf{ev});[\![K]\!] \\ [\![K\ \mathtt{where}\ \mathtt{x} :: \mathtt{nil}\ \mathtt{is}\ L]\!] & = & (\mathsf{q}^{\to\mathsf{k}}[\![K]\!]);[\![L]\!] \end{array}$$

We shall see in Sect. 11.3.4 that the naturality style description of an exponential from A to $\underline{B}$ is the vertex $\underline{V}$ together with isomorphisms

$$\begin{array}{rcll} \mathcal{D}_X(\underline{Y},\underline{V}) & \cong & \mathcal{D}_{X\times A}(\pi^*\underline{Y},\underline{B}) & \text{natural in } X \text{ and } \underline{Y} \\ \mathcal{O}^p_X\underline{V} & \cong & \mathcal{O}^p_{X\times A}\underline{B} & \text{natural in } X \end{array}$$

mutually natural in $\underline{Y}$. This mutual naturality is defined using the $\mathcal{O}^p\mathcal{D}_X(g,\underline{Z})$ notation as in Sect. 9.6.4.

9.6.6 Distributive Coproducts—Interpreting $\sum$

We now extend the notion of distributive coproduct to a situation where there are some left modules.

Definition 9.29 Let $\mathcal{C}$ be a cartesian category, and let $\{\mathcal{N}^p\}_{p\in P}$ be a family of left $\mathcal{C}$-modules. A *distributive coproduct* within $(\mathcal{C},\{\mathcal{N}_p\}_{p\in P})$ for a family of $\mathcal{C}$-objects $\{A_i\}_{i\in I}$ consists of a $\mathcal{C}$-object V (the vertex) together with a $\mathcal{C}$-morphism $A_i \xrightarrow{\mathsf{in}_i} V$, for each $i \in I$, such that the functions

$$\begin{array}{rcll} \mathcal{C}_{X\times V}Y & \longrightarrow & \prod_{i\in I}\mathcal{C}_{X\times A_i}Y & \text{for all } X,Y \\ \mathcal{N}^p_{X\times V} & \longrightarrow & \prod_{i\in I}\mathcal{N}^p_{X\times A_i} & \text{for all } X,p \\ f & \longmapsto & \lambda i.((X\times\mathsf{in}_i)^*f) & \end{array}$$

are isomorphisms—we write $\mathsf{q}^{\Sigma}{}_X{}^{\mathsf{v}}Y$ and $\mathsf{q}^{\Sigma}{}_X{}^{\not\mathsf{v}}I$ for the inverses. □

There are 3 cases of interest.

cartesian category $\mathcal{C}$ Here we take the module family to be empty. This gives us exactly Def. 9.17(3).

JWA judgement model $(\mathcal{C},\mathcal{N})$ Here we take the module family to be the singleton $\{\mathcal{N}\}$. The isomorphism condition thus states that the functions

$$\begin{array}{lrcll} \textbf{values} & \mathcal{C}_{X\times V}Y & \longrightarrow & \prod_{i\in I}\mathcal{C}_{X\times A_i}Y & \text{for all } X,Y \\ \textbf{nonret. commands} & \mathcal{N}_{X\times V} & \longrightarrow & \prod_{i\in I}\mathcal{N}_{X\times A_i} & \text{for all } X \\ & f & \longmapsto & \lambda i.((X\times\mathsf{in}_i)^*f) & \end{array}$$

are isomorphisms. This corresponds to the reversible derivations for $\sum$ in JWA:

$$\frac{\cdots \quad \Gamma, A_i \vdash^{\mathsf{v}} B \quad \cdots}{\Gamma, \sum_{i \in I} A_i \vdash^{\mathsf{v}} B}$$

$$\frac{\cdots \quad \Gamma, A_i \vdash^{\mathsf{n}} \quad \cdots \; {}_{i \in I}}{\Gamma, \sum_{i \in I} A_i \vdash^{\mathsf{n}}}$$

CBPV judgement model $(\mathcal{C}, \mathcal{D}, \mathcal{O})$ Here we take the module family to be $\mathcal{N}_{\mathsf{c}} + \mathcal{N}_{\mathsf{k}}$ where

- $\mathcal{N}_{\mathsf{c}}$ is the module family containing $\mathcal{O}_{-}\underline{B}$ for each comp-object $\underline{B}$
- $\mathcal{N}_{\mathsf{k}}$ for the module family containing $\mathcal{D}_{-}(\underline{A}, \underline{B})$ for each pair of comp-objects $\underline{A}$ and $\underline{B}$.

The isomorphism condition thus states that the functions

values	$\mathcal{C}_{X \times V} Y$	$\longrightarrow$	$\prod_{i \in I} \mathcal{C}_{X \times A_i} Y$	for all X, Y
stacks	$\mathcal{D}_{X \times V}(\underline{Y}, \underline{Z})$	$\longrightarrow$	$\prod_{i \in I} \mathcal{D}_{X \times A_i}(\underline{Y}, \underline{Z})$	for all $X, \underline{Y}, \underline{Z}$
computations	$\mathcal{O}_{X \times V} \underline{Z}$	$\longrightarrow$	$\prod_{i \in I} \mathcal{O}_{X \times A_i} \underline{Z}$	for all $X, \underline{Z}$
	f	$\longmapsto$	$\lambda i.((X \times \mathsf{in}_i)^* f)$	

are isomorphisms. This corresponds to the reversible derivations for $\sum$ in JWA:

$$\frac{\cdots \quad \Gamma, A_i \vdash^{\mathsf{v}} B \quad \cdots \; {}_{i \in I}}{\Gamma, \sum_{i \in I} A_i \vdash^{\mathsf{v}} B}$$

$$\frac{\cdots \quad \Gamma, A_i | \underline{B} \vdash^{\mathsf{k}} \underline{C} \quad \cdots \; {}_{i \in I}}{\Gamma, \sum_{i \in I} A_i | \underline{B} \vdash^{\mathsf{k}} \underline{C}}$$

$$\frac{\cdots \quad \Gamma, A_i \vdash^{\mathsf{c}} \underline{C} \quad \cdots \; {}_{i \in I}}{\Gamma, \sum_{i \in I} A_i \vdash^{\mathsf{c}} \underline{C}}$$

In all 3 cases, the semantic equations for $\sum$ type using a distributive coproduct are the same:

$$\begin{aligned} [\![(\hat{\imath}, V)]\!] &= [\![V]\!]; \mathsf{in}_{\hat{\imath}} \\ [\![\texttt{pm}\ V\ \texttt{as}\ \{\ldots, (i, \texttt{x}).P_i, \ldots\}]\!] &= (\mathsf{id}, [\![V]\!])^* \mathsf{q}^{\sum} \lambda i.[\![P_i]\!] \end{aligned}$$

We shall see in Sect. 11.3.5 that the naturality style description of a distributive coproduct for $\{A_i\}_{i \in I}$ within $(\mathcal{C}, \{\mathcal{N}_p\}_{p \in P})$ consists of the vertex V together with isomorphisms

$$\begin{aligned} \mathcal{C}_{X \times V} \pi^* Y &\cong \prod_{i \in I} \mathcal{C}_{X \times A_i} \pi^* Y \quad \text{natural in } X \text{ and } Y \\ \mathcal{N}^p_{X \times V} &\cong \prod_{i \in I} \mathcal{N}^p_{X \times A_i} \quad \text{natural in } X \end{aligned}$$

mutually natural in Y. The "natural in Y" condition uses the $\mathcal{C}_X h$ notation of Sect. 9.4.2, whilst the "mutually natural in Y" condition uses the following notation. Given a $\mathcal{N}$-morphism $X \times Y \xrightarrow{h}$, we write

$$\mathcal{C}_X Y \xrightarrow{\mathcal{C}\mathcal{N}^p_X h} \mathcal{N}^p_X$$

for the function taking g to $(\mathsf{id}, g)^* h$.

As with products and exponentials, certain isomorphism conditions can be redundant.

Proposition 99 Let $(\mathcal{C}, \mathcal{N})$ be a JWA judgement model with all jump-withs. Then any distributive coproduct in $\mathcal{C}$ is automatically a distributive coproduct in $(\mathcal{C}, \mathcal{N})$. (The converse is trivial.) □

Proof The isomorphism

$$\mathcal{C}_X \neg 1 \xrightarrow{\cong} \mathcal{N}_{X \times 1} \xrightarrow{\cong} \mathcal{N}_X$$

maps f to $(f, ())^* \mathsf{jump}$, by calculation. So the diagram

$$\begin{array}{ccc} \mathcal{C}_{X \times V} \neg 1 & \longrightarrow & \prod_{i \in I} \mathcal{C}_{X \times A_i} \neg 1 \\ \cong \downarrow & & \downarrow \cong \\ \mathcal{N}_{X \times V} & \longrightarrow & \prod_{i \in I} \mathcal{N}_{X \times A_i} \end{array}$$

commutes by calculation. So if the top arrow is an isomorphism, the bottom must be too. □

Proposition 100 Let $(\mathcal{C}, \mathcal{D}, \mathcal{O})$ be a CBPV judgement model equipped with all right adjunctives. Then any distributive coproduct in $(\mathcal{C}, \mathcal{N}_{\mathsf{k}})$ is automatically a distributive coproduct in $(\mathcal{C}, \mathcal{D}, \mathcal{O})$ i.e. in $(\mathcal{C}, \mathcal{N}_{\mathsf{c}} \cup \mathcal{N}_{\mathsf{k}})$. (The converse is trivial.) □

Proof The diagram

$$\begin{array}{ccc} \mathcal{C}_{X\times V}U\underline{Y} & \longrightarrow & \prod_{i\in I}\mathcal{C}_{X\times A_i}U\underline{Y} \\ \downarrow \cong & & \downarrow \cong \\ \mathcal{O}_{X\times V}\underline{Y} & \longrightarrow & \prod_{i\in I}\mathcal{O}_{X\times A_i}\underline{Y} \end{array}$$

commutes by calculation. So if the top arrow is an isomorphism, the bottom arrow must be too. □

9.6.7 Tying It All Together

We can now give our categorical definitions

Definition 9.30 A *strong adjunction* consists of a CBPV judgement model $(\mathcal{C}, \mathcal{D}, \mathcal{O})$ with all right adjunctives and left adjunctives. □

Definition 9.31 A *CBPV adjunction model* consists of a strong adjunction with all countable distributive coproducts, all countable products and all exponentials. □

Definition 9.32 A *JWA module model* consists of a JWA judgement model $(\mathcal{C}, \mathcal{N})$ with all countable distributive coproducts and all jumpwiths. □

Proposition 101 Any CBPV adjunction model, and indeed any strong adjunction, is isomorphic to one in which the isomorphism

$$\mathcal{C}_X U\underline{Y} \cong \mathcal{O}_X\underline{Y} \tag{9.1}$$

is the identity, and in which therefore

$$\begin{aligned} [\![\texttt{thunk}\ M]\!] &= [\![M]\!] \\ [\![\texttt{force}\ V]\!] &= [\![V]\!] \end{aligned}$$

□

Despite Prop. 101, we do not incorporate this constraint into our definition of model. For while this constraint is reasonable for certain models (e.g. Eilenberg-Moore models, as we shall see in Sect. 9.7.2), for others, such as continuations, games and possible worlds, requiring (9.1) to be the identity would detract from the intuitiveness of the presentation.

9.6.8 Adjunctions

This section is **abstract material**. It provides some mathematical context to Def. 9.30.

To explain the "strong adjunction" terminology, recall the following.

Definition 9.33 An *adjunction* from a category $\mathcal{B}$ to a category $\mathcal{D}$ consists of

- two functors

$$\mathcal{B} \underset{U}{\overset{F}{\rightleftarrows}} \mathcal{D}$$

  (U is called the *right adjoint*, F is called the *left adjoint*);

- natural transformations $1 \xrightarrow{\eta} UF$ (the *unit*) and $FU \xrightarrow{\epsilon} 1$ (the *counit*)

satisfying the *triangle laws*

$$\begin{array}{ccc} U & & \\ {\scriptstyle \eta U}\downarrow & \searrow^{id} & \\ UFU & \xrightarrow[U\epsilon]{} & U \end{array} \qquad \begin{array}{ccc} F & \xrightarrow{F\eta} & FUF \\ & {\scriptstyle id}\searrow & \downarrow{\scriptstyle \epsilon F} \\ & & F \end{array}$$

□

This definition makes sense in any 2-categories, so we can speak about adjunctions in $\mathcal{C}$**LICat**. The following result is proved in Sect. 11.6.1.

Proposition 102 Let $\mathcal{C}$ be a cartesian category and let $\mathcal{D}$ be a locally $\mathcal{C}$-indexed category. A strong adjunction from $\mathcal{C}$ to $\mathcal{D}$ corresponds to adjunction from $\mathsf{self}\,\mathcal{C}$ to $\mathcal{D}$ in $\mathcal{C}$**LICat**. □

Now, in any 2-category, an adjunction $(\mathcal{D}, U, F, \eta, \epsilon)$ from $\mathcal{B}$ gives rise to a monad $(UF, \eta, U\epsilon F)$ on $\mathcal{B}$. So by Prop. 102 and Prop. 96, a strong adjunction from $\mathcal{C}$ gives rise to a strong monad on $\mathcal{C}$—that is the reason for the terminology "strong adjunction".

Conversely, if we are given a monad (T, η, μ) on $\mathcal{B}$ we can look for *resolutions* of this monad i.e. adjunctions that give rise to it. As explained in e.g. [Lambek and Scott, 1986], these resolutions form a category $\mathsf{Res}(T)$, when we say that a *comparison functor* from the resolution $(\mathcal{D}, U, F, \epsilon)$

to the resolution $(\mathcal{D}', U', F', \epsilon)$ is a functor $\mathcal{D} \xrightarrow{K} \mathcal{D}'$ such that

$$\begin{aligned} U'K &= U \\ KF &= F' \\ K\epsilon &= \epsilon' K \end{aligned}$$

It is shown in [Mac Lane, 1971] that there are two canonical ways of resolving a monad in **Cat**: the *Eilenberg-Moore resolution*, which is terminal in $\mathsf{Res}(T)$, and the *Kleisli resolution*, which is initial in $\mathsf{Res}(T)$ (and every comparison functor from it is fully faithful). These methods exist also in $\mathcal{C}$**LICat**, where $\mathcal{C}$ is any cartesian category. We look at Eilenberg-Moore resolutions in Sect. 9.7.2 and Sect. 11.6.2, and at Kleisli resolutions in Sect. 11.6.3.

9.7 Examples

9.7.1 Trivial Models Of CBPV

We saw in Sect. 3.11 the *trivialization transform* from CBPV+stacks to the $\times\sum\prod \rightarrow$-calculus. This gives us the following construction on models.

Definition 9.34 Let $\mathcal{C}$ be a countably bicartesian closed category. We obtain a CBPV adjunction model, called a *trivial model*, by setting the stack category to be $\mathsf{self}\,\mathcal{C}$ and $\mathcal{O}_X Y$ to be $\mathcal{C}_X Y$. Both U and F (on objects) are the identity. □

Indeed, CBPV adjunction model can be thought of as a generalization of countably bicartesian closed category, where we separate into different categories the distributive coproduct requirement and the countable product/exponential requirement.

9.7.2 Eilenberg-Moore Models Of CBPV

Suppose we have a cartesian category $\mathcal{C}$ and a strong monad (T, η, μ, t). We define the *Eilenberg-Moore resolution* $\mathsf{EM}(T)$ of this strong monad (discussed further in Sect. 11.6.2) to be the following strong adjunction:

- the value category is $\mathcal{C}$
- the stack category $\mathcal{D}$ is the locally $\mathcal{C}$-indexed category of T-algebras and T-algebra homomorphisms
- the computation homset $\mathcal{O}_X(Y, \theta)$ is set to $\mathcal{C}_X Y$
- the right adjunctive $U(X, \theta)$ is the carrier X

- the left adjunctive FX is the free algebra $(TX, \mu X)$.

Under what conditions does it possess distributive coproducts, products and exponentials?

Proposition 103 Let (T, η, μ, t) be a strong monad on a cartesian category $\mathcal{C}$.

1 A distributive coproduct of $\{A_i\}_{i \in I}$ in $\mathsf{EM}(T)$ corresponds to a distributive coproduct of $\{A_i\}_{i \in I}$ in $\mathcal{C}$.

2 A product of $\{(A_i, \theta_i)\}_{i \in I}$ in $\mathsf{EM}(T)$ corresponds to a product of $\{A_i\}_{i \in I}$ in $\mathcal{C}$.

3 An exponential from A to (B, θ) in $\mathsf{EM}(T)$ corresponds to an exponential from A to B in $\mathcal{C}$.

□

The proof of this is straightforward, and uses the construction of product and exponential algebras given in Def. 1.17.

Corollary 104 The Eilenberg-Moore resolution $\mathsf{EM}(T)$ of the strong monad (T, η, μ, t) on $\mathcal{C}$ is a CBPV adjunction model iff

- $\mathcal{C}$ is countably distributive
- $\mathcal{C}$ possesses all countable products of, and exponentials to, carriers of T-algebras.

□

Abstract material here is a special case.

Corollary 105 Let $\mathcal{C}$ be a countably distributive category in which all idempotents split. Let (T, η, μ, t) be a strong monad on $\mathcal{C}$, and suppose $\mathcal{C}$ possesses all countable Kleisli products (i.e. every product of a countable family $\{TA_i\}_{i \in I}$) and all Kleisli exponentials (i.e. every exponential from A to TB). Then the Eilenberg-Moore resolution $\mathsf{EM}(T)$ is a CBPV adjunction model. □

Proof This follows from Cor. 104, because every algebra carrier is a retract of a free algebra carrier [Mac Lane, 1971]. □

9.7.3 Between JWA and CBPV

We saw in Fig. 7.1 that JWA can be seen as embedded in CBPV+<u>`Ans`</u>; this corresponds to the following construction. Suppose $(\mathcal{C}, \mathcal{D}, \mathcal{O})$ is a ($\mathcal{A}$-set/cppo/domain enriched) CBPV adjunction model, and we are given a comp-object $\underline{\mathsf{Ans}} \in \mathsf{ob}\ \mathcal{D}$. We obtain a ($\mathcal{A}$-set/cppo/domain enriched) JWA module model where

- the value category is $\mathcal{C}$
- the nonreturning command homsets are given by

$$\mathcal{N}_X = \mathcal{O}_X \underline{\mathsf{Ans}}$$

Here are the required isomorphisms.

$$\begin{array}{lrcl}
\textbf{jumpwith} & \mathcal{C}_X U(A \to \underline{\mathsf{Ans}}) & \cong & \mathcal{O}_{X \times A}\underline{\mathsf{Ans}} \\
\textbf{distr. coproduct} & & & \\
\quad \textbf{values} & \mathcal{C}_{X \times \sum_{i \in I} A_i} Y & \cong & \prod_{i \in I} \mathcal{C}_{X \times A_i} Y \\
\quad \textbf{nonret. commands} & \mathcal{O}_{X \times \sum_{i \in I} A_i} \underline{\mathsf{Ans}} & \cong & \prod_{i \in I} \mathcal{O}_{X \times A_i} \underline{\mathsf{Ans}}
\end{array}$$

A special case of this construction is where the CBPV model we begin with is trivial in the sense of Def. 9.34. Thus, we are given a countably bicartesian closed category $\mathcal{C}$ and an object Ans, and we obtain a JWA module model with value category $\mathcal{C}$, where $\mathcal{N}_X$ is set to be $\mathcal{C}_X \mathsf{Ans}$. In that case, $\neg A$ is set to be $A \to \mathsf{Ans}$, since U is identity (on objects).

Some authors e.g. [Hofmann, 1995] have generalized this by noticing that $\mathcal{C}$ does not need to be countably bicartesian closed. It is sufficient that $\mathcal{C}$ be countably distributive, and equipped with an exponential from A to Ans (whose vertex we write $\neg A$) for every object A. Moreover, every JWA module model $(\mathcal{C}, \mathcal{N})$ is isomorphic to one arising from such a $(\mathcal{C}, \mathsf{Ans})$—to show this, put Ans to be $\neg 1$. (More technically, there is a 2-equivalence between the 2-category of such $(\mathcal{C}, \mathsf{Ans})$ and the 2-category of JWA models.)

In the opposite direction, we have a construction corresponding to the StkPS transform from CBPV+stacks to JWA. Suppose we have a ($\mathcal{A}$-set/cppo/domain enriched) JWA module model $(\mathcal{C}, \mathcal{N})$. We obtain a ($\mathcal{A}$-set/cppo/domain enriched) CBPV adjunction model where

- the value category is $\mathcal{C}$
- the stack category is the locally $\mathcal{C}$-indexed category $(\mathsf{self}\,\mathcal{C})^{\mathsf{op}}$, so a morphism over A from B to C is a $\mathcal{C}$-morphism $A \times C \xrightarrow{h} B$
- the computation homsets are given by

$$\mathcal{O}_X Y = \mathcal{N}_{X \times Y}$$

Here are the required isomorphisms.

$$
\begin{array}{lrcl}
\textbf{right adjunctive} & \mathcal{C}_X \neg B & \cong & \mathcal{N}_{X \times B} \\
\textbf{left adjunctive} & (\mathsf{self}\,\mathcal{C})^{\mathsf{op}}_X(\neg A, Z) & \cong & \mathcal{N}_{(X \times A) \times Z} \\
\textbf{distr. coproduct} & & & \\
\quad \textbf{values} & \mathcal{C}_{X \times \sum_{i \in I} A_i} Y & \cong & \prod_{i \in I} \mathcal{C}_{X \times A_i} Y \\
\quad \textbf{computations} & \mathcal{N}_{(X \times \sum_{i \in I} A_i) \times Z} & \cong & \prod_{i \in I} \mathcal{N}_{(X \times A_i) \times Z} \\
\quad \textbf{stacks} & (\mathsf{self}\,\mathcal{C})^{\mathsf{op}}_{X \times \sum_{i \in I} A_i}(Y, Z) & \cong & \prod_{i \in I} (\mathsf{self}\,\mathcal{C})^{\mathsf{op}}_{X \times A_i}(Y, Z) \\
\textbf{product} & & & \\
\quad \textbf{computations} & \prod_{i \in I} \mathcal{N}_{X \times \sum_{i \in I} B_i} & \cong & \prod_{i \in I} \mathcal{N}_{X \times B_i} \\
\quad \textbf{stacks} & (\mathsf{self}\,\mathcal{C})^{\mathsf{op}}_X(Y, \sum_{i \in I} B_i) & \cong & \prod_{i \in I} (\mathsf{self}\,\mathcal{C})^{\mathsf{op}}_X(Y, B_i) \\
\textbf{exponential} & & & \\
\quad \textbf{computations} & \mathcal{N}_{X \times (A \times B)} & \cong & \mathcal{N}_{(X \times A) \times B} \\
\quad \textbf{stacks} & (\mathsf{self}\,\mathcal{C})^{\mathsf{op}}_X(Y, A \times B) & \cong & (\mathsf{self}\,\mathcal{C})^{\mathsf{op}}_{X \times A}(Y, B)
\end{array}
$$

A model obtained by this construction is called a *continuation model.* ("Stack-passing model" would be a better name in the CBPV setting, but we use the traditional term.) It is worth noting that in every CBPV adjunction model we have an *adjunction* between values and stacks, but in a continuation model we have, in addition, a *duality* between values and stacks.

Putting these constructions together, we see that, if we are given a ($\mathcal{A}$-set/cppo/domain enriched) CBPV adjunction model $(\mathcal{C}, \mathcal{D}, \mathcal{O})$ and a comp-object $\underline{\mathsf{Ans}} \in \mathsf{ob}\ \mathcal{D}$, we obtain a ($\mathcal{A}$-set/cppo/domain enriched) continuation model $(\mathcal{C}, (\mathsf{self}\,\mathcal{C})^{\mathsf{op}}, \mathcal{O}')$ where

$$
\begin{array}{rcl}
\mathcal{O}'_X Y & = & \mathcal{O}_{X \times Y} \underline{\mathsf{Ans}} \\
U'B & = & U(B \to \underline{\mathsf{Ans}}) \\
F'A & = & U(A \to \underline{\mathsf{Ans}})
\end{array}
$$

9.7.4 Erratic Choice

The relation model for erratic choice presented in Sect. 5.5.2 is a CBPV adjunction model, as follows.

- The value category $\mathcal{C}$ is **Set**.
- The stack category **Rel** is the locally **Set**-indexed category of relations i.e. an object is a set and a morphism over A from B to C is a relation from $A \times B$ to C. Usual composition of relations is used.

- The computation morphisms from A to B are relations from A to B, and we write $\mathbf{Rel}_A B$ for the set of all such. Again, usual composition of relations is used.

Here are the required isomorphisms.

$$\begin{array}{lrcl}
\textbf{right adjunctive} & \mathbf{Set}_X \mathcal{P}B & \cong & \mathbf{Rel}_X B \\
\textbf{left adjunctive} & \mathbf{Rel}_X(A, Z) & \cong & \mathbf{Rel}_{X \times A} Z \\
\textbf{distr. coproduct} & & & \\
\quad \textbf{values} & \mathbf{Set}_{X \times \sum_{i \in I} A_i} Y & \cong & \prod_{i \in I} \mathbf{Set}_{X \times A_i} Y \\
\quad \textbf{computations} & \mathbf{Rel}_{X \times \sum_{i \in I} A_i} Z & \cong & \prod_{i \in I} \mathbf{Rel}_{X \times A_i} Z \\
\quad \textbf{stacks} & \mathbf{Rel}_{X \times \sum_{i \in I} A_i}(Y, Z) & \cong & \prod_{i \in I} \mathbf{Rel}_{X \times A_i}(Y, Z) \\
\textbf{product} & & & \\
\quad \textbf{computations} & \mathbf{Rel}_X \sum_{i \in I} B_i & \cong & \prod_{i \in I} \mathbf{Rel}_X B_i \\
\quad \textbf{stacks} & \mathbf{Rel}_X(Y, \sum_{i \in I} B_i) & \cong & \prod_{i \in I} \mathbf{Rel}_X(Y, B_i) \\
\textbf{exponential} & & & \\
\quad \textbf{computations} & \mathbf{Rel}_X A \times B & \cong & \mathbf{Rel}_{X \times A} B \\
\quad \textbf{stacks} & \mathbf{Rel}_X(Y, A \times B) & \cong & \mathbf{Rel}_{X \times A}(Y, B)
\end{array}$$

The model for erratic choice + divergence with may-testing equivalence presented in Sect. 5.5.4 is also a CBPV adjunction model.

- In the value category $\mathcal{C}$, an object is a poset and a morphism from A to B is a pointwise ideal opbimodule.
- In the locally $\mathcal{C}$-indexed stack category, an object is a poset and a morphism over A from B to C is an opbimodule from $A \times B$ to C. The identity over A on B relates (a, b) to b' when $b \geqslant b'$. Usual composition of relations is used.
- The computation morphisms from A to B are opbimodules from A to B. Usual composition of relations is used.

9.7.5 Global Store

In Sect. 5.3.3 we saw how to translate CBPV+stacks+global store into CBPV+stacks. This corresponds to the following construction on models. Suppose $(\mathcal{C}, \mathcal{D}, \mathcal{O})$ is a ($\mathcal{A}$-set/cppo/domain enriched) CBPV adjunction model and we are given a val-object $S \in \mathsf{ob}\, \mathcal{C}$. We construct another ($\mathcal{A}$-set/cppo/domain enriched) CBPV adjunction model with the same value category $\mathcal{C}$ and the same stack category $\mathcal{D}$ by writing $\mathcal{O}^S$

for the right $\mathcal{D}$-module given by

$$\mathcal{O}^S_X \underline{Y} = \mathcal{O}_{S \times X} \underline{Y}$$

We see that $(\mathcal{C}, \mathcal{D}, \mathcal{O}^S)$ has all the required isomorphisms.

right adjunctive	$\mathcal{C}_X U(S \to \underline{B})$	$\cong$	$\mathcal{O}^S_X \underline{B}$
left adjunctive	$\mathcal{D}_X(F(S \times A), Z)$	$\cong$	$\mathcal{O}^S_{X \times A} Z$
distr. coproduct			
values	$\mathcal{C}_{X \times \sum_{i \in I} A_i} Y$	$\cong$	$\prod_{i \in I} \mathcal{C}_{X \times A_i} Y$
computations	$\mathcal{O}^S_{X \times \sum_{i \in I} A_i} \underline{Z}$	$\cong$	$\prod_{i \in I} \mathcal{O}^S_{X \times A_i} \underline{Z}$
stacks	$\mathcal{D}_{X \times \sum_{i \in I} A_i}(\underline{Y}, \underline{Z})$	$\cong$	$\prod_{i \in I} \mathcal{D}_{X \times A_i}(\underline{Y}, \underline{Z})$
product			
computations	$\mathcal{O}^S_X \prod_{i \in I} \underline{B}_i$	$\cong$	$\prod_{i \in I} \mathcal{O}^S_X \underline{B}_i$
stacks	$\mathcal{D}_X(Y, \prod_{i \in I} \underline{B}_i)$	$\cong$	$\prod_{i \in I} \mathcal{D}_X(Y, \underline{B}_i)$
exponential			
computations	$\mathcal{O}^S_X A \to \underline{B}$	$\cong$	$\mathcal{O}^S_{X \times A} \underline{B}$
stacks	$\mathcal{D}_X(Y, A \to \underline{B})$	$\cong$	$\mathcal{D}_{X \times A}(Y, \underline{B})$

We have a similar construction for JWA module models, corresponding to a translation from JWA + global store to JWA. Given a JWA module model $(\mathcal{C}, \mathcal{N})$ and an object $S \in \mathsf{ob}\, \mathcal{C}$ we obtain a JWA module model with the same value category $\mathcal{C}$ by setting $\mathcal{N}^S_X = \mathcal{N}_{S \times X}$. We see that $(\mathcal{C}, \mathcal{N}^S)$ has all the required isomorphisms.

jumpwith	$\mathcal{C}_X \neg(S \times A)$	$\cong$	$\mathcal{N}^S_{X \times A}$
distr. coproduct			
values	$\mathcal{C}_{X \times \sum_{i \in I} A_i} Y$	$\cong$	$\prod_{i \in I} \mathcal{C}_{X \times A_i} Y$
nonret. commands	$\mathcal{N}^S_{X \times \sum_{i \in I} A_i}$	$\cong$	$\prod_{i \in I} \mathcal{N}^S_{X \times A_i}$

9.7.6 Possible Worlds: Part 1

We give the first part of the possible worlds construction here, but we finish it in Sect. 11.5, after we have treated parametrized representability.

Let $\mathcal{W}$ be a countable category (i.e. countably many objects and morphisms) whose objects we call "worlds". For each world w we write $w/\mathcal{W}$ for the countable set $\sum_{w' \in \mathcal{W}} \mathcal{C}(w, w')$. In the semantics of Chap. 6, $\mathcal{W}$ is a poset, so $w/\mathcal{W}$ is the set of worlds $w' \geqslant w$. In the special case of global store, $\mathcal{W}$ is the singleton poset.

$\mathcal{W}$ gives rise to the following construction on models. Suppose that $(\mathcal{C}, \mathcal{D}, \mathcal{O})$ is a ($\mathcal{A}$-set/cppo/domain enriched) CBPV adjunction model, and we are given a val-object $Sw \in \mathsf{ob}\,\mathcal{C}$ for each world w. Then we obtain a new ($\mathcal{A}$-set/cppo/domain enriched) adjunction model as follows.

- The value category is the functor category $[\mathcal{W}, \mathcal{C}]$, which is cartesian (finite products defined pointwise).
- In the stack category, an object is a contravariant functor from $\mathcal{W}^{\mathsf{op}}$ to $\mathcal{D}_1$, and a morphism $\underline{Y} \xrightarrow[X]{h} \underline{Z}$ provides for each world w a morphism $\underline{Y}w \xrightarrow[Xw]{hw} \underline{Z}w$, such that the diagram over Xw

$$\begin{array}{ccc} \underline{Y}w & \xrightarrow{hw} & \underline{Z}w \\ {\scriptstyle ()^*\underline{Y}f}\uparrow & & \uparrow{\scriptstyle ()^*\underline{Z}f} \\ \underline{Y}w' & \xrightarrow[(Xf)^*(hw')]{} & \underline{Z}w' \end{array}$$

  commutes for each $w \xrightarrow{f} w'$
- The computation homsets are given by

$$\mathcal{O}^S_X \underline{Y} = \prod_{w \in \mathcal{W}} \mathcal{O}_{Sw \times Xw} \underline{Y}w$$

  Notice that if h is an element of this homset, there is no naturality constraint between the various hw—indeed S is not even a functor.

The definition of the connectives at a world w is given as in Sect. 6.8.2, viz.

$$\begin{array}{rcl} (U\underline{B})w & = & U\prod_{(w',f)\in w/\mathcal{W}}(Sw' \to \underline{B}w') \\ (FA)w & = & F\sum_{(w',f)\in w/\mathcal{W}}(Sw' \times Aw') \\ (\sum_{i\in I} A_i)w & = & \sum_{i\in I} A_i w \\ (A \times A')w & = & Aw \times A'w \\ (\prod_{i\in I} \underline{B}_i)w & = & \prod_{i\in I} \underline{B}_i w \\ (A \to \underline{B})w & = & Aw \to \underline{B}w \end{array}$$

The definition of the connectives at a morphism $w \xrightarrow{g} x$ makes use of parametrized representability (Sect. 11.4) which tells us how to treat all the connectives as functors. So we defer it to Sect. 11.5.

The category $\mathcal{W}$ gives rise similarly to a construction on JWA module models, corresponding to a translation from JWA + cell generation to JWA. Given a ($\mathcal{A}$-set/cppo/domain enriched) JWA module model $(\mathcal{C}, \mathcal{N})$, and an object $Sw \in \mathsf{ob}\ \mathcal{C}$ for each world w, we obtain the following ($\mathcal{A}$-set/cppo/domain enriched) JWA module model.

- The value category is $[\mathcal{W}, \mathcal{C}]$, which is cartesian (finite products defined pointwise).
- The nonreturning command homsets are given by

$$\mathcal{N}_X^S = \prod_{w \in \mathcal{W}} \mathcal{N}_{Sw \times Xw}$$

  Again, there is no naturality constraint on elements of this homset.

The definition of the connectives at a world w is given by

$$\begin{array}{rcl} (\neg A)w & = & \neg \sum_{(w',f) \in w/\mathcal{W}} (Sw' \times Aw') \\ (\sum_{i \in I} A_i)w & = & \sum_{i \in I} A_i w \\ (A \times A')w & = & Aw \times A'w \end{array}$$

The definition of the connectives at a morphism $w \xrightarrow{g} x$ again uses parametrized representability and we defer it to Sect. 11.5.

9.8 Families

9.8.1 Composite Connectives

Certain composite connectives have reversible derivations, and it is generally possible to describe these connectives directly in categorical terms. We shall treat those examples which are necessary for describing families constructions.

The most important example is in JWA, where the type $\neg \sum_{i \in I} A_i$ satisfies the following reversible derivation

$$\frac{\cdots\ \Gamma, A_i \vdash^{\mathsf{n}}\ \cdots_{\ i \in I}}{\Gamma \vdash^{\mathsf{v}} \neg \sum_{i \in I} A_i}$$

This suggests the following definition.

Definition 9.35 A *coproduct-jumpwith* for a family of objects $\{A_i\}_{i \in I}$ in a JWA judgement model $(\mathcal{C}, \mathcal{N})$ consists of an object V (the vertex) together with a $\mathcal{N}$-morphism $V \times A_i \xrightarrow{\mathsf{jump}_i}$, for each $i \in I$, such that the functions

$$\begin{array}{rcll} \mathcal{C}_X V & \longrightarrow & \prod_{i\in I} \mathcal{N}_{X\times A_i} & \text{for all } X \\ f & \longmapsto & \lambda i.((f \times A_i)^* \mathsf{jump}_i) & \end{array}$$

are isomorphisms. □

Clearly, a JWA module model will have all countable coproduct-jumpwiths. We shall see in Sect. 11.3.1 that, in naturality style, a coproduct-jumpwith for $\{A_i\}_{i\in I}$ is the vertex V together with an isomorphism

$$\mathcal{C}_X V \cong \prod_{i\in I} \mathcal{N}_{X\times A_i} \qquad \text{natural in } X.$$

Similarly, in CBPV, the type $F\sum_{i\in I} A_i$ has the following reversible derivation

$$\frac{\cdots \quad \Gamma, A_i \vdash^{\mathsf{c}} \underline{B} \quad \cdots \; {}_{i\in I}}{\Gamma | F\sum_{i\in I} A_i \vdash^{\mathsf{k}} \underline{B}}$$

This suggests the following definition.

Definition 9.36 A *left coproduct-adjunctive* for a family of val-objects $\{A_i\}_{i\in I}$ in a CBPV judgement model $(\mathcal{C}, \mathcal{D}, \mathcal{O})$ consists of a comp-object $\underline{V}$ (the vertex) together with an $\mathcal{O}$-morphism $\xrightarrow[A_i]{\mathsf{return}_i} \underline{V}$, for each $i \in I$, such that the functions

$$\begin{array}{rcll} \mathcal{D}_X(\underline{V}, \underline{Y}) & \longrightarrow & \prod_{i\in I} \mathcal{O}_{X\times A_i} \underline{Y} & \text{for all } X, \underline{Y} \\ h & \longmapsto & \lambda i.((\pi'^* \mathsf{return}_i); (\pi^* h)) & \end{array}$$

are isomorphisms. □

Clearly a CBPV adjunction model will have all countable left coproduct-adjunctives. We shall see in Sect. 11.3.2 that, in naturality style, a left coproduct-adjunctive for $\{A_i\}_{i\in I}$ is the vertex $\underline{V}$ together with an isomorphism

$$\mathcal{D}_X(\underline{V}, \underline{Y}) \cong \prod_{i\in I} \mathcal{O}_{X\times A_i} \pi^* \underline{Y} \qquad \text{natural in } X \text{ and } \underline{Y}$$

Likewise the types $U\prod_{i\in I}\underline{B}_i$, $U(A \to \underline{B})$ and $U\prod_{i\in I}(A_i \to \underline{B}_i)$ satisfy respectively the following reversible derivations

$$\frac{\cdots\ \Gamma \vdash^{\mathsf{c}} \underline{B}_i\ \cdots\ _{i\in I}}{\Gamma \vdash^{\mathsf{v}} U\prod_{i\in I}\underline{B}_i}$$

$$\frac{\Gamma, A \vdash^{\mathsf{c}} \underline{B}}{\Gamma \vdash^{\mathsf{v}} U(A \to \underline{B})}$$

$$\frac{\cdots\ \Gamma, A_i \vdash^{\mathsf{c}} \underline{B}_i\ \cdots\ _{i\in I}}{\Gamma \vdash^{\mathsf{v}} U\prod_{i\in I}(A_i \to \underline{B}_i)}$$

These suggest, respectively, the following definitions.

Definition 9.37 Let $(\mathcal{C}, \mathcal{D}, \mathcal{O})$ be a CBPV judgement model.

1 A *right product-adjunctive* for a family of comp-objects $\{\underline{B}_i\}_{i\in I}$ consists of a val-object V (the vertex) together with a morphism $\xrightarrow[V]{\mathsf{force}_i} \underline{B}_i$, for each $i \in I$, such that the functions

$$\begin{array}{rcll} \mathcal{C}_X V & \longrightarrow & \prod_{i\in I} \mathcal{O}_X \underline{B}_i & \qquad \text{for all } X \\ f & \longmapsto & \lambda i.(f^*\mathsf{force}_i) & \end{array}$$

are isomorphisms.

2 A *right exponential-adjunctive* from a val-object A to a comp-object $\underline{B}$ consists of a val-object V (the vertex) together with a morphism $\xrightarrow[V\times A]{\mathsf{ev}} \underline{B}$, such that the functions

$$\begin{array}{rcll} \mathcal{C}_X V & \longrightarrow & \mathcal{O}_{X\times A} \underline{B} & \qquad \text{for all } X \\ f & \longmapsto & (f \times A)^*\mathsf{ev} & \end{array}$$

are isomorphisms.

3 A *right exponential-product-adjunctive* for a family of (val-object, comp-object) pairs $\{(A_i, \underline{B}_i)\}_{i\in I}$ consists of a val-object V (the vertex) together with a morphism $\xrightarrow[V\times A_i]{\mathsf{ev}_i} \underline{B}_i$, for each $i \in I$, such that the functions

$$\begin{array}{rcll} \mathcal{C}_X V & \longrightarrow & \prod_{i\in I} \mathcal{O}_{X\times A_i} \underline{B}_i & \qquad \text{for all } X \\ f & \longmapsto & \lambda i.((f \times A_i)^*\mathsf{ev}_i) & \end{array}$$

are isomorphisms.

□

Of course a CBPV model will have all countable right exponential-product-adjunctives—and hence all countable right product-adjunctives and all right exponential-adjunctives. We shall see in Sect. 11.3.1 that, in naturality style, these structures are described by the vertex V together with an isomorphism from $\mathcal{C}_X V$ to

right product-adjunctive $\prod_{i \in I} \mathcal{O}_X \underline{B}_i$

right exponential-adjunctive $\mathcal{O}_{X \times A} \pi^* \underline{B}$

right exponential-product-adjunctive $\prod_{i \in I} \mathcal{O}_{X \times A_i} \pi^* \underline{B}_i$

natural in X.

9.8.2 Families Construction For JWA

All the pointer-game models in Chap. 8 are given by families constructions, whose use for denotational semantics was introduced in [Abramsky and McCusker, 1998]. The most important is the JWA model. We recapitulate Def. 8.11.

Definition 9.38 A *JWA pre-families model* is a JWA judgement model $(\mathcal{C}, \mathcal{N})$ equipped with all countable coproduct-jumpwiths. We write $\neg_{i \in I} A_i$ for the vertex of the coproduct-jumpwith of the family $\{A_i\}_{i \in I}$. □

Definition 9.39 Let $(\mathcal{C}, \mathcal{N})$ be a ($\mathcal{A}$-set/cppo/domain enriched) JWA pre-families model. Then we obtain a ($\mathcal{A}$-set/cppo/domain enriched) JWA module model $(\mathcal{C}', \mathcal{N}')$ as follows. An object of $\mathcal{C}'$ is a *countable family* of $\mathcal{C}$-objects. To describe a $\mathcal{C}'$-morphism from $\{A_i\}_{i \in I}$ to $\{B_j\}_{j \in J}$, we must first give a function between indexing sets $I \xrightarrow{f} J$. Then, for each $i \in I$, we must provide a $\mathcal{C}$-morphism from A_i to B_{fi}. To describe a $\mathcal{N}'$-morphism from $\{A_i\}_{i \in I}$, we provide, for each $i \in I$, a $\mathcal{N}$-morphism from A_i. In summary:

values	$\mathcal{C}'_{\{A_i\}_{i \in I}} \{B_j\}_{j \in J}$	$=$	$\prod_{i \in I} \sum_{j \in J} \mathcal{C}_{A_i} B_j$
nonret. commands	$\mathcal{N}'_{\{A_i\}_{i \in I}}$	$=$	$\prod_{i \in I} \mathcal{N}_{A_i}$

Identities and composition are defined in the obvious way. The connectives are given by

$$
\begin{aligned}
\{A_i\}_{i \in I} \times \{B_j\}_{j \in J} &= \{A_i \times B_j\}_{(i,j) \in I \times J} \\
1 &= \{1\} \\
\textstyle\sum_{i \in I} \{A_{ij}\}_{j \in J_i} &= \{A_{ij}\}_{(i,j) \in \sum_{i \in I} J_i} \\
\neg \{A_i\}_{i \in I} &= \{\neg_{i \in I} A_i\}
\end{aligned}
$$

□

The key examples is the pointer game of model of JWA in Sect. 8.3. An object is an arena, and the homsets are given by

$$\mathcal{C}_R S = \mathsf{vStrat}(R, S) \qquad\qquad \mathcal{N}_R = \mathsf{nStrat}(R)$$

The connectives are given by

$$\frac{1 \quad \times \quad \neg_{i \in I}}{\emptyset \quad \uplus \quad \mathsf{pt}_{i \in I}}$$

9.8.3 Families Construction For CBPV

If we want to build a CBPV model where both value objects and computation objects are families, then we require the following.

Definition 9.40 A *pre-families CBPV model* is a CBPV judgement model $(\mathcal{C}, \mathcal{D}, \mathcal{O})$, with

- all exponentials—we write $A \to \underline{B}$ for the vertex
- all countable right product-adjunctives—we write $U_{i \in I}\underline{B}_i$ for the vertex
- all countable left coproduct-adjunctives—we write $F_{i \in I}A_i$ for the vertex

□

Definition 9.41 Let $(\mathcal{C}, \mathcal{D}, \mathcal{O})$ be a ($\mathcal{A}$-set/cppo/domain enriched) pre-families CBPV model. We define a ($\mathcal{A}$-set/cppo/domain enriched) CBPV adjunction model $(\mathcal{C}', \mathcal{D}', \mathcal{O}')$ as follows.

- A val-object is a countable family of $\mathcal{C}$-objects.
- A comp-object is a countable family of $\mathcal{D}$-objects.

The homsets are given by

values	$\mathcal{C}'_{\{A_i\}_{i \in I}}\{B_j\}_{j \in J}$	$=$	$\prod_{i \in I} \sum_{j \in J} \mathcal{C}_{A_i} B_j$
stacks	$\mathcal{D}'_{\{A_i\}_{i \in I}}(\{\underline{B}_j\}_{j \in J}, \{\underline{C}_k\}_{k \in K})$	$=$	$\prod_{i \in I} \prod_{k \in K} \sum_{j \in J} \mathcal{D}_{A_i}(\underline{B}_j, \underline{C}_k)$
computations	$\mathcal{O}'_{\{A_i\}_{i \in I}}\{\underline{B}_j\}_{j \in J}$	$=$	$\prod_{i \in I} \prod_{j \in J} \mathcal{O}_{A_i} \underline{B}_j$

with the obvious identities, composition and reindexing. The connectives are given by

$$\begin{array}{rcl} 1 & = & \{1\} \\ \{A_i\}_{i\in I} \times \{B_j\}_{j\in J} & = & \{A_i \times B_j\}_{(i,j)\in I\times J} \\ \sum_{i\in I}\{A_{ij}\}_{j\in J_i} & = & \{A_{ij}\}_{(i,j)\in \sum_{i\in I} J_i} \\ \{A_i\}_{i\in I} \to \{\underline{B}_j\}_{j\in J} & = & \{A_i \to \underline{B}_j\}_{(i,j)\in I\times J} \\ \prod_{i\in I}\{\underline{B}_{ij}\}_{j\in J_i} & = & \{\underline{B}_{ij}\}_{(i,j)\in \sum_{i\in I} J_i} \\ U\{\underline{B}_i\}_{i\in I} & = & \{U_{i\in I}\underline{B}_i\} \\ F\{A_i\}_{i\in I} & = & \{F_{i\in I}A_i\} \end{array}$$

□

We have 4 examples of this.

1 The model in Sect. 8.5.2 is an example. Objects of $\mathcal{C}$ and $\mathcal{D}$ are Q/A-labelled arenas. Homsets are given by

$$\begin{array}{rcl} \mathcal{C}_R S & = & \mathsf{vStrat}(R, S) \\ \mathcal{D}_R(S,T) & = & \mathsf{vStrat}(R \uplus T, S) \\ \mathcal{O}_R S & = & \mathsf{nStrat}(R \uplus S) \end{array}$$

and connectives are given by

1	$\times$	$\to$	$U_{i\in I}$	$F_{i\in I}$
$\emptyset$	$\uplus$	$\uplus$	$\mathsf{pt}^{\mathsf{Q}}_{i\in I}$	$\mathsf{pt}^{\mathsf{A}}_{i\in I}$

2 Take example (1), restrict ob $\mathcal{C}$ to *stackfree-in-value* arenas (no root and no successor of an A-token is an A-token), and restrict ob $\mathcal{D}$ to *stackfree-in-stack* arenas (no successor of an A-token is an A-token).

3 Take example (2), and restrict to P-bracketed strategies. This is the model of Sect. 8.6.

4 Take example (2), and use instead OP-bracketed strategies, i.e. strategies for the game where both Player and Opponent are constrained by bracketing.

In Sect. 11.6.4 we shall see more families constructions, for building CBPV models out of CBN models and out of CBV models.

Chapter 10

ALL MODELS ARE CATEGORICAL MODELS

10.1 Introduction

We have seen that any cartesian model gives a sound semantics for $\times$-calculus, any CBPV adjunction model gives a sound semantics for CBPV, etc. But it is useful to know that *every* model of $\times$-calculus is a cartesian category, so that when we search for cartesian categories, we are not excluding a potentially interesting model.

To express such a statement, we have to say what we mean by a "model of $\times$-calculus", and then our categorical semantics gives a functor from a category of cartesian categories to a category of models. Our required result will assert that this functor is an equivalence.

In this chapter, we go through this theory for cartesian categories and $\times$-calculus. We then briefly give analogous accounts for CBPV and for JWA; they are essentially the same story.

We shall see in Sect. 10.2.4 that a serious drawback of our account is that we have to fix object structure. So, in a situation where the operations on objects (in the categorical structure) are not the same as the operations on types (in the language), our account would not be suitable. Fortunately, this problem does not arise for our categorical semantics. But further work is needed to address it.

10.2 All Models Of $\times$-Calculus Are Cartesian Categories

10.2.1 The Theorem, Vaguely Stated

Recall that $\times$-calculus is the fragment of Fig. 3.8 whose only connectives are 1 and $\times$.

We will begin by stating a plausible result, and then consider what definitions are required to make it true. Here is the result:

Proposition 106 Models of $\times$-calculus and cartesian categories are equivalent. □

10.2.2 Fixing The Object Structure

Our first step is to describe the semantics of types:

Definition 10.1 A *semantics of types for* $\times$*-calculus* , also called a $\times$ *object structure* is a tuple $\tau = (\tau_{\mathsf{ob}}, 1, \times)$ consisting of

- a class τ_{ob}, whose elements we call *objects*;
- an object 1;
- a binary operation $\times : \tau_{\mathsf{ob}} \times \tau_{\mathsf{ob}} \longrightarrow \tau_{\mathsf{ob}}$

□

We can now replace Prop. 106 by the following stronger statement; like Prop. 106 it is not precise because we have not yet defined "model of $\times$-calculus".

Proposition 107 Let τ be a $\times$ object structure. Then

- models of $\times$-calculus with semantics of types given by τ, and
- cartesian categories with object structure τ.

are equivalent. □

This statement will be easier to make precise than Prop. 106. We will form two categories and they will be equivalent.

For the purposes of Prop. 107, we need a very specific definition of cartesian category.

Definition 10.2 1 A *cartesian category* is a category $\mathcal{C}$ with a distinguished terminal object and distinguished binary products.

2 The *object structure* of a cartesian category $\mathcal{C}$ is the $\times$ object structure $(\mathsf{ob}\ \mathcal{C}, 1, \times)$ where

- 1 is the distinguished terminal object of $\mathcal{C}$;
- $A \times A'$ is the vertex of the distinguished product of the $\mathcal{C}$-objects A and A'.

□

As we mention at the end of Sect. 10.2.4, our approach would not work if we defined a cartesian category to be "a category with distinguished n-ary products for all $n \in \mathbb{N}$"

We can now make precise one half of Prop. 107.

Definition 10.3 Let τ be a $\times$ object structure. The category $\mathbf{CartCat}_\tau$ is defined as follows:

- an object is a cartesian category with object structure τ.
- a morphism is an identity-on-objects functor preserving all structure.

□

Notice that the fixing of τ has allowed us to sidestep the question of whether a general cartesian functor should preserve structure on the nose or up to isomorphism, because the only cartesian functors we use are identity-on-objects.

10.2.3 What Is A Model Of ×-Calculus?

The real problem in making Prop. 107 precise is to give, direct from the equational theory, an *a priori* notion of "model for ×-calculus". The approach that we use was developed independently in [Jeffrey, 1999] and in [Levy, 1996].

Definition 10.4 Let τ be a $\times$ object structure (as in Sect. 10.2).

1. A τ-sequent is a sequence of objects $A_0, \ldots, A_{n-1}$ together with an object B. It is written $A_0, \ldots, A_{n-1} \vdash B$. Thus, it is a judgement without a term.
2. A *τ-signature* (or *τ-multigraph*) s provides, for each τ-sequent $A_0, \ldots, A_{n-1} \vdash B$, a set $s(A_0, \ldots, A_{n-1} \vdash B)$. The elements are called *operation symbols* with *arity* $A_0, \ldots, A_{n-1}$ and *result type* B, or *edges* from $A_0, \ldots, A_{n-1}$ to B.
3. We write Sig_τ for the category of τ-signatures, where a morphism from s to s' provides a function from $s(Q)$ to $s'(Q)$ for each τ-sequent Q.

□

The term "multigraph" is sometimes used because, whereas each edge in a graph has one source object and one target object, each edge in a multigraph has several source objects and one target object.

It is clear that, given semantics of types τ, a semantics of the $\vdash$ judgement for $\times$-calculus is precisely a τ-signature s. For s tells us that a term

$A_0, \ldots, A_{n-1} \vdash M : B$ will denote an edge from $[\![A_0]\!], \ldots, [\![A_{n-1}]\!]$ to $[\![B]\!]$, but does not tell us *which* edge—that will depend on the particular term M.

We still need a way of characterizing a semantics of terms for the $\times$-calculus. The key fact is that, for each object structure τ, *the terms and equations of $\times$-calculus define a monad $\mathcal{T}$ on* Sig_τ. To see this, suppose that s is a τ-signature; we will build from s another τ-multigraph $\mathcal{T}s$ as follows. First, we inductively define the *terms built from the signature* s, using the rules of $\times$-calculus, together with the rule

$$\frac{\Gamma \vdash M_0 : a_0 \quad \cdots \quad \Gamma \vdash M_{n-1} : A_{n-1}}{\Gamma \vdash f(M_0, \ldots, M_{n-1}) : B}$$

for each operation symbol f with arity $A_0, \ldots, A_{n-1}$ and result type B in s. We require the following.

Definition 10.5 A *hypercongruence* on terms generated from s is a congruence $\sim$ closed under substitution. □

We write $\sim_s$ for the least congruence on these terms containing the equational laws of $\times$-calculus.

Proposition 108 The congruence $\sim_s$ is a hypercongruence. □

We define $\mathcal{T}s$ to be the τ-multigraph in which an edge from Γ to B is an equivalence class under $\sim_s$ of terms $\Gamma \vdash M : B$ built from the signature s. This new multigraph can be thought of as the "free $\times$-calculus model generated by s", keeping τ fixed throughout. Because of Prop. 108, we can perform substitution on these equivalence classes.

The rest of the monad structure is given as follows.

- ηs takes an edge f in s to the term $f(\mathtt{x}_0, \ldots, \mathtt{x}_{n-1})$.
- μs is the "flattening transform": it takes $t(M_0, \ldots, M_{n-1})$, where t is an edge in $\mathcal{T}s$, to $t[\overrightarrow{(\mu s)M_i/\mathtt{x}_i}]$. It preserves all other term constructors; for example, it maps (M, M') to $((\mu s)M, (\mu s)M')$.
- If $s \xrightarrow{\alpha} s'$ is a multigraph homomorphism, then $\mathcal{T}\alpha$ replaces each occurrence of a function symbol f in a term built from s by αf. Thus it takes $f(M_0, \ldots, M_{n-1})$, where f is an edge in s, to $\alpha(f)((\mathcal{T}\alpha)M_0, \ldots, (\mathcal{T}\alpha)M_{n-1})$, and it preserves all other term constructors.

We omit the proof that $\mathcal{T}$ preserves identity and composition, η and μ are natural and $(\mathcal{T}, \eta, \mu)$ satisfies all the monad laws. These are all straightforward inductions.

Now let us think what information a direct model for $\times$-calculus should provide.

semantics of types It should provide a $\times$ object structure τ.

semantics of judgement It should provide a τ-multigraph s. Then we know that a term $A_0, \ldots, A_{n-1} \vdash M : B$ is going to denote an edge from $[\![A_0]\!], \ldots, [\![A_{n-1}]\!]$ to $[\![B]\!]$.

semantics of terms Given a term $A_0, \ldots, A_{n-1} \vdash M : B$ generated from signature s, it should specify the edge from $[\![A_0]\!], \ldots, [\![A_{n-1}]\!]$ to $[\![B]\!]$ in s that M denotes. Furthermore, provably equal terms should denote the same edge. Thus the model should provide a multigraph homomorphism θ from $\mathcal{T}s$ to s.

In fact, (s, θ) should be an algebra for the $\mathcal{T}$ monad. We summarize:

Definition 10.6 1 A *direct model for $\times$-calculus* consists of

- a $\times$ object structure τ;
- an algebra (s, θ) for the $\mathcal{T}$ monad on Sig_τ.

2 We write $\mathbf{Direct}_\tau$ for the category of direct models for $\times$-calculus with object structure τ. Morphisms are algebra homomorphisms.

□

10.2.4 The Theorem, Precisely Stated

We can now formulate Prop. 107 precisely. We write $A_0 \times \cdots \times A_{n-1}$ for the object $(\cdots(1 \times A_0)\cdots) \times A_{n-1}$.

Definition 10.7 Let τ be a $\times$ object structure. Define a functor

$$\mathbf{CartCat}_\tau \xrightarrow{\mathcal{I}_\tau} \mathbf{Direct}_\tau$$

describing our categorical semantics. Explicitly, if $\mathcal{C}$ is a cartesian category based on τ, then define a τ-multigraph s by setting $s(A_0, \ldots, A_{n-1} \vdash B)$ to be $\mathcal{C}_{A_0 \times \cdots \times A_{n-1}} B$. We define, by induction, a function $[\![-]\!]$ from the multigraph of terms generated by s to s. This is given by the semantic equations from Sect. 9.4.1, together with an additional equation for operation symbols:

$$[\![f(M_0, \ldots, M_{n-1})]\!] = ([\![M_0]\!], \ldots, [\![M_{n-1}]\!])^* f$$

We define θ from $\mathcal{T}s$ to s to take the $\sim_s$ equivalence class of M to $[\![M]\!]$. That this is well defined is precisely the content of the soundness theorem: terms related by $\sim_s$ (i.e. provably equal terms in the equational

theory) have the same denotation. Finally we define $\mathcal{I}_\tau(\mathcal{C})$ to be (s, θ).
□

Proposition 109 Let τ be a $\times$ object structure. Then the functor $\mathcal{I}_\tau$ is an equivalence. □

Proof (outline)

- Let (s, θ) be a direct model based on τ. Then we construct a cartesian category $\mathcal{C}$ based on τ by setting $\mathcal{C}_A B$ to be $s(A \vdash^{\mathsf{v}} B)$. The structure is defined in the evident way:

$$\begin{aligned} \mathsf{id} &= \theta \mathtt{x}_0 \\ f^* g &= \theta g(f(\mathtt{x}_0)) \\ \pi &= \theta \mathbf{pm}\ \mathtt{x}_0\ \mathtt{as}\ (\mathtt{x}_1, \mathtt{x}_2).\ \mathtt{x}_1 \\ \pi' &= \theta \mathbf{pm}\ \mathtt{x}_0\ \mathtt{as}\ (\mathtt{x}_1, \mathtt{x}_2).\ \mathtt{x}_2 \\ (f, g) &= \theta(f(\mathtt{x}_0), g(\mathtt{x}_0)) \end{aligned}$$

  and all required equations verified.

- This operation is inverse to $\mathcal{I}_\tau$, up to isomorphism.

□

Our fixing of object structure in Prop. 109 sidesteps some subtle coherence issues. Observe, for example, that if instead of Def. 10.2 we had defined "cartesian category" to be "category with distinguished n-ary products for every $n \in \mathbb{N}$"—which clearly ought to be an acceptable definition—then our approach would not work. Further work is certainly required to remedy this problem, and we will not consider it further.

10.3 All Models Of CBPV Are CBPV Adjunction Models

The direct model/categorical model equivalence result for CBPV proceeds through exactly the same steps as that for $\times$-calculus in Sect. 10.2. The first step is to define "object structure" (meaning "semantics of types") in the obvious way.

Definition 10.8 A *CBPV object structure*, is a tuple

$$\tau = (\tau_{\mathsf{valob}}, \tau_{\mathsf{compob}}, U, \textstyle\sum, 1, \times, F, \prod, \to) \quad \text{consisting of}$$

- a class τ_{valob}, whose elements we call *val-objects*
- a class τ_{compob}, whose elements we call *comp-objects*

- a function $U : \tau_{\mathsf{compob}} \longrightarrow \tau_{\mathsf{valob}}$
- a function $\sum_{i \in I} : \tau^I_{\mathsf{valob}} \longrightarrow \tau_{\mathsf{valob}}$ for every countable set I
- a value object 1
- a binary operation $\times : \tau_{\mathsf{valob}} \times \tau_{\mathsf{valob}} \longrightarrow \tau_{\mathsf{valob}}$
- a function $F : \tau_{\mathsf{valob}} \longrightarrow \tau_{\mathsf{compob}}$
- a function $\prod_{i \in I} : \tau^I_{\mathsf{compob}} \longrightarrow \tau_{\mathsf{compob}}$ for every countable set I
- a binary operation $\to: \tau_{\mathsf{valob}} \times \tau_{\mathsf{compob}} \longrightarrow \tau_{\mathsf{compob}}$.

□

Clearly any CBPV adjunction model gives us a CBPV object structure. Next we need to look at signatures for a given object structure.

Definition 10.9 Let τ be a CBPV type structure.

1 A *τ-sequent* is $\Gamma \vdash^{\mathsf{v}} B$, or $\Gamma \vdash^{\mathsf{c}} \underline{C}$ or $\Gamma|\underline{B} \vdash^{\mathsf{k}} \underline{C}$ where Γ is a sequence of τ-val-objects, B is a τ-val-object and $\underline{B}$ and $\underline{C}$ are τ-comp-objects.

2 A *τ-signature* s provides for each sequent Q a set $s(Q)$.

3 We write Sig_τ for the category of τ-signatures, where a morphism from s to s' provides a function from $s(Q)$ to $s'(Q)$ for each τ-sequent Q.

□

As in Sect. 10.2.3, we need to define a monad $\mathcal{T}$ on Sig_τ. Suppose s is a τ-signature. We inductively define the terms built from s, using all the rules of CBPV+stacks+complex values/stacks, together with rules for the operation symbols in s appearing in Fig. —reffig:sigrules

The reader may find the rules for a stack operation symbol f to be rather strange. The intention is that $f(V_0, \ldots, V_{r-1}|M)$ is a computation whose execution begins by executing M, and so the context $f(V_0, \ldots, V_{r-1}|[\cdot])$ is placed onto the stack for future use. This is represented by the CK-machine transition

	Γ	$f(V_0, \ldots, V_{r-1}\|M)$	$\underline{B}'$	K	$\underline{C}$
$\leadsto$	Γ	M	$\underline{B}$	$f(V_0, \ldots, V_{r-1}\|[\cdot]) :: K$	$\underline{C}$

It is helpful, at this stage, to give a formal definition of dismantling and concatenation. This is done in Fig. 10.2.

$$\frac{\Gamma \vdash^{\mathsf{v}} V_0 : A_0 \quad \cdots \quad \Gamma \vdash^{\mathsf{v}} V_{r-1} : A_{r-1}}{\Gamma \vdash^{\mathsf{v}} f(V_0, \ldots, V_{r-1}) : B} \quad f \in s\,(A_0, \ldots, A_{r-1} \vdash^{\mathsf{v}} B)$$

$$\frac{\Gamma \vdash^{\mathsf{v}} V_0 : A_0 \quad \cdots \quad \Gamma \vdash^{\mathsf{v}} V_{r-1} : A_{r-1}}{\Gamma \vdash^{\mathsf{c}} f(V_0, \ldots, V_{r-1}) : \underline{B}} \quad f \in s\,(A_0, \ldots, A_{r-1} \vdash^{\mathsf{c}} \underline{B})$$

$$\frac{\Gamma \vdash^{\mathsf{v}} V_0 : A_0 \quad \cdots \quad \Gamma \vdash^{\mathsf{v}} V_{r-1} : A_{r-1} \quad \Gamma \vdash^{\mathsf{c}} M : \underline{B}}{\Gamma \vdash^{\mathsf{c}} f(V_0, \ldots, V_{r-1} | M) : \underline{B}'} \quad f \in s\,(A_0, \ldots, A_{r-1} | \underline{B} \vdash^{\mathsf{k}} \underline{B}')$$

$$\frac{\Gamma \vdash^{\mathsf{v}} V_0 : A_0 \quad \cdots \quad \Gamma \vdash^{\mathsf{v}} V_{r-1} : A_{r-1} \quad \Gamma | \underline{B}' \vdash^{\mathsf{k}} K : \underline{C}}{\Gamma | \underline{B} \vdash^{\mathsf{k}} f(V_0, \ldots, V_{r-1} | [\cdot]) :: K : \underline{C}} \quad f \in s\,(A_0, \ldots, A_{r-1} | \underline{B} \vdash^{\mathsf{k}} \underline{B}')$$

Figure 10.1. Rules For A Signature s

K	$M \bullet K$	$K + + L$
$\mathtt{nil}$	M	L
$[\cdot]\ \mathtt{to}\ \mathtt{x}.\ M :: K$	$(M\ \mathtt{to}\ \mathtt{x}.\ N) \bullet K$	$[\cdot]\ \mathtt{to}\ \mathtt{x}.\ M :: (K + + L)$
$\hat{\imath} :: K$	$(\hat{\imath}\text{`}M) \bullet K$	$\hat{\imath} :: (K + + L)$
$V :: K$	$(V\text{`}M) \bullet K$	$V :: (K + + L)$
$\mathtt{let}\ V\ \mathtt{be}\ \mathtt{x}.\ K$	$\mathtt{let}\ V\ \mathtt{be}\ \mathtt{x}.\ (M \bullet K)$	$\mathtt{let}\ V\ \mathtt{be}\ \mathtt{x}.\ (K + + L)$
$\mathtt{pm}\ V\ \mathtt{as}\ \{\ldots, (i, \mathtt{x}).K_i, \ldots\}$	$\mathtt{pm}\ V\ \mathtt{as}\ \{\ldots, (i, \mathtt{x}).(M \bullet K_i), \ldots\}$	$\mathtt{pm}\ V\ \mathtt{as}\ \{\ldots, (i, \mathtt{x}).K_i + + L, \ldots\}$
$\mathtt{pm}\ V\ \mathtt{as}\ (\mathtt{x}, \mathtt{y}).K$	$\mathtt{pm}\ V\ \mathtt{as}\ (\mathtt{x}, \mathtt{y}).(M \bullet K)$	$\mathtt{pm}\ V\ \mathtt{as}\ (\mathtt{x}, \mathtt{y}).(K + + L)$
$K\ \mathtt{where\ nil\ is}\ K'$	$(M \bullet K) \bullet K'$	$K\ \mathtt{where\ nil\ is}\ (K' + + L)$
$\{\ldots, K_i\ \mathtt{where}\ i :: \mathtt{nil}, \ldots\}\ \mathtt{is}\ K'$	$(\lambda\{\ldots, i.(M \bullet K_i), \ldots\}) \bullet K'$	$\{\ldots, K_i\ \mathtt{where}\ i :: \mathtt{nil}, \ldots\}\ \mathtt{is}\ (K' + + L)$
$K\ \mathtt{where}\ \mathtt{x} :: \mathtt{nil}\ \mathtt{is}\ K'$	$(\lambda \mathtt{x}.(M \bullet K)) \bullet K'$	$K\ \mathtt{where}\ \mathtt{x} :: \mathtt{nil}\ \mathtt{is}\ (K' + + L)$
$f(V_0, \ldots, V_{r-1} \| [\cdot]) :: K$	$f(V_0, \ldots, V_{r-1} \| M) \bullet K$	$f(V_0, \ldots, V_{r-1} \| [\cdot]) :: (K + + L)$

Figure 10.2. Dismantling and Concatenation (Including Complex Stacks And Operation Symbols)

Definition 10.10 A *hypercongruence* on the terms generated from a s is a congruence closed under substitution, dismantling and concatenation. □

Our next step is to write $\sim_s$ for the least congruence on the terms built from s containing the equational laws of $\times$-calculus.

Proposition 110 The congruence $\sim$ is a hypercongruence. □

We define $\mathcal{T}s$ to be the τ-signature associating to each τ-sequent Q the set of equivalence classes under $\sim_s$ of terms in judgement Q using the signature s. This new multigraph can be thought of as the "free $\times$-calculus model generated by s", keeping τ fixed throughout. Because of Prop. 110, we can perform substitution, dismantling and concatenation on these equivalence classes.

The rest of the monad structure is given as follows.

- ηs takes
  - f in $s(A_0, \ldots, A_{n-1} \vdash^{\mathsf{v}} B)$ to the value $f(\mathtt{x}_0, \ldots, \mathtt{x}_{n-1})$
  - f in $s(A_0, \ldots, A_{n-1} \vdash^{\mathsf{v}} \underline{C})$ to the computation $f(\mathtt{x}_0, \ldots, \mathtt{x}_{n-1})$
  - f in $s(A_0, \ldots, A_{n-1}|\underline{B} \vdash^{\mathsf{k}} \underline{C})$ to the stack $f(\mathtt{x}_0, \ldots, \mathtt{x}_{n-1}|[-]) ::$ `nil` .
- μs is the "flattening transform": it takes
  - a value $t(V_0, \ldots, V_{n-1})$, where t is an value-edge in $\mathcal{T}s$, to the value $t[\overrightarrow{(\mu s)V_i/\mathtt{x}_i}]$
  - a computation $t(V_0, \ldots, V_{n-1})$, where t is an computation-edge in $\mathcal{T}s$, to the computation $t[\overrightarrow{(\mu s)V_i/\mathtt{x}_i}]$
  - a computation $t(V_0, \ldots, V_{n-1}|M)$, where t is a stack-edge in $\mathcal{T}s$, to $(\mu s)M \bullet t[\overrightarrow{(\mu s)V_i/\mathtt{x}_i}]$
  - a stack $t(V_0, \ldots, V_{n-1}|[\cdot]) :: K$, where t is a stack-edge in $\mathcal{T}s$, to $t[\overrightarrow{(\mu s)V_i/\mathtt{x}_i}] \mathbin{+\!\!+} (\mu s)K$.

  It preserves all other term constructors; for example, it maps (M, M') to $((\mu s)M, (\mu s)M')$.
- If $s \xrightarrow{\alpha} s'$ is a multigraph homomorphism, then $\mathcal{T}\alpha$ replaces each operation symbol f by αf.

We can now formulate our direct model/categorical model equivalence.

Proposition 111 Let τ be a CBPV object structure. The "categorical semantics" functor from CBPV adjunction models on τ to the category of algebras for $\mathcal{T}$ on Sig_τ is an equivalence. □

Proof (outline)

- The categorical semantics functor is defined as follows. Let $(\mathcal{C}, \mathcal{D}, \mathcal{O})$ be a CBPV adjunction model based on τ. Then we construct a τ-multigraph s by setting

  - $s(A_0, \ldots, A_{n-1} \vdash^{\mathsf{v}} B)$ to be $\mathcal{C}_{A_0 \times \cdots \times A_{n-1}} B$
  - $s(A_0, \ldots, A_{n-1} \vdash^{\mathsf{c}} \underline{B})$ to be $\mathcal{O}_{A_0 \times \cdots \times A_{n-1}} \underline{B}$
  - $s(A_0, \ldots, A_{n-1} | \underline{B} \vdash^{\mathsf{k}} \underline{C})$ to be $\mathcal{D}_{A_0 \times \cdots \times A_{n-1}}(\underline{B}, \underline{C})$

  . We define, by induction, a function $[\![-]\!]$ from the multigraph of terms generated by s to s. This is given by the semantic equations from Sect. 9.5–9.6, together with additional equations for operation symbols:

$$\begin{aligned} [\![f(V_0, \ldots, V_{n-1})]\!] &= ([\![V_0]\!], \ldots, [\![V_{n-1}]\!])^* f \\ [\![f(V_0, \ldots, V_{r-1}|M)]\!] &= [\![M]\!]; ([\![V_0]\!], \ldots, [\![V_{n-1}]\!])^* f \\ [\![f(V_0, \ldots, V_{r-1}|[\cdot]) :: K]\!] &= ([\![V_0]\!], \ldots, [\![V_{n-1}]\!])^* f; [\![K]\!] \end{aligned}$$

  We define θ from $\mathcal{T}s$ to s to take the $\sim_s$ equivalence class of P to $[\![P]\!]$. That this is well defined is precisely the content of the soundness theorem: terms related by $\sim_s$ (i.e. provably equal terms in the equational theory) have the same denotation.

- Let (s, θ) be a direct model based on τ. Then we construct a CBPV adjunction model $\mathcal{C}$ based on τ by setting

  - $\mathcal{C}_A B$ to be $s(A \vdash^{\mathsf{v}} B)$
  - $\mathcal{D}_A(\underline{B}, \underline{C})$ to be $s(A | \underline{B} \vdash^{\mathsf{k}} \underline{C})$
  - $\mathcal{O}_A \underline{B}$ to be $s(A \vdash^{\mathsf{c}} \underline{B})$

  The combinators are defined following Fig. 10.3, and all the required equations verified.

- These operations are inverse up to isomorphism.

□

id (in $\mathcal{C}$)	$=$	$\theta\mathtt{x}_0$
f^*g (g a $\mathcal{C}$- or $\mathcal{O}$-morphism)	$=$	$\theta g(f(\mathtt{x}_0))$
f^*h (h a $\mathcal{D}$-morphism)	$=$	$\theta h(f(\mathtt{x}_0)\|[\cdot]) :: \mathtt{nil}$
$g;h$ (g an $\mathcal{O}$-morphism)	$=$	$\theta h(\mathtt{x}_0\|g(\mathtt{x}_0))$
id (in $\mathcal{D}$)	$=$	$\theta\mathtt{nil}$
$h;k$ (h a $\mathcal{D}$-morphism)	$=$	$\theta h(\mathtt{x}_0\|[\cdot]) :: k(\mathtt{x}_0\|[\cdot]) :: \mathtt{nil}$
π	$=$	$\theta\mathtt{pm}\ \mathtt{x}_0\ \mathtt{as}\ (\mathtt{x}_1,\mathtt{x}_2).\ \mathtt{x}_1$
π'	$=$	$\theta\mathtt{pm}\ \mathtt{x}_0\ \mathtt{as}\ (\mathtt{x}_1,\mathtt{x}_2).\ \mathtt{x}_2$
(f,g)	$=$	$\theta(f(\mathtt{x}_0), g(\mathtt{x}_0))$
force	$=$	$\theta\mathtt{force}\ \mathtt{x}_0$
$\mathsf{q}^U f$	$=$	$\theta\mathtt{thunk}\ f(\mathtt{x}_0)$
return	$=$	$\theta\mathtt{return}\ \mathtt{x}_0$
$\mathsf{q}^F f$	$=$	$\theta[\cdot]\ \mathtt{to}\ \mathtt{x}_1.\ f((\mathtt{x}_0,\mathtt{x}_1)) :: \mathtt{nil}$
$\mathsf{in}_{\hat{\imath}}$	$=$	$\theta(\hat{\imath},\mathtt{x}_0)$
$\mathsf{q}^{\Sigma}\lambda i.g_i$	$=$	$\theta\mathtt{pm}\ \mathtt{x}_0\ \mathtt{as}\ \{\ldots,(\mathtt{x}_1,(i,\mathtt{x}_2)).f_i((\mathtt{x}_1,\mathtt{x}_2)),\ldots\}$
$\pi_{\hat{\imath}}$	$=$	$\theta\hat{\imath} :: \mathtt{nil}$
$\mathsf{q}^{\Pi}\lambda i.g_i$ (g_i an $\mathcal{O}$-morphism)	$=$	$\theta\lambda\{\ldots,i.g_i(\mathtt{x}_0),\ldots\}$
$\mathsf{q}^{\Pi}\lambda i.h_i$ (h_i a $\mathcal{D}$-morphism)	$=$	$\theta\{\ldots,(h_i(\mathtt{x}_0\|[\cdot]) :: \mathtt{nil})\ \mathtt{where}\ i :: \mathtt{nil},\ldots\}\ \mathtt{is}\ \mathtt{nil}$
ev	$=$	$\theta\mathtt{x}_0 :: \mathtt{nil}$
$\mathsf{q}^{\to}g$ (g an $\mathcal{O}$-morphism)	$=$	$\theta\lambda\mathtt{x}_1.g((\mathtt{x}_0,\mathtt{x}_1))$
$\mathsf{q}^{\to}h$ (h a $\mathcal{D}$-morphism)	$=$	$\theta(h((\mathtt{x}_0,\mathtt{x}_1)\|[\cdot]) :: \mathtt{nil})\ \mathtt{where}\ \mathtt{x}_1 :: \mathtt{nil}\ \mathtt{is}\ \mathtt{nil}$

Figure 10.3. Adjunction Model From A Direct Model (s,θ)

10.4 All Models Of JWA Are JWA Module Models

The direct model/categorical model equivalence result for JWA is organized in exactly the same way as that for ×-calculus in Sect. 10.2 and that for CBPV in Sect. 10.3. We define *JWA object structure* in the obvious way; clearly each JWA module model gives us a JWA object structure. If τ is a JWA object structure, a τ*-sequent* is either $A_0,\ldots,A_{n-1} \vdash^{\mathsf{v}} B$ or $A_0,\ldots,A_{n-1} \vdash^{\mathsf{n}}$, where $A_0,\ldots,A_{n-1}$ and B are τ-objects. A τ*-signature* s provides a set $s(Q)$ for each τ-sequent Q.

Given a τ-signature s, we define the set of terms built from s. These are inductively defined by the rules of JWA+complex values, together with the rules

$$\frac{\Gamma \vdash^{\mathsf{v}} V_0 : A_0 \quad \cdots \quad \Gamma \vdash^{\mathsf{v}} V_{r-1} : A_{r-1}}{\Gamma \vdash^{\mathsf{v}} f(V_0,\ldots,V_{r-1}) : B} \ f \in s\,(A_0,\ldots,A_{r-1} \vdash^{\mathsf{v}} B)$$

$$\frac{\Gamma \vdash^{\mathsf{v}} V_0 : A_0 \quad \cdots \quad \Gamma \vdash^{\mathsf{v}} V_{r-1} : A_{r-1}}{\Gamma \vdash^{\mathsf{n}} f(V_0,\ldots,V_{r-1})} \ f \in s\,(A_0,\ldots,A_{r-1} \vdash^{\mathsf{n}})$$

We take the least congruence $\sim_s$ on these terms containing the equational laws of JWA, which as in Prop. 108 and Prop. 110 is a hypercongruence (i.e. closed under substitution), Quotienting by this congruence we obtain a new signature called $\mathcal{T}s$. This gives a monad $\mathcal{T}$ on the category of τ-signatures.

Proposition 112 Let τ be a CBPV object structure. The "categorical semantics" functor from JWA module models with object structure τ to the category of algebras for $\mathcal{T}$ on Sig_τ is an equivalence. □

This is proved in the same way as Prop. 109 and Prop. 111.

10.5 Enrichment

In a similar manner to Prop. 111, we can show that our categorical semantics of CBPV + stacks + `print` in an $\mathcal{A}$-set enriched CBPV adjunction model is an equivalence. And in a similar manner to Prop. 112, we can show that our categorical semantics of JWA + `print` in an $\mathcal{A}$-set enriched JWA module model is an equivalence.

It appears that these 2 theorems could be generalized from printing to other strong monads specified by *algebraic operations and equations* in the manner of [Plotkin and Power, 2002]; but we do not treat this here.

Chapter 11

REPRESENTING OBJECTS

This chapter is independent of Chap. 10.

11.1 Introduction

A key concept of category theory is that of *representable object* (or *representation*, or *universal element*). This concept can be defined in both element style and naturality style—the equivalence of the two definitions follows from the Yoneda Lemma. So by describing something as a representable object, we automatically obtain both an element-style and a naturality-style description.

The main goal of this chapter is to describe each of our connectives as a representing object for a suitable functor. In each case we recover the element style definition given in Chap. 9 as well as the naturality style definition briefly mentioned there.

It is because our connectives are representing objects that they are automatically unique up to unique isomorphism, and there is no need for any coherence conditions such as are required in [Selinger, 2001; Thielecke, 1997b] and in models for linear logic. c

The functors that our connectives represent are built out of homset functors. We described the homset functor for ordinary categories (Def. 9.4) and locally indexed categories (Def. 9.12), and these suffice for most of our connectives. But for distributive coproducts and CBPV products and exponentials, we require other homset functors.

- In Sect. 11.3.3, we define a special homset functor for a cartesian category $\mathcal{C}$, and use it to define distributive coproduct in $\mathcal{C}$ as a representing object.

- For CBPV products and exponentials, we will see in Sect. 11.3.4 how to glue right modules to a category (in the manner of [Carboni et al., 1998]) and give an appropriate homset functor, enabling us to characterize these structures as representing objects.
- The most difficult case is in Sect. 11.3.5, where we treat a cartesian category with some left modules. This was the structure used in Sect. 9.6.6 to define distributive coproducts in JWA and CBPV. Again by glueing modules, we define a suitable homset functor, and then we can characterize a distributive coproduct as a representing object.

We then look at *parametrized representability*, which show us how to make each connective into a functor. We use this firstly to complete our definition of possible world semantics that we began in Sect. 9.7.6, and secondly to show that a strong adjunction is indeed an adjunction (Prop. 102).

Finally, we look at CBV and CBN. We give a categorical structure for each, and see how, for a given CBPV adjunction model, we interpret CBV inside the Kleisli part and CBN inside the co-Kleisli part.

Note Some authors use the term "representation" instead of "representing object", reserving the latter term for what we have called the *vertex*. However, "representation" has many other meanings.

11.2 Categorical Structures

11.2.1 Joint vs. Separate Naturality

We frequently wish to use the idiom "natural in X and Y". This appears to be ambiguous, because it could refer to either joint naturality or separate naturality. However, there is no ambiguity, because it is well known that the two are equivalent:

Proposition 113 Let $\mathcal{D}$ and $\mathcal{D}'$ be categories. Let F and G be functors from $\mathcal{D}\times\mathcal{D}'$ to $\mathcal{E}$. Suppose we are given a function $F(X,Y) \xrightarrow{\alpha(X,Y)} G(X,Y)$ for each $X \in \mathsf{ob}\ \mathcal{D}$ and $Y \in \mathsf{ob}\ \mathcal{D}'$. Then α is a natural transformation from F to G iff

- $\alpha(X,Y)$ is natural in $X \in \mathcal{D}$ for each $Y \in \mathsf{ob}\ \mathcal{D}'$
- $\alpha(X,Y)$ is natural in $Y \in \mathcal{D}'$ for each $X \in \mathsf{ob}\ \mathcal{D}$.

□

Sometimes, however, we want to use the idiom "natural in X and $\underline{Y}$" where ${}_X\underline{Y}$ ranges over not a product category but an opGr category. In that situation, we require the following variant (in fact it is a generalization) of Prop. 113.

Proposition 114 Let $\mathcal{D}$ be a locally $\mathcal{C}$-indexed category, and let F and G be functors from $\mathsf{opGr}\,\mathcal{D}$ to $\mathcal{E}$. Suppose we are given a function $F_X\underline{Y} \xrightarrow{\alpha_X\underline{Y}} G_X\underline{Y}$, for each $X \in \mathsf{ob}\,\mathcal{C}$ and $\underline{Y} \in \mathsf{ob}\,\mathcal{D}$. Then α is a natural transformation from F to G iff

- $\alpha_X\underline{Y}$ is natural in $X \in \mathcal{C}^{\mathsf{op}}$ for each $\underline{Y} \in \mathsf{ob}\,\mathcal{D}$
- $\alpha_X\underline{Y}$ is natural in $\underline{Y} \in \mathcal{D}_X$ for each $X \in \mathsf{ob}\,\mathcal{C}$.

□

Finally, we can adapt Prop. 113 and Prop. 114 to allow us to use the idiom "$\alpha_X(\underline{Y}, \underline{Z})$ is natural in X, $\underline{Y}$ and $\underline{Z}$".

Proposition 115 Let $\mathcal{D}$ and $\mathcal{D}'$ be locally $\mathcal{C}$-indexed categories, and let F and G be functors from $\mathsf{opGr}(\mathcal{D} \times \mathcal{D}')$ to $\mathcal{E}$. Suppose we are given a function $F_X(\underline{Y}, \underline{Z}) \xrightarrow{\alpha_X(\underline{Y},\underline{Z})} G_X(\underline{Y}, \underline{Z})$, for each $X \in \mathsf{ob}\,\mathcal{C}$, $\underline{Y} \in \mathsf{ob}\,\mathcal{D}$ and $\underline{Z} \in \mathsf{ob}\,\mathcal{D}'$. Then α is a natural transformation from F to G iff

- $\alpha_X(\underline{Y}, \underline{Z})$ is natural in $X \in \mathcal{C}^{\mathsf{op}}$ for each $\underline{Y} \in \mathsf{ob}\,\mathcal{D}$ and $\underline{Z} \in \mathsf{ob}\,\mathcal{D}'$
- $\alpha_X(\underline{Y}, \underline{Z})$ is natural in $\underline{Y} \in \mathcal{D}_X$ for each $X \in \mathsf{ob}\,\mathcal{C}$ and $\underline{Z} \in \mathsf{ob}\,\mathcal{D}'$
- $\alpha_X(\underline{Y}, \underline{Z})$ is natural in $\underline{Z} \in \mathcal{D}'_X$ for each $X \in \mathsf{ob}\,\mathcal{C}$ and $\underline{Y} \in \mathsf{ob}\,\mathcal{D}$.

□

11.2.2 Context Extension Functors

In type theory, the operation written as a comma in Γ, A is of great importance and is often called *context extension*. It is closely related to the following functors, as we shall see in Sect. 11.3. Let $\mathcal{C}$ be a cartesian category.

Firstly, we know that if $X' \xrightarrow{k} X$ and $Y' \xrightarrow{f} Y$, then we can define a morphism $X' \times Y' \xrightarrow{k \times f} X \times Y$. But suppose that, instead of f, we have $X' \times Y' \xrightarrow{g} Y$. Despite the extra dependency, we *still* obtain a morphism $X' \times Y' \longrightarrow X \times Y$ viz. $((\pi; k), g)$. We write $k \cdot g$ for this morphism, and $\cdot$ defines a functor from $\mathsf{opGr}((\mathsf{self}\,\mathcal{C})^{\mathsf{op}})$ to $\mathcal{C}^{\mathsf{op}}$, given by $\times$ on objects. From this we can recover the usual functor $\times$

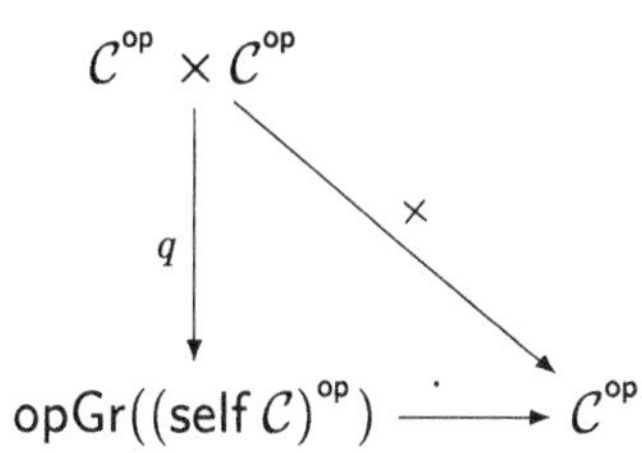

Here, the functor q takes (X, Y) to ${}_X Y$ and (k, f) to ${}_k \pi'^* f$. (We could remove all these $-^{\mathsf{op}}$ marks by using the *Grothendieck construction* rather than opGrothendieck here.)

We make use of $\cdot$ in defining the following most important functor.

Definition 11.1 Let $\mathcal{D}$ be a locally $\mathcal{C}$-indexed category, where $\mathcal{C}$ is cartesian. We write $\mathsf{growall}$ for the functor

$$\mathsf{opGr}((\mathsf{self}\,\mathcal{C})^{\mathsf{op}} \times \mathcal{D}) \longrightarrow \mathsf{opGr}\ \mathcal{D}$$

$${}_X(Y, \underline{Z}) \longmapsto {}_{X \times Y}\underline{Z}$$

$${}_k(f, h) \longmapsto {}_{k \cdot f}\pi^* h$$

Given a $\mathcal{C}$-object A, we write $\mathsf{grow}A$, for the functor

$$\mathsf{opGr}\ \mathcal{D} \xrightarrow{\mathsf{opGr}\ i} \mathsf{opGr}(1 \times \mathcal{D}) \xrightarrow{\mathsf{opGr}(\mathsf{obj}\,A \times \mathcal{D})} \mathsf{opGr}((\mathsf{self}\,\mathcal{C})^{\mathsf{op}} \times \mathcal{D}) \xrightarrow{\mathsf{growall}} \mathsf{opGr}\ \mathcal{D}$$

Explicitly, this is given by

$$\mathsf{opGr}\ \mathcal{D} \longrightarrow \mathsf{opGr}\ \mathcal{D}$$

$${}_X\underline{Z} \longmapsto {}_{X \times A}\underline{Z}$$

$${}_k h \longmapsto {}_{k \times A}\pi^* h$$

□

In the case that $\mathcal{D} = 1$, it is evident that $\mathsf{opGr}\mathcal{D} \cong \mathcal{C}^{\mathsf{op}}$. Modulo this isomorphism, $\mathsf{growall}$ corresponds to *cdot*, and $\mathsf{grow}A$ corresponds to $- \times A$. So we will sometimes write $\mathsf{grow}A$ instead of $- \times A$.

The following is often used in conjunction with context extension functors.

Definition 11.2 If, for each $i \in I$, we have a functor $\mathcal{B} \xrightarrow{\mathcal{F}_i} \mathbf{Set}$, then we write $\mathcal{B} \xrightarrow{\prod_{i \in I} \mathcal{F}_i} \mathbf{Set}$ for their pointwise product. □

11.3 Representable Functors

11.3.1 Ordinary Category

Firstly, we recall from Sect. 9.3.2 the homset functor $\mathsf{hom}_{\mathcal{D}}$ for an ordinary category $\mathcal{D}$.

We present our key definition [Mac Lane, 1971] in both element style and naturality style.

Definition 11.3 Let $F : \mathcal{D} \longrightarrow \mathbf{Set}$ be a functor. Let $\iota : \mathcal{B} \longrightarrow \mathcal{D}$ be a fully faithful functor (it will be $\mathsf{id}_{\mathcal{D}}$ in many examples). A *representing object* for F within ι consists of a $\mathcal{B}$-object V (the *vertex*) together with

element style an element $v \in F\iota V$ such that the functions

$$\begin{array}{rcll} \mathcal{D}(\iota V, X) & \longrightarrow & FX & \quad\text{for all } X \\ f & \longmapsto & (Ff)v & \end{array}$$

are isomorphisms.

naturality style an isomorphism

$$\mathcal{D}(\iota V, X) \xrightarrow[\cong]{\alpha X} FX \quad \text{natural in } X.$$

We omit the phrase "within ι" when $\iota = \mathsf{id}_{\mathcal{D}}$. □

The equivalence of these two definitions is essentially the Yoneda Lemma:

Proposition 116 The naturality style and element style formulations of Def. 11.3 are equivalent.

naturality to element style the element v is $\alpha(\iota V)\mathsf{id}$.

element to naturality style the isomorphism α is $\lambda X.\lambda f.(Ff)v$.

□

By defining a categorical structure to be a representing object, we automatically obtain both an element style and naturality style description, and we know that they are equivalent. Here are examples, for the structures we have seen so far.

Definition 11.4 Let $\mathcal{C}$ be a cartesian category.

- A *product* for $\{B_i\}_{i \in I}$ is a representing object for $\prod_{i\in I} \mathcal{C}\_B_i$.
- An *exponential* from A to B is a representing object for $\mathsf{grow}A; \mathcal{C}\_B$.
- A *exponential-product* for $\{(A_i, B_i)\}_{i\in I}$ is a representing object for $\prod_{i\in I}(\mathsf{grow}A_i; \mathcal{C}\_B_i)$.

Let $(\mathcal{C}, \mathcal{N})$ be a JWA judgement model.

- A *jumpwith* for A is a representing object for $\mathsf{grow}A; \mathcal{N}$.
- A *coproduct-jumpwith* for $\{A_i\}_{i\in I}$ is a representing object for $\prod_{i\in I}(\mathsf{grow}A_i; \mathcal{N})$.

Let $(\mathcal{C}, \mathcal{D}, \mathcal{O})$ be a CBPV judgement model.

- A *right adjunctive* for $\underline{B}$ is a representing object for $\mathcal{O}_{-}\underline{B}$.
- A *right product-adjunctive* for $\{\underline{B}_i\}_{i \in I}$ is a representing object for $\prod_{i \in I} \mathcal{O}_{-}\underline{B}_i$.
- A *right exponential-adjunctive* from A to $\underline{B}$ is a representing object for $\mathsf{grow}A; \mathcal{O}_{-}\underline{B}$.
- A *right exponential-product-adjunctive* for $\{(A_i, \underline{B}_i)\}_{i \in I}$ is a representing object for $\prod_{i \in I}(\mathsf{grow}A_i; \mathcal{O}_{-}\underline{B}_i)$.

□

Notice, and this will be apparent throughout all our examples, how the functor being represented is a description of the upper part of the reversible derivation; it is a product, and each factor describes one of the judgements above the double line. Each A on the left of the turnstile in a judgement corresponds to $\mathsf{grow}A$ in the corresponding factor.

When we define a categorical structure as a representing object of a functor, we know that is unique up to unique isomorphism, in the following sense.

Proposition 117 Let F be a functor from $\mathcal{D}$ to **Set**, and let $\iota : \mathcal{B} \longrightarrow \mathcal{D}$ be a fully faithful functor. Let (V, α) and (V', α') be naturality-style representing objects for F within ι. Then there is a unique isomorphism $V \xrightarrow{f} V'$ in $\mathcal{B}$ such that

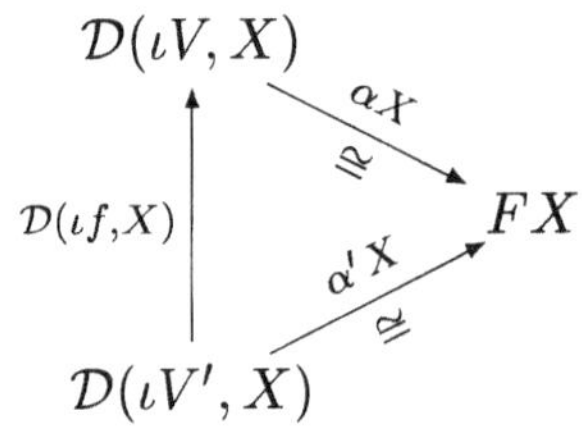

commutes for all X.

If we write the representing objects in element style as (V, v) and (V', v') then f is characterized by $(F\iota f)v = v'$. □

Proof This is proved the same way as Prop. 121 below; in fact it is a special case. □

11.3.2 Locally Indexed Category

Firstly, we recall from Sect. 9.3.5 the homset functor $\mathsf{hom}_{\mathcal{D}}$ for an locally $\mathcal{C}$-indexed category $\mathcal{D}$, where $\mathcal{C}$ is cartesian. We repeat the definitions of Sect. 11.3.1 in the locally $\mathcal{C}$-indexed setting.

Definition 11.5 Let $F : \mathsf{opGr}\ \mathcal{D} \longrightarrow \mathbf{Set}$ be a functor. Let $\iota : \mathcal{B} \longrightarrow \mathcal{D}$ be a fully faithful functor. A *global representing object* for F within ι consists of a $\mathcal{B}$-object $\underline{V}$ (the *vertex*) together with

naturality style an isomorphism

$$\mathcal{D}_X(\iota\underline{V}, \underline{Y}) \xrightarrow[\cong]{\alpha_X \underline{Y}} F_X\underline{Y} \quad \text{natural in } X \text{ and } \underline{Y}.$$

element style an element $v \in F_1\iota\underline{V}$ such that the functions

$$\begin{array}{rcl} \mathcal{D}_X(\iota\underline{V}, \underline{Y}) & \longrightarrow & F_X\underline{Y} \\ f & \longmapsto & (F_{()}f)v \end{array} \qquad \text{for all } X \text{ and } \underline{Y}$$

are isomorphisms.

□

Proposition 118 The naturality style and element style formulations of Def. 11.5 are equivalent, and a global representing object is unique up to unique isomorphism. □

Proof A global representing object for F within ι can be seen as a representing object for F within the composite

$$\mathcal{B}_1 \xrightarrow{\iota_1} \mathcal{D}_1 \xrightarrow{1-} \mathsf{opGr}\ \mathcal{D}$$

(This is the reason for the "global" terminology.) □

Definition 11.6 Let $(\mathcal{C}, \mathcal{D}, \mathcal{O})$ be a CBPV judgement model.

- A *left adjunctive* for a val-object A is a global representing object for $\mathsf{grow} A; \mathcal{O}$.
- A *left coproduct-adjunctive* for $\{A_i\}_{i \in I}$ is a global representing object for $\prod_{i \in I}(\mathsf{grow} A_i; \mathcal{O})$.

□

11.3.3 Cartesian Category

Let $\mathcal{C}$ be a cartesian category. So far, we have seen two homset functors associated that $\mathcal{C}$ gives rise to:

- the functor $\mathsf{hom}_{\mathcal{C}}$ obtained by regarding $\mathcal{C}$ as an ordinary category, ignoring its cartesian structure
- the functor $\mathsf{hom}_{\mathsf{self}\,\mathcal{C}}$ obtained by converting $\mathcal{C}$ into the locally $\mathcal{C}$-indexed category $\mathsf{self}\,\mathcal{C}$.

But, in fact, there is a third homset functor $\mathsf{hom}_{\times\mathcal{C}}$ special to cartesian categories:

$$\mathsf{opGr}\ \mathsf{self}\,\mathcal{C} \longrightarrow \mathbf{Set}$$

$$_{X}Y \longmapsto \mathcal{C}_X Y$$

$$\begin{array}{ccccc}
X & X' \times Y & & \mathcal{C}_X Y & g \\
{\scriptstyle k}\uparrow & \downarrow{\scriptstyle h} & \longmapsto & \downarrow{\scriptstyle \mathcal{C}_k h} & \downarrow \\
X' & Y' & & \mathcal{C}_{X'} Y' & (\mathsf{id}, (k; g))^* h
\end{array}$$

This functor provides all the homset information we need. From it, we can recover $\mathsf{hom}_{\mathcal{C}}$

$$\begin{array}{ccc}
\mathcal{C}^{\mathsf{op}} \times \mathcal{C} & & \\
{\scriptstyle q}\downarrow & \searrow^{\mathsf{hom}_{\mathcal{C}}} & \\
\mathsf{opGr}\ \mathsf{self}\,\mathcal{C} & \xrightarrow[\mathsf{hom}_{\times\mathcal{C}}]{} & \mathbf{Set}
\end{array}$$

where q is as in Sect. 11.2.2. And we can also recover $\mathsf{hom}_{\mathsf{self}\,\mathcal{C}}$

$$\begin{array}{ccc}
\mathsf{opGr}((\mathsf{self}\,\mathcal{C})^{\mathsf{op}} \times \mathsf{self}\,\mathcal{C}) & & \\
{\scriptstyle \mathsf{growall}}\downarrow & \searrow^{\mathsf{hom}_{\mathsf{self}\,\mathcal{C}}} & \\
\mathsf{opGr}\ \mathsf{self}\,\mathcal{C} & \xrightarrow[\mathsf{hom}_{\times\mathcal{C}}]{} & \mathbf{Set}
\end{array} \qquad (11.1)$$

where growall is as in Sect. 11.2.2. As a consequence of (11.1) we have

$$\mathsf{self}\,\mathcal{C}_{-}(A, -) = \mathsf{grow}A; \mathsf{hom}_{\times\mathcal{C}} \qquad (11.2)$$

and we make the following definition:

Definition 11.7 A *distributive coproduct* for a family of objects $\{A_i\}_{i \in I}$ in a cartesian category $\mathcal{C}$ is a representing object for $\prod_{i \in I}(\mathsf{grow}A_i; \mathsf{hom}_{\times\mathcal{C}})$. □

Because of (11.2), we can equivalently define a distributive coproduct to be a locally indexed coproduct in $\mathsf{self}\,\mathcal{C}$ [Jacobs, 1999]. We prefer the formulation of Def. 11.7, because it continues the pattern that we noted in Sect. 11.3.1: a connective is a representing object for a functor that describes the upper line of its reversible derivation.

11.3.4 Category With Right Modules

Ordinary Category With Right Modules

Suppose, as in Sect. 9.6.4, that we have a category $\mathcal{D}$ with a family of right $\mathcal{D}$-modules $\{\mathcal{O}^p\}_{p \in P}$. Our aim is to make a homset functor that describes each homset $\mathcal{D}(Y, Z)$ as well as each homset $\mathcal{O}^p Z$.

Our first step is to "glue" the modules to the category in the following manner [Carboni et al., 1998]:

Definition 11.8 We write $\mathcal{O} \bullet \mathcal{D}$ for the category in which

- an object is either ${}^{\mathsf{c}}p$, where $p \in P$, or ${}^{\mathsf{k}}Y$, where Y is an object of $\mathcal{D}$
- a morphism ${}^{\mathsf{k}}Y \longrightarrow {}^{\mathsf{k}}Z$ is a $\mathcal{D}$-morphism $Y \longrightarrow Z$
- a morphism ${}^{\mathsf{c}}p \longrightarrow {}^{\mathsf{k}}Z$ is an $\mathcal{O}^p$-morphism $\longrightarrow Z$
- the only morphism to ${}^{\mathsf{c}}p$ is the identity
- identities and composition are defined as in $\mathcal{D}$ and the various $\mathcal{O}^p$.

We write ${}^{\mathsf{k}}-$ for the fully faithful functor from $\mathcal{D}$ to $\mathcal{O} \bullet \mathcal{D}$. □

Now we are in a position to define the required homset functor.

Definition 11.9 We write $\mathsf{hom}^{\mathcal{O}\mathcal{D}}$ for the functor

$$(\mathcal{O} \bullet \mathcal{D})^{\mathsf{op}} \times \mathcal{D} \xrightarrow{(\mathcal{O} \bullet \mathcal{D})^{\mathsf{op}} \times {}^{\mathsf{k}}-} (\mathcal{O} \bullet \mathcal{D})^{\mathsf{op}} \times (\mathcal{O} \bullet \mathcal{D}) \xrightarrow{\mathsf{hom}_{\mathcal{O} \bullet \mathcal{D}}} \mathbf{Set}$$

□

We can see that this achieves our goal, because

$$\begin{aligned} \mathsf{hom}^{\mathcal{O}\mathcal{D}}({}^{\mathsf{k}}Y, Z) &= \mathcal{D}(Y, Z) \\ \mathsf{hom}^{\mathcal{O}\mathcal{D}}({}^{\mathsf{c}}p, Z) &= \mathcal{O}^p Z \end{aligned}$$

Furthermore we have

$$\mathsf{hom}^{\mathcal{O}\mathcal{D}}(-, \underline{B}) = \mathcal{O} \bullet \mathcal{D}(-, {}^{\mathsf{k}}\underline{B}) \tag{11.3}$$

We can use Def. 11.9 to turn Def. 9.26 into naturality style.

Definition 11.10 Let $\mathcal{D}$ be a category and $\{\mathcal{O}^p\}_{p \in P}$ a family of right $\mathcal{D}$-modules. A *product* for an object family $\{B_i\}_{i \in I}$ is a representing object within ${}^{\mathsf{k}}-$ for $\prod_{i \in I} \mathsf{hom}^{\mathcal{O}\mathcal{D}}(-, B_i)$. □

Locally Indexed Category With Right Modules

We now repeat the above in the locally indexed setting. Suppose, as in Sect. 9.6.4–9.6.5, that $\mathcal{C}$ is cartesian and we have a locally $\mathcal{C}$-indexed category $\mathcal{D}$ with a family of right $\mathcal{D}$-modules $\{\mathcal{O}^p\}_{p\in P}$. Our aim is to make a homset functor that describes each homset $\mathcal{D}_X(\underline{Y}, \underline{Z})$ as well as each homset $\mathcal{O}^p_X \underline{Z}$.

Our first step is to "glue" the modules to the category in the following manner [Carboni et al., 1998]:

Definition 11.11 We write $\mathcal{O}\bullet\mathcal{D}$ for the locally $\mathcal{C}$-indexed category in which

- an object is either ${}^{\mathsf{c}}p$, where $p \in P$, or ${}^{\mathsf{k}}\underline{Y}$, where $\underline{Y}$ is an object of $\mathcal{D}$
- a morphism ${}^{\mathsf{k}}\underline{Y} \xrightarrow[X]{} {}^{\mathsf{k}}\underline{Z}$ is a $\mathcal{D}$-morphism $\underline{Y} \xrightarrow[X]{} \underline{Z}$
- a morphism ${}^{\mathsf{c}}p \xrightarrow[X]{} {}^{\mathsf{k}}\underline{Z}$ is an $\mathcal{O}^p$-morphism $\xrightarrow[X]{} \underline{Z}$
- the only morphisms from ${}^{\mathsf{c}}p$ are identities
- identities, composition and reindexing are defined as in $\mathcal{D}$ and $\mathcal{O}^p$.

We write ${}^{\mathsf{k}}-$ for the fully faithful functor from $\mathcal{D}$ to $\mathcal{O}\bullet\mathcal{D}$. □

Now we are in a position to define the required homset functor.

Definition 11.12 We write $\mathsf{hom}^{\mathcal{OD}}$ for the functor

$$\mathsf{opGr}((\mathcal{O}\bullet\mathcal{D})^{\mathsf{op}} \times \mathcal{D}) \xrightarrow{\mathsf{opGr}((\mathcal{O}\bullet\mathcal{D})^{\mathsf{op}} \times {}^{\mathsf{k}}-)} \mathsf{opGr}((\mathcal{O}\bullet\mathcal{D})^{\mathsf{op}} \times (\mathcal{O}\bullet\mathcal{D})) \xrightarrow{\mathsf{hom}_{\mathcal{O}\bullet\mathcal{D}}} \mathbf{Set}$$

□

We can see that this achieves our goal, because

$$\begin{aligned} \mathsf{hom}^{\mathcal{OD}}_X({}^{\mathsf{k}}\underline{Y}, \underline{Z}) &= \mathcal{D}_X(\underline{Y}, \underline{Z}) \\ \mathsf{hom}^{\mathcal{OD}}_X({}^{\mathsf{c}}p, \underline{Z}) &= \mathcal{O}^p_X \underline{Z} \end{aligned}$$

Furthermore we have

$$\mathsf{hom}^{\mathcal{OD}}_{-}(-, \underline{B}) = (\mathcal{O}\bullet\mathcal{D})_{-}(-, {}^{\mathsf{k}}\underline{B}) \tag{11.4}$$

We can use this to adapt Def. 9.27 and Def. 9.28 to naturality style.

Definition 11.13 Let $\mathcal{C}$ be a cartesian category. Let $\mathcal{D}$ be a locally $\mathcal{C}$-indexed category (whose objects we underline) and $\{\mathcal{O}^p\}_{p\in P}$ a family of right $\mathcal{D}$-modules.

- A *product* for a family of $\mathcal{D}$-objects $\{\underline{B}_i\}_{i \in I}$ is a global representing object within ${}^{\mathsf{k}}-$ for $\prod_{i \in I} \mathsf{hom}_{-}^{\mathcal{OD}}(-, \underline{B}_i)$.
- An *exponential* from a $\mathcal{C}$-object A to a $\mathcal{D}$-object $\underline{B}$ is a global representing object within ${}^{\mathsf{k}}-$ for $\mathsf{grow}A; \mathsf{hom}_{-}^{\mathcal{OD}}(-, \underline{B})$.
- A *exponential-product* for $\{(A_i, \underline{B}_i)\}_{i \in I}$ is a global representing object within ${}^{\mathsf{k}}-$ for $\prod_{i \in I}(\mathsf{grow}A_i; \mathsf{hom}_{-}^{\mathcal{OD}}(-, \underline{B}_i))$.

□

11.3.5 Cartesian Category With Left Modules

Now suppose, as in Sect. 9.6.6, that we have a cartesian category $\mathcal{C}$ with a family of left modules $\mathcal{N} = \{\mathcal{N}^p\}_{p \in P}$. We have already seen in Sect. 11.3.3 a homset functor $\mathsf{hom}_{\times \mathcal{C}}$ for $\mathcal{C}$. But now we want to extend this to a homset functor giving all the homsets $\mathcal{N}_X^p$ as well as the homsets $\mathcal{C}_X Y$.

To do this, like in Sect. 11.3.4, we must first glue the modules to the category.

Definition 11.14 We write $\mathsf{self}\,\mathcal{CN}$ for the locally $\mathcal{C}$-indexed category in which

- an object is either ${}^{\mathsf{v}}Y$, where $Y \in \mathsf{ob}\,\mathcal{C}$, or ${}^{\not\vee}p$, where $P \in P$
- a morphism ${}^{\mathsf{v}}Y \xrightarrow[X]{} {}^{\mathsf{v}}Z$ is a $\mathcal{C}$-morphism $X \times Y \longrightarrow Z$
- a morphism ${}^{\mathsf{v}}Y \xrightarrow[X]{} {}^{\not\vee}p$ is a $\mathcal{N}^p$-morphism $X \times Y \longrightarrow$
- the only morphisms from ${}^{\not\vee}p$ are the identities
- the identity over X on ${}^{\mathsf{v}}Y$ is π'
- the composite of ${}^{\mathsf{v}}Y \xrightarrow[X]{f} {}^{\mathsf{v}}Z \xrightarrow[X]{g} {}^{\mathsf{v}}W$ is $(\pi, f)^* g$
- the composite of ${}^{\mathsf{v}}Y \xrightarrow[X]{f} {}^{\mathsf{v}}Z \xrightarrow[X]{g} {}^{\not\vee}p$ is $(\pi, f)^* g$
- the reindexing along $X' \xrightarrow{k} X$ of ${}^{\mathsf{v}}Y \xrightarrow[X]{f} {}^{\mathsf{v}}Z$ is $(k \times Y)^* f$
- the reindexing along $X' \xrightarrow{k} X$ of ${}^{\mathsf{v}}Y \xrightarrow[X]{f} {}^{\not\vee}p$ is $(k \times Y)^* f$.

We write ${}^{\mathsf{v}}-$ for the fully faithful functor from $\mathsf{self}\,\mathcal{C}$ to $\mathsf{self}\,\mathcal{CN}$. □

Now we are in a position to define the required homset functor.

Definition 11.15 We write $\mathsf{hom}_{\times\mathcal{CN}}$ for the homset functor

$$\mathsf{opGr}\ \mathsf{self}\,\mathcal{CN} \longrightarrow \mathbf{Set}$$

$$\overset{\mathsf{v}}{{}_X}Y \longmapsto \mathcal{C}_X Y$$

$$\overset{\not\mathsf{v}}{{}_X}p \longmapsto \mathcal{N}^p_X$$

$$\begin{array}{cccc}
X & X' \times Y & \mathcal{C}_X Y & g \\
\uparrow k & \downarrow h \quad \longmapsto & \downarrow \mathcal{C}_k h & \downarrow \\
X' & Y' & \mathcal{C}_{X'} Y' & (\mathsf{id}, (k; g))^* h
\end{array}$$

$$\begin{array}{cccc}
X & X' \times Y & \mathcal{C}_X Y & g \\
\uparrow k & h \downarrow (p) \quad \longmapsto & \downarrow \mathcal{C}_k h & \downarrow \\
X' & & \mathcal{N}^p_{X'} & (\mathsf{id}, (k; g))^* h
\end{array}$$

□

It is clear that this achieves our goal. Furthermore, the diagram (11.1) can be generalized:

Proposition 119 The following diagram commutes

$$\begin{array}{ccc}
\mathsf{opGr}((\mathsf{self}\,\mathcal{C})^{\mathsf{op}} \times (\mathsf{self}\,\mathcal{CN})) & \xrightarrow{\mathsf{opGr}((^{\mathsf{v}}-)^{\mathsf{op}} \times)\mathsf{self}\,\mathcal{CN}))} & \mathsf{opGr}(\mathsf{self}\,\mathcal{CN}^{\mathsf{op}} \times (\mathsf{self}\,\mathcal{CN})) \\
\downarrow \mathsf{growall} & & \downarrow \mathsf{hom}_{\mathsf{self}\,\mathcal{CN}} \\
\mathsf{opGr}\ \mathsf{self}\,\mathcal{CN} & \xrightarrow{\mathsf{hom}_{\times\mathcal{CN}}} & \mathbf{Set}
\end{array}$$

and we write $\mathsf{growhom}_{\times\mathcal{CN}}$ for the composite of this. For each $\mathcal{C}$-object A, we therefore have

$$\begin{array}{rcl}
\mathsf{growhom}_{\times\mathcal{CN}_}(A, -) & = & \mathsf{hom}_{\mathsf{self}\,\mathcal{CN}}\ {}_{-}(^{\mathsf{v}}A, -) \\
 & = & \mathsf{grow}\,A; \mathsf{hom}_{\times\mathcal{CN}}
\end{array}$$

□

This enables us to write Def. 9.29 in naturality style:

Definition 11.16 Let $\mathcal{C}$ be a cartesian category and let $\{\mathcal{N}^p\}_{p\in P}$ be a family of left $\mathcal{C}$-modules. A *distributive coproduct* for an object family $\{A_i\}_{i\in I}$ is a global representing object within $^{\mathsf{v}}-$ for $\prod_{i\in I}\mathsf{grow}A_i;\mathsf{hom}_{\times\mathcal{C}\mathcal{N}}$. □

This completes our naturality style description of the connectives.

Proposition 120 Let V be the vertex of a distributive coproduct for $\{A_i\}_{i\in I}$ in a CBPV judgement model $(\mathcal{C},\mathcal{D},\mathcal{O})$—recall from Sect. 9.6.6 that this means a distributive coproduct in $(\mathcal{C},\mathcal{N}_{\mathsf{c}}+\mathcal{N}_{\mathsf{k}})$. Then the associated isomorphism

$$\mathsf{hom}^{\mathcal{O}\mathcal{D}}_{X\times V}(\underline{Y},\underline{Z})\cong\prod_{i\in I}\mathsf{hom}^{\mathcal{O}\mathcal{D}}_{X\times A_i}(\underline{Y},\underline{Z})\qquad\text{natural in } X$$

is natural also in $\underline{Y}$ and $\underline{Z}$. □

Proof Immediate from the element-style description of this isomorphism, viz. $f\mapsto\lambda i.((X\times\mathsf{in}_i)^*f)$. □

11.4 Parametrized Representability

Note The results in this section are related to the Yoneda Lemma. However, we omit this relationship, because it appears to be easier to prove them directly than to derive them from the Yoneda Lemma. This is especially true of Prop. 124.

We commonly wish to build functors out of representing objects. Suppose, for example, that a category $\mathcal{C}$ has a product for each pair of objects A and A', whose vertex we write $A\times A'$. We would like to make $\times$ into a functor so that the isomorphism

$$\mathcal{C}(X,A)\times\mathcal{C}(X,A')\cong\mathcal{C}(X,A\times A')\qquad\text{natural in } X$$

becomes natural in A and A'. There is a unique way of doing this, by a theorem called "parametrized representability" (the name is apparently due to Hyland). We need such results for two purposes:

- to complete our definition of the possible world construction (Sect. 9.7.6)
- to compare different definitions of "adjunction", and proving Prop. 102.

Here is the basic theorem.

Proposition 121 Let $F:\mathfrak{I}\times\mathcal{D}\longrightarrow\mathbf{Set}$ be a functor, and let $\iota:\mathcal{B}\longrightarrow\mathcal{D}$ be a fully faithful functor. Suppose that for each $I\in\mathfrak{I}$ there is a representing object within ι for $F(I,-)$, which we write in naturality style as an isomorphism

$$\mathcal{D}(\iota V(I), X) \xrightarrow[\cong]{\alpha(I,X)} F(I,X) \quad \text{natural in } X.$$

Then V extends uniquely to a functor from $\mathfrak{J}$ to $\mathcal{B}^{\mathsf{op}}$ so as to make $\alpha(I,X)$ natural in I as well as X.

If we write the representing object of $F(I,-)$ in element style as $(V(I), v_I)$, then the functor V is characterized by

$$F(J, \iota V(f))v_J = F(f, \iota V(I))v_I \text{ for } I \xrightarrow[A]{f} J$$

□

Proof Similar to the proof of Prop. 122 below. □

We adapt Prop. 121 to the locally indexed setting as follows.

Proposition 122 Let $F : \mathsf{opGr}(\mathfrak{J} \times \mathcal{D}) \longrightarrow \mathbf{Set}$ be a functor, and let $\iota : \mathcal{B} \longrightarrow \mathcal{D}$ be a fully faithful functor. Suppose that for each $\underline{I} \in \mathfrak{J}$ there is a global representing object within ι for $F_{-}(\underline{I},-)$, written in naturality style as an isomorphism

$$\mathcal{D}_X(\iota \underline{V}(\underline{I}), \underline{Y}) \xrightarrow[\cong]{\alpha_X(\underline{I},\underline{Y})} F_X(\underline{I}, \underline{Y}) \quad \text{natural in } X \text{ and } \underline{Y}.$$

Then $\underline{V}$ extends uniquely to a functor from $\mathfrak{J}$ to $\mathcal{B}^{\mathsf{op}}$ so as to make $\alpha_X(\underline{I},\underline{Y})$ natural in $\underline{I}$ as well as X and $\underline{Y}$.

If we write the representing object of $F(\underline{I},-)$ in element style as $(\underline{V}(\underline{I}), v_{\underline{I}})$, then the functor $\underline{V}$ is characterized by

$$F_{()}(\underline{J}, \iota \underline{V}(f))v_{\underline{J}} = F_{()}(f, \iota \underline{V}(\underline{I}))v_{\underline{I}} \text{ for } \underline{I} \xrightarrow{f} \underline{J} \tag{11.5}$$

□

Proof Using (11.5) it is easy but tedious to check that V is a functor and that α is natural. We do this by making diagrams of homset operations, representing an element of a homset as a morphism from 1. Conversely, suppose we have a functor V making α natural. Then for each $\underline{I} \xrightarrow[A]{f} \underline{J}$ we have

$$\begin{array}{ccc}
\mathcal{D}_A(\iota \underline{VI}, \iota \underline{VI}) & \xrightarrow{\alpha_A(\underline{I}, \iota \underline{VI})} & F_A(\underline{I}, \iota \underline{VI}) \\
{\scriptstyle \mathcal{D}_A(\iota \underline{V}f, \iota \underline{VI})} \downarrow & & \downarrow {\scriptstyle F_A(f, \iota \underline{VI})} \\
\mathcal{D}_A(\iota \underline{VJ}, \iota \underline{VI}) & \xrightarrow[\alpha_A(\underline{J}, \iota \underline{VI})]{} & F_A(\underline{J}, \iota \underline{VI})
\end{array}$$

Applying this to the identity over A on $\iota \underline{VI}$ gives (11.5). □

Corollary 123 1 If $(\mathcal{C}, \mathcal{D}, \mathcal{O})$ is a CBPV judgement model with all left adjunctives, then F extends to a unique functor $\mathsf{self}\,\mathcal{C} \xrightarrow{F} \mathcal{D}$ making

$$\mathcal{D}_X(FA, \underline{Y}) \cong \mathcal{O}_{X \cdot A} \pi^* \underline{Y}$$

natural in A as well as X and $\underline{Y}$.

2 If $\mathcal{C}$ is cartesian, $\mathcal{D}$ is a locally $\mathcal{C}$-indexed category and $\{\mathcal{O}^p\}_{p \in P}$ a family of right $\mathcal{D}$-modules with all I-indexed products, then $\prod_{i \in I}$ extends to a unique functor $\mathcal{D}^I \xrightarrow{\prod_{i \in I}} \mathcal{D}$ making

$$\mathsf{hom}_X^{\mathcal{O}\mathcal{D}}(\underline{Y}, \textstyle\prod_{i \in I} \underline{B}_i) \cong \prod_{i \in I} \mathsf{hom}_X^{\mathcal{O}\mathcal{D}}(\underline{Y}, \underline{B}_i)$$

natural in all $\{\underline{B}_i\}_{i \in I}$ as well as in X and $\underline{Y}$.

3 If $\mathcal{C}$ is cartesian, $\mathcal{D}$ is a locally $\mathcal{C}$-indexed category and $\{\mathcal{O}^p\}_{p \in P}$ a family of right $\mathcal{D}$-modules with all exponentials, then $\rightarrow$ extends to a unique functor $(\mathsf{self}\,\mathcal{C})^{\mathsf{op}} \times \mathcal{D} \xrightarrow{\rightarrow} \mathcal{D}$ making

$$\mathsf{hom}_X^{\mathcal{O}\mathcal{D}}(\underline{Y}, A \rightarrow \underline{B}) \cong \mathsf{hom}_{X \cdot A}^{\mathcal{O}\mathcal{D}}(\pi^* \underline{Y}, \underline{B})$$

natural in A and $\underline{B}$ as well as in X and $\underline{Y}$.

4 If $\mathcal{C}$ is cartesian and $\mathcal{N} = \{\mathcal{N}^p\}_{p \in P}$ is a family of left $\mathcal{C}$-modules with all I-indexed distributive coproducts, then $\sum_{i \in I}$ extends to a unique functor $\mathsf{self}\,\mathcal{C}^I \xrightarrow{\sum} \mathsf{self}\,\mathcal{C}$ making

$$(\mathsf{hom}_{\times \mathcal{C}\mathcal{N}})_{X \cdot \sum_{i \in I} A_i} \underline{Y} \cong \prod_{i \in I} (\mathsf{hom}_{\times \mathcal{C}\mathcal{N}})_{X \cdot A_i} \underline{Y}$$

natural in $\{A_i\}_{i \in I}$ as well as in X and $\underline{Y}$.

□

There is one more parametrized representability theorem that we require, involving the special homset functor for cartesian categories that we defined in Sect. 11.3.3. Unfortunately, it does not appear to be a special case of Prop. 121 or Prop. 122; it would be interesting to find a common generalization of these results.

Proposition 124 Suppose $F : \mathsf{opGr}\, \mathfrak{I} \longrightarrow \mathbf{Set}$ is a functor. Suppose that for each $\underline{I} \in \mathfrak{I}$ there is a representing object for $F_{-}\underline{I}$, which we write in naturality style as an isomorphism

$$\mathcal{C}_X V(\underline{I}) \underset{\cong}{\xrightarrow{\alpha X \underline{I}}} F_X \underline{I} \quad \text{natural in } X$$

Then V extends uniquely to a functor from $\mathfrak{I}$ to $\mathsf{self}\,\mathcal{C}$ so as to make $\alpha_X \underline{I}$ natural in $\underline{I}$ as well as in X.

If we write the representing object of $F_{\underline{I}}$ in element style as $(V(\underline{I}), v_{\underline{I}})$ then the functor V is characterized by

$$(F_{V(f)}\underline{J})v_{\underline{J}} = (F_{\pi'}\pi^* f)v_{\underline{I}} \text{ for } \underline{I} \xrightarrow[A]{f} \underline{J} \tag{11.6}$$

□

Proof Using (11.6), it is easy but tedious to check that V is a functor and that α is natural. Conversely, suppose we have a functor V making α natural. Then for each $\underline{I} \xrightarrow[A]{f} \underline{J}$ we have

$$\begin{array}{ccc}
\mathcal{C}_{V\underline{I}} V\underline{I} & \xrightarrow{\alpha_{V\underline{I}}\underline{I}} & F_{V\underline{I}}\underline{I} \\
{\scriptstyle \mathcal{C}_{\pi'}\pi^* Vf}\downarrow & & \downarrow{\scriptstyle F_{\pi'}\pi^* f} \\
\mathcal{C}_{A\times V\underline{I}}\underline{J} & \xrightarrow[\alpha_{A\times V\underline{I}}\underline{J}]{} & F_{A\times V\underline{I}}\underline{J}
\end{array}$$

Applying this to the identity on $V\underline{I}$ gives the element-style characterization of Vf. □

Corollary 125 1 If $\mathcal{C}$ is a cartesian category with all I-indexed products, then $\prod_{i\in I}$ extends to a unique functor $\mathsf{self}\,\mathcal{C}^I \xrightarrow{\prod_{i\in I}} \mathsf{self}\,\mathcal{C}$ making

$$\mathcal{C}_X \textstyle\prod_{i\in I} B_i \cong \prod_{i\in I} \mathcal{C}_X B_i$$

natural in $\{B_i\}_{i\in I}$ as well as in X.

2 If $\mathcal{C}$ is a cartesian category with all exponentials, then $\to$ extends to a unique functor $(\mathsf{self}\,\mathcal{C})^{\mathsf{op}} \times \mathcal{C} \xrightarrow{\to} \mathcal{C}$ making

$$\mathcal{C}_X A \to B \cong \mathcal{C}_{X\cdot A} B$$

natural in A and B as well as in X.

3 If $(\mathcal{C}, \mathcal{D}, \mathcal{O})$ is a CBPV judgement model with all right adjunctives, then U extends to a unique functor $\mathcal{D} \xrightarrow{U} \mathsf{self}\,\mathcal{C}$ making

$$\mathcal{C}_X U\underline{B} \cong \mathcal{O}_X \underline{B}$$

natural in $\underline{B}$ as well as in X.

4 If $(\mathcal{C}, \mathcal{N})$ is a JWA judgement model with all jumpwiths, then $\neg$ extends to a unique functor $(\mathsf{self}\, \mathcal{C})^{\mathsf{op}} \xrightarrow{\neg} \mathsf{self}\, \mathcal{C}$ making

$$\mathcal{C}_X \neg A \cong \mathcal{N}_{X \cdot A}$$

natural in A as well as X.

□

11.5 Possible Worlds: Part 2

In Sect. 9.7.6 we built various locally $\mathcal{C}$-indexed functors between $\mathsf{self}\, \mathcal{C}$ and $\mathcal{D}$. They restrict to functors between $\mathsf{self}\, \mathcal{C}_1$ and $\mathcal{D}_1$. Modulo the isomorphism j from $\mathcal{C}$ to $\mathsf{self}\, \mathcal{C}_1$, these are functors between $\mathcal{C}$ and $\mathcal{D}_1$, which is what we require for possible worlds.

For the CBPV possible world construction, we define the connectives at a morphism $w \xrightarrow{g} x$ as follows.

- $(U\underline{B})g$ is obtained by obtained by applying the functor $\mathcal{D}_1 \xrightarrow{U} \mathcal{C}$ to the $\mathcal{D}_1$-morphism

$$\prod_{(w',f) \in w/\mathcal{W}} (Sw' \to \underline{B}w') \xrightarrow{h} \prod_{(x',f) \in x/\mathcal{W}} (Sx' \to \underline{B}x')$$

  such that $h; \pi_{(x',f)} = \pi_{(x',(g;f))}$.

- $(FA)g$ is obtained by applying the functor $\mathcal{C} \xrightarrow{F} \mathcal{D}_1$ to the $\mathcal{C}$-morphism

$$\sum_{(x',f) \in x/\mathcal{W}} (Sx' \times Ax') \xrightarrow{h} \sum_{(w',f) \in w/\mathcal{W}} (Sw' \times Aw')$$

  such that $\mathsf{in}_{(x',f)}; h = \mathsf{in}_{(x',(g;f))}$.

- $(\sum_{i \in I} A_i)g = \sum_{i \in I} A_i g$ using $\mathcal{C}^I \xrightarrow{\sum_{i \in I}} \mathcal{C}$.

- $(A \times A')g$ is defined to be $Ag \times A'g$ using $\mathcal{C} \times \mathcal{C} \xrightarrow{\times} \mathcal{C}$.

- $(\prod_{i \in I} \underline{B}_i)g$ is defined to be $\prod_{i \in I} \underline{B}_i g$ using $\mathcal{D}_1^I \xrightarrow{\prod_{i \in I}} \mathcal{D}_1$.

- $(A \to \underline{B})g$ is defined to be $Ag \to \underline{B}g$ using $\mathcal{C}^{\mathsf{op}} \times \mathcal{D}_1 \xrightarrow{\to} \mathcal{D}_1$

For the JWA possible world construction, we define the connectives at a morphism $w \xrightarrow{g} x$ as follows.

- $\neg A$ at $w \xrightarrow{g} x$ is obtained by applying $\mathcal{C}^{\mathsf{op}} \xrightarrow{\neg} \mathcal{C}$ to the morphism

$$\textstyle\sum_{(x',f)\in x/\mathcal{W}}(Sx' \times Ax') \xrightarrow{h} \sum_{(w',f)\in w/\mathcal{W}}(Sw' \times Aw')$$

such that $\mathsf{in}_{(x',f)}; h = \mathsf{in}_{(x',(g;f))}$.

- $(\sum_{i\in I} A_i)g$ is defined to be $\sum_{i\in I} A_i g$ using $\mathcal{C}^I \xrightarrow{\sum_{i\in I}} \mathcal{C}$.
- $(A \times A')g$ is defined to be $Ag \times A'g$ using $\mathcal{C} \times \mathcal{C} \xrightarrow{\times} \mathcal{C}$.

Defining all the homset operations in the possible world model and proving that all the required equations hold can be done by expressing them as diagrams of homsets in the original model. We use all the naturalities in Prop. 120, Cor. 123 and Cor. 125 together with the equations

$$\begin{aligned} \mathcal{C}_f()^* jg &= \mathcal{C}(f,g) \\ f \cdot ()^* jg &= f \times g \end{aligned}$$

An alternative approach is to work in the term calculus, exploiting our knowledge that all the equations of CBPV + stacks are validated in a CBPV adjunction model, and all the equations of JWA are validated in a JWA module model.

11.6 Adjunctions and Monads

11.6.1 Definitions of Adjunction

Ordinary Adjunctions

Before we prove Prop. 102, we look at a variety of definitions of adjunction between ordinary categories, all of them equivalent to Def. 9.33.

Definition 11.17 Let $\mathcal{B}$ and $\mathcal{D}$ be categories. We underline objects of $\mathcal{D}$.

1 An *adjunction* from $\mathcal{B}$ to $\mathcal{D}$ consists of

- two functors

$$\mathcal{B} \underset{U}{\overset{F}{\rightleftarrows}} \mathcal{D}$$

- an isomorphism

$$\mathcal{B}(Y, U\underline{Z}) \cong \mathcal{D}(FY, \underline{Z}) \text{ natural in } Y \text{ and } \underline{Z}.$$

2 An *adjunction* from $\mathcal{B}$ to $\mathcal{D}$ consists of

- a functor $\mathcal{D} \xrightarrow{U} \mathcal{B}$
- for each $Y \in \mathsf{ob}\ \mathcal{B}$, a representing object for $\mathcal{B}(Y, U-)$, whose vertex we call FY.

3 An *adjunction* from $\mathcal{B}$ to $\mathcal{D}$ consists of

- a functor $\mathcal{B} \xrightarrow{F} \mathcal{D}$
- for each $\underline{Z} \in \mathsf{ob}\ \mathcal{D}$ a representing object for $\mathcal{D}(F-, \underline{Z})$, whose vertex we call $U\underline{Z}$.

4 An *adjunction* from $\mathcal{B}$ to $\mathcal{D}$ consists of

- two functions

$$\mathsf{ob}\ \mathcal{B} \underset{U}{\overset{F}{\rightleftarrows}} \mathsf{ob}\ \mathcal{D}$$

- for each $X \in \mathsf{ob}\ \mathcal{B}$ and $\underline{Z} \in \mathsf{ob}\ \mathcal{D}$ a bijection

$$\mathcal{B}(Y, U\underline{Z}) \cong \mathcal{D}(FY, \underline{Z})$$

such that, for each $\mathcal{B}$-morphism $Y' \xrightarrow{f} Y$ and each $\mathcal{C}$-morphism $\underline{Z} \xrightarrow{g} \underline{Z}'$, the following commutes:

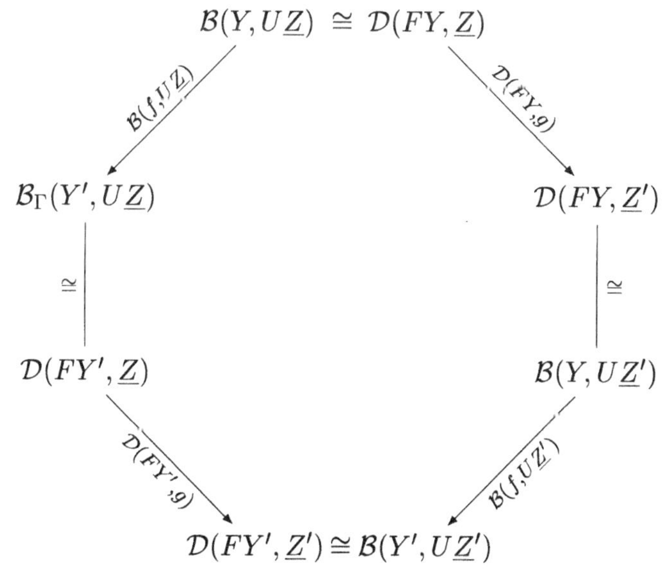

5 An *adjunction* from $\mathcal{B}$ to $\mathcal{D}$ consists of

- a $(\mathcal{B},\mathcal{D})$-bimodule i.e. a functor

$$\mathcal{M} : \mathcal{B}^{\mathsf{op}} \times \mathcal{D} \to \mathbf{Set}$$

- for each object $\underline{Z} \in \mathcal{D}$, a representing object for $\mathcal{M}(-, \underline{Z})$, whose vertex we call $U\underline{Z}$
- for each object $Y \in \mathcal{B}$, a representing object for $\mathcal{M}(Y, -)$, whose vertex we call FY.

□

We compare these definitions as follows.

- In Def. 11.17(1)—like in Def. 9.33—both F and U are given on both objects and morphisms. Accordingly, the isomorphism

$$\mathcal{B}(Y, U\underline{Z}) \cong \mathcal{D}(FY, \underline{Z}) \tag{11.7}$$

  is required to be natural in both Y and $\underline{Z}$.

- In Def. 11.17(2), U is given on both objects and morphisms but F is given only on objects. Accordingly, the isomorphism (11.7) is required to be natural in $\underline{Z}$ but not in Y.
- In Def. 11.17(3), F is given on both objects and morphisms but U is given only on objects. Accordingly, the isomorphism (11.7) is required to be natural in Y but not in $\underline{Z}$.
- in Def. 11.17(4)—(5), both U and F are given only on objects. Accordingly, in Def. 11.17(4) the isomorphism (11.7) is not required to be natural in either Y or $\underline{Z}$, while in Def. 11.17(5), it is divided into two isomorphisms:

$$\begin{array}{rcll} \mathcal{B}(Y, U\underline{Z}) & \cong & \mathcal{M}(Y, \underline{Z}) & \text{natural in } Y \\ \mathcal{M}(Y, \underline{Z}) & \cong & \mathcal{D}(FY, \underline{Z}) & \text{natural in } \underline{Z} \end{array}$$

Proposition 126 Def. 11.17(1)–(5) and Def. 9.33 are all equivalent. □

Proof The equivalence of Def.9.33) and Def. 11.17(1) is standard. The equivalence of (1)–(3) and (5) is a consequence of parametrized representability (Prop. 121). In moving from (1) to (5) we can either set $\mathcal{M}(Y, \underline{Z})$ to be $\mathcal{B}(Y, U\underline{Z})$ or we can set it to be $\mathcal{D}(FY, \underline{Z})$. The equivalence of (4) and (5) is straightforward. □

Locally Indexed Adjunctions

Each of these definitions can be adapted to locally $\mathcal{C}$-indexed adjunctions.

Definition 11.18 (cf.11.17) Let $\mathcal{B}$ and $\mathcal{D}$ be locally $\mathcal{C}$-indexed categories. We underline objects of $\mathcal{D}$.

1 An *adjunction* from $\mathcal{B}$ to $\mathcal{D}$ consists of

- two functors

$$\mathcal{B} \underset{U}{\overset{F}{\rightleftarrows}} \mathcal{D}$$

- a bijection

$$\mathcal{B}_\Gamma(Y, U\underline{Z}) \cong \mathcal{D}_\Gamma(FY, \underline{Z}) \quad \text{natural in } \Gamma, Y \text{ and } \underline{Z}.$$

Note In the literature, the condition of naturality in Γ is usually replaced by the *Beck-Chevalley condition*. The two conditions are equivalent (assuming naturality in Y and $\underline{Z}$), but we use the former because we consider it to be more intuitive.

2 An *adjunction* from $\mathcal{B}$ to $\mathcal{D}$ consists of

- a functor $\mathcal{D} \xrightarrow{U} \mathcal{B}$
- for each $Y \in \mathsf{ob}\ \mathcal{B}$, a representing object for $\mathcal{B}_-(Y, U-)$, whose vertex we call FY.

3 An *adjunction* from $\mathcal{B}$ to $\mathcal{D}$ consists of

- a functor $\mathcal{B} \xrightarrow{F} \mathcal{D}$
- for each $\underline{Z} \in \mathsf{ob}\ \mathcal{D}$ a representing object for $\mathcal{D}_-(F-, \underline{Z})$, whose vertex we call $U\underline{Z}$.

4 An *adjunction* from $\mathcal{B}$ to $\mathcal{D}$ consists of

- two functions

$$\mathsf{ob}\ \mathcal{B} \underset{U}{\overset{F}{\rightleftarrows}} \mathsf{ob}\ \mathcal{D}$$

- for each $Y \in \mathsf{ob}\ \mathcal{B}$ and $\underline{Z} \in \mathsf{ob}\ \mathcal{D}$ a bijection

$$\mathcal{B}_X(Y, U\underline{Z}) \cong \mathcal{D}_X(FY, \underline{Z}) \text{ natural in } X$$

such that, for each $\mathcal{B}$-morphism $Y' \xrightarrow[X]{f} Y$ and each $\mathcal{C}$-morphism $\underline{Z} \xrightarrow[X]{g} \underline{Z}'$, the following commutes:

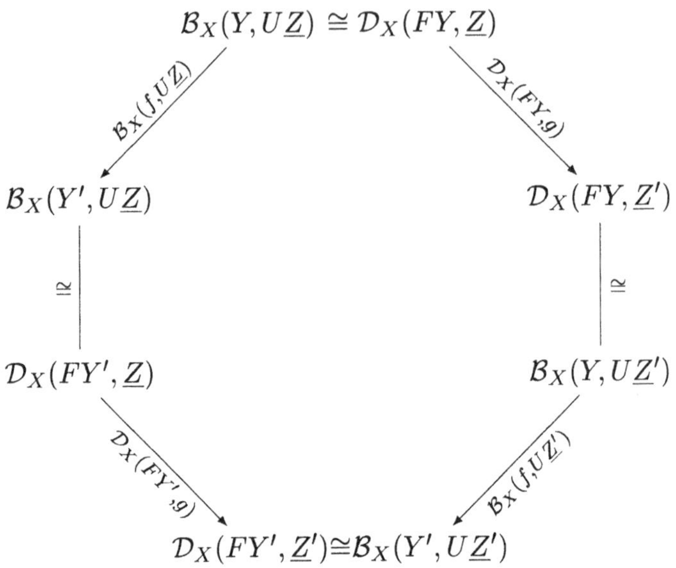

5 An *adjunction* from $\mathcal{B}$ to $\mathcal{D}$ consists of

- a ($\mathcal{B}$,$\mathcal{D}$)-bimodule i.e. a functor

$$\mathcal{M} : \mathsf{opGr}(\mathcal{B}^{\mathsf{op}} \times \mathcal{D}) \to \mathbf{Set}$$

- for each object $\underline{Z} \in \mathcal{D}$, a representing object for $\mathcal{M}_{-}(-, \underline{Z})$, whose vertex we call $U\underline{Z}$
- for each object $Y \in \mathcal{B}$, a representing object for $\mathcal{M}_{-}(Y, -)$, whose vertex we call FY.

□

Strong Adjunctions

We now prove Prop. 102, using Def. 11.18(2) to define adjunction between $\mathsf{self}\,\mathcal{C}$ and $\mathcal{D}$.

Proof Given a strong adjunction from $\mathcal{C}$ to $\mathcal{D}$, we construct U as in Cor. 125(3). Given a $\mathcal{C}$-object A, we construct an isomorphism

$$\mathcal{D}_X(FA, \underline{Y}) \cong \mathsf{self}\,\mathcal{C}_X(A, \underline{Y}) \quad \text{natural in } X \text{ and } \underline{Y}$$

as the composite

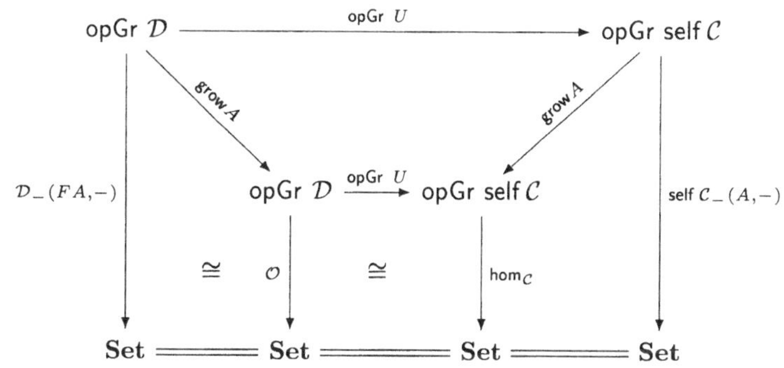

The right trapezium commutes by Prop. 119.

Conversely, given an adjunction in the sense of Def. 11.18(2) from self $\mathcal{C}$ to $\mathcal{D}$, define $\mathcal{O}$ to be opGr U; hom$_\mathcal{C}$. This immediately gives the required structure.

It is clear that these two transformations are inverse up to isomorphism. □

11.6.2 Terminality of Eilenberg-Moore

In Sect. 9.7.2 we saw that, given a strong monad (T, η, μ, t) on a cartesian category $\mathcal{C}$, we can construct a strong adjunction of T-algebras and algebra homomorphisms. We write EM(T) for this strong adjunction; it has the following properties.

Proposition 127 1 Let (T, η, μ, t) be a strong monad on $\mathcal{C}$. Then EM(T) is a resolution of it.

2 Let $(\mathcal{C}, \mathcal{D}, \mathcal{O})$ be a strong adjunction. Write (T, η, μ, t) for the strong monad it gives rise to.

- There is a unique comparison functor K (in the sense of Sect. 9.6.8) from $(\mathcal{C}, \mathcal{D}, \mathcal{O})$ to EM(T). It takes
  - a $\mathcal{D}$-object $\underline{Y}$ to the algebra whose carrier is $U\underline{Y}$ and whose structure map is

$$UFU\underline{Y} \xrightarrow{\cong} 1 \times UFU\underline{Y} \xrightarrow{U\epsilon_{\underline{Y}}} U\underline{Y}$$

  - a $\mathcal{D}$-morphism $\underline{Y} \xrightarrow[X]{h} \underline{Z}$ to $X \times U\underline{Y} \xrightarrow{Uh} U\underline{Z}$
  - an $\mathcal{O}$-morphism $\xrightarrow[X]{g} \underline{Z}$ to the corresponding morphism $X \longrightarrow U\underline{Z}$.

Therefore $\mathsf{EM}(T)$ is terminal in the category of resolutions for (T, η, μ, t).

- Any distributive coproduct in $(\mathcal{D}, \mathcal{O})$ of $\{A_i\}_{i \in I}$ is mapped by K to a distributive coproduct in $\mathsf{EM}(T)$ of $\{A_i\}_{i \in I}$.
- Any product in $(\mathcal{D}, \mathcal{O})$ of $\{\underline{B}_i\}_{i \in I}$ is mapped by K to a product in $\mathsf{EM}(T)$ of $\{K\underline{B}_i\}_{i \in I}$.
- Any exponential in $(\mathcal{D}, \mathcal{O})$ from A to $\underline{B}$ is mapped by K to an exponential in $\mathsf{EM}(T)$ from A to $K\underline{B}$.

□

Proof This is straightforward if we assume, without loss of generality by Prop. 101, that the U isomorphism

$$\mathcal{C}_X U \underline{Y} \cong \mathcal{O}_X \underline{Y}$$

is the identity. □

It follows from this that if $(\mathcal{D}, \mathcal{O})$ is a CBPV adjunction model, then $\mathsf{EM}(T)$ will have all countable distributive coproducts, and it will have all finite products of comp-objects by Prop. 103(2). If all idempotents split in $\mathcal{C}$, then Cor. 105 tells us that $\mathsf{EM}(T)$ must be a CBPV adjunction model. Otherwise, it might not be.

11.6.3 Kleisli and Co-Kleisli Adjunctions

Ordinary And Locally Indexed Adjunctions

Definition 11.19 An (ordinary or locally $\mathcal{C}$-indexed) adjunction $(\mathcal{B}, \mathcal{D}, \mathcal{M}, U, F)$ is *Kleisli* iff $\mathsf{ob}\ \mathcal{B} = \mathsf{ob}\ \mathcal{D}$ and the left adjoint F is identity-on-objects. It is *co-Kleisli* iff $\mathsf{ob}\ \mathcal{B} = \mathsf{ob}\ \mathcal{D}$ and the right adjoint U is identity-on-objects. □

Proposition 128 A Kleisli (co-Kleisli) adjunction is initial in the category of resolutions for its monad (comonad), and all comparison functors from it are fully faithful. Furthermore, every monad (comonad) has a Kleisli (co-Kleisli) resolution. □

This is standard e.g. [Mac Lane, 1971].

Definition 11.20 Let $(\mathcal{B}, \mathcal{D}, \mathcal{M}, U, F)$ be an (ordinary or locally $\mathcal{C}$-indexed) adjunction in the manner of Def. 11.17(5) or Def. 11.18(5). We define its *Kleisli part* $(\mathcal{B}, \mathcal{D}', \mathcal{M}', U', F')$ to be the Kleisli adjunction given by

$$\begin{aligned} \mathsf{ob}\,\mathcal{D}' &= \mathsf{ob}\,\mathcal{B} \\ \mathcal{D}'(A,B) &= \mathcal{D}(FA,FB) \\ \mathcal{M}'(A,B) &= \mathcal{M}(A,FB) \\ F'A &= A \\ U'B &= UFB \end{aligned}$$

with its identities, composition, isomorphisms etc. all inherited from $(\mathcal{B},\mathcal{D},\mathcal{M},U,F)$. Dually, we define its *co-Kleisli part* $(\mathcal{B}',\mathcal{D},\mathcal{M}',U',F')$ to the co-Kleisli adjunction given by

$$\begin{aligned} \mathsf{ob}\,\mathcal{B}' &= \mathsf{ob}\,\mathcal{D} \\ \mathcal{B}'(C,D) &= \mathcal{B}(UC,UD) \\ \mathcal{M}'(C,D) &= \mathcal{M}(UC,D) \\ U'D &= D \\ F'C &= FUC \end{aligned}$$

with all structure inherited. □

Proposition 129 An adjunction $(\mathcal{B},\mathcal{D},\mathcal{M},U,F)$ and its Kleisli part $(\mathcal{C},\mathcal{D},\mathcal{M},U',F')$ give rise to the same monad. The unique comparison functor is given by F on objects, and the identity on morphisms. Similarly for the co-Kleisli part. □

Given an adjunction $(\mathcal{B},\mathcal{D},\mathcal{M},U,F)$ we thus obtain the following diagram showing the Kleisli and co-Kleisli parts, and their embeddings.

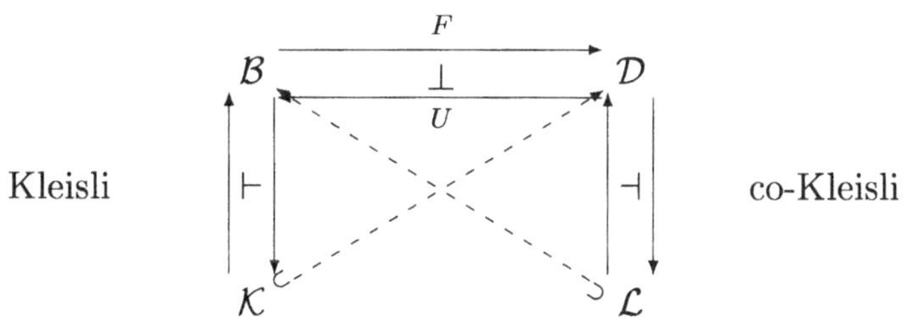

Strong Adjunctions

We shall see that the notions of Kleisli adjunction, co-Kleisli adjunction and Kleisli part apply straightforwardly to strong adjunctions, but not the notion of co-Kleisli part.

Definition 11.21 1 A strong adjunction $(\mathcal{C},\mathcal{D},\mathcal{O},U,F)$ from $\mathcal{C}$ to $\mathcal{D}$ is *Kleisli* iff $\mathsf{ob}\,\mathcal{C} = \mathsf{ob}\,\mathcal{D}$ and the left adjoint F is identity-on-objects.

It is *co-Kleisli* iff $\mathsf{ob}\ \mathcal{C} = \mathsf{ob}\ \mathcal{D}$ and the right adjoint U is identity-on-objects.

2 Let $(\mathcal{C}, \mathcal{D}, \mathcal{O}, U, F)$ be a strong adjunction. We define its *Kleisli part* $(\mathcal{C}, \mathcal{D}', \mathcal{O}', U', F')$ to be the Kleisli strong adjunction given by

$$\begin{array}{rcl} \mathsf{ob}\ \mathcal{D}' & = & \mathsf{ob}\ \mathcal{C} \\ \mathcal{D}'_A(B, C) & = & \mathcal{D}_A(FB, FC) \\ \mathcal{O}'_A B & = & \mathcal{O}_A FB \\ F'A & = & A \\ U'B & = & UFB \end{array}$$

with identities, composition, reindexing, isomorphisms etc. inherited from $(\mathcal{C}, \mathcal{D}, \mathcal{O}, U, F)$.

□

These agree with the corresponding notions for locally indexed adjunctions.

Proposition 130 1 A strong adjunction from $\mathcal{C}$ is Kleisli (resp. co-Kleisli) if the corresponding strong adjunction from $\mathsf{self}\,\mathcal{C}$ is Kleisli (resp. co-Kleisli).

2 If the strong adjunction $(\mathcal{C}, \mathcal{D}, \mathcal{O})$ corresponds to the locally $\mathcal{C}$-indexed adjunction $(\mathsf{self}\,\mathcal{C}, \mathcal{D}, \mathcal{M})$ then their Kleisli parts correspond too.

□

But in the co-Kleisli part $(\mathcal{B}', \mathcal{D}, \mathcal{M}')$ of an adjunction $(\mathsf{self}\,\mathcal{C}, \mathcal{D}, \mathcal{M})$, the category $\mathcal{B}'$ is not of the form $\mathsf{self}\,\mathcal{C}'$, and so this part does not correspond to a strong adjunction. We therefore must define the co-Kleisli part of a strong adjunction somewhat differently.

Definition 11.22 Let $(\mathcal{C}, \mathcal{D}, \mathcal{O}, U, F)$ be a strong adjunction, where $\mathcal{D}$—and hence $(\mathcal{D}, \mathcal{O})$ by Prop. 97—has finite products. We define the *co-Kleisli part* of this strong adjunction to be the co-Kleisli strong adjunction $(\mathcal{C}', \mathcal{D}', \mathcal{O}', U', F')$ given by

$$\begin{array}{rcl} \mathsf{ob}\ \mathcal{C}' & = & \mathsf{ob}\ \mathcal{D} \\ \mathcal{C}'_{\underline{A}}\underline{B} & = & \mathcal{C}_{U\underline{A}}U\underline{B} \\ \mathcal{D}'_{\underline{A}}(\underline{B}, \underline{C}) & = & \mathcal{D}_{U\underline{A}}(\underline{B}, \underline{C}) \\ \mathcal{O}'_{\underline{A}}\underline{B} & = & \mathcal{O}_{U\underline{A}}\underline{B} \\ U'\underline{B} & = & \underline{B} \\ F'\underline{A} & = & FU\underline{A} \end{array}$$

with the cartesian structure on $\mathcal{C}'$ given by the finite products on $\mathcal{D}$, and all structure inherited. □

11.6.4 CBV Is Kleisli, CBN Is Co-Kleisli

We recall that the translation of a CBV term into CBPV always has the form $\Gamma \vdash^{\mathsf{c}} M : FA$. So, in a given CBPV model, it denotes an $\mathcal{O}$-morphism g over $[\![\Gamma]\!]$ to $F[\![A]\!]$, which lies in the *Kleisli part*.

On the other hand, the translation of a CBN term into CBPV always has the form $U\underline{A}_0, \ldots, U\underline{A}_{n-1} \vdash^{\mathsf{c}} M : \underline{B}$. In a given CBPV model, it denotes an $\mathcal{O}$-morphism g over $U[\![\underline{A}_0]\!] \times \cdots \times U[\![\underline{A}_{n-1}]\!]$ to $[\![\underline{B}]\!]$. Using the isomorphism $U\underline{A} \times U\underline{A}' \cong U(\underline{A} \sqcap \underline{A}')$, this morphism g corresponds to a morphism g' over $U[\![\Gamma]\!]$ to $[\![\underline{B}]\!]$, where we write $[\![\Gamma]\!]$ for $[\![\underline{A}_0]\!] \sqcap \cdots \sqcap [\![\underline{A}_{n-1}]\!]$. This morphism g' lies in the *co-Kleisli part*.

This suggests that we can formulate categorical semantics for CBV and CBN by trying to characterize the Kleisli part and the co-Kleisli part of a CBPV adjunction model. Of course, to see how these are genuinely suitable for CBV and CBN, we would have to introduce stack terms just as we did for CBPV.

Definition 11.23 A ($\mathcal{A}$-set/cppo/domain enriched) *CBV adjunction model* is a ($\mathcal{A}$-set/cppo/domain enriched) CBPV judgement model $(\mathcal{C}, \mathcal{D}, \mathcal{O})$ where $\mathcal{D}$ and $\mathcal{C}$ have the same objects (which we do not underline), equipped with

- a left adjunctive for each object A, whose vertex is A
- all countable[1] right exponential-product-adjunctives, whose vertex we write as $\prod^{\rightarrow}_{i \in I} A_i B_i$
- all countable distributive coproducts.

□

In such a model, the distributive coproducts model sum types in CBV, while the countable right exponential-product-adjunctives model a less familiar connective, a special case of the general function type considered in Sect. A.1.3. Briefly, $\prod^{\rightarrow}_{i \in I} A_i B_i$ is a type in which a value is a λ-abstraction that takes two operands—a tag $i \in I$ and a value of type A_i—and returns a value of type B_i.

Definition 11.24 A ($\mathcal{A}$-set/cppo/domain enriched) *CBN adjunction model* is a ($\mathcal{A}$-set/cppo/domain enriched) CBPV judgement model $(\mathcal{C}, \mathcal{D}, \mathcal{O})$ where $\mathcal{D}$ and $\mathcal{C}$ have the same objects (which we underline), equipped with

[1] If we are modelling finitely wide syntax only, it suffices to require all right exponential-adjunctives, as in [Levy et al., 2003]. This automatically gives us all finite right exponential-product-adjunctives, using the finite products in $\mathcal{C}$.

- a right adjunctive for each object $\underline{A}$, whose vertex is $\underline{A}$
- all countable left coproduct-adjunctives, whose vertex we write as $\sum_{i \in I} \underline{A}_i$
- all countable products and exponentials.

□

In such a model, the left coproduct-adjunctives model sum types in CBN.

We now see that the Kleisli and co-Kleisli parts of a CBPV model do indeed have these structures.

Proposition 131 1 A CBV adjunction model is a Kleisli strong adjunction. The Kleisli part of any ($\mathcal{A}$-set/cppo/domain enriched) CBPV adjunction model $(\mathcal{C}, \mathcal{D}, \mathcal{O})$ is a ($\mathcal{A}$-set/cppo/domain enriched) CBV adjunction model.

2 A CBN adjunction model is a co-Kleisli strong adjunction. The co-Kleisli part of any ($\mathcal{A}$-set/cppo/domain enriched) CBPV adjunction model $(\mathcal{C}, \mathcal{D}, \mathcal{O})$—which is well defined because $\mathcal{D}$ must have finite products—is a ($\mathcal{A}$-set/cppo/domain enriched) CBN adjunction model.

□

This is summarized by the slogan "CBV is Kleisli, CBN is co-Kleisli". A special case is where the CBPV adjunction model is a continuation model, obtained from the JWA module model $(\mathcal{C}, \mathcal{N})$. In this case,

- the denotation of a CBV term is a $\mathcal{N}$-morphism from $[\![\Gamma]\!] \times \neg[\![A]\!]$
- the denotation of a CBN term is a $\mathcal{N}$-morphism from $\neg[\![\underline{\Gamma}]\!] \times [\![\underline{B}]\!]$.

This is described in [Streicher and Reus, 1998] as "duality between CBV and CBN, with control operators", and the theme is further developed in [Curien and Herbelin, 2000; Selinger, 2001; Wadler, 2003]. It is a consequence of the duality between values and stacks, and between F and U, in a continuation model.

Families Constructions From CBV And CBN Models

Given a CBN model or CBV model, it is possible to construct a CBPV model using an appropriate families construction, as we now see. We treat the CBN case first [Abramsky and McCusker, 1998], as it is simpler and more common. The idea is that, since every value type in CBPV is isomorphic to one of the form $\sum_{i \in I} U\underline{B}_i$, we can simulate a val-object as a family of comp-objects.

Definition 11.25 Let $(\mathcal{C}, \mathcal{D}, \mathcal{O})$ be a ($\mathcal{A}$-set/cppo/domain enriched) CBN adjunction model. We define a ($\mathcal{A}$-set/cppo/domain enriched) CBPV adjunction model $(\mathcal{C}', \mathcal{D}', \mathcal{O}')$ as follows.

- A val-object is a countable family $\{\underline{A}\}_{i \in I}$ of objects of the CBN model—to be thought of as $\sum_{i \in I} U\underline{A}_i$.
- A comp-object is an object $\underline{B}$ of the CBN model.

The homsets are given by

$$\begin{array}{rcl} \mathcal{C}'_{\{\underline{A}_i\}_{i \in I}} \{\underline{B}_j\}_{j \in J} & = & \prod_{i \in I} \sum_{j \in J} \mathcal{C}_{\underline{A}_i} \underline{B}_j \\ \mathcal{D}'_{\{\underline{A}_i\}_{i \in I}} (\underline{B}, \underline{C}) & = & \prod_{i \in I} \mathcal{D}_{\underline{A}_i} (\underline{B}, \underline{C}) \\ \mathcal{O}'_{\{\underline{A}_i\}_{i \in I}} \underline{B} & = & \prod_{i \in I} \mathcal{O}_{\underline{A}_i} \underline{B}_j \end{array}$$

with the evident identities, composition and reindexing. The connectives are given by

$$\begin{array}{rcl} 1 & = & \{1\} \\ \{\underline{A}_i\}_{i \in I} \times \{\underline{B}_j\}_{j \in J} & = & \{\underline{A}_i \times \underline{B}_j\}_{(i,j) \in I \times J} \\ \sum_{i \in I} \{\underline{A}_{ij}\}_{j \in J_i} & = & \{\underline{A}_{ij}\}_{(i,j) \in \sum_{i \in I} J_i} \\ \{\underline{A}_i\}_{i \in I} \to \underline{B} & = & \prod_{i \in I} (\underline{A}_i \to \underline{B}) \\ U\underline{B} & = & \{\underline{B}\} \\ F\{\underline{A}_i\}_{i \in I} & = & \sum_{i \in I} \underline{A}_i \end{array}$$

and comp-object products as in the CBN model. □

Clearly the co-Kleisli part of such a CBPV model is isomorphic to our original CBN adjunction model.

The predomain/domain CBPV model is an instance of this construction, because we defined a predomain to be any cpo that is a countable disjoint union of domains. This construction is used also in [Abramsky and McCusker, 1998] to build Q/A-labelled pointer games. Indeed, given a CBPV pre-families model, we can first obtain a CBN adjunction model and then a CBPV adjunction model, and this composite construction is the same as Def. 9.41.

The construction from a CBV model is more complicated. The idea is that, since every computation type in CBPV is isomorphic to one of the form $\prod_{i \in I}(A_i \to FB_i)$—albeit not uniquely—we can simulate a comp-object as a family of pairs of val-objects.

Definition 11.26 Let $(\mathcal{C}, \mathcal{D}, \mathcal{O})$ be a ($\mathcal{A}$-set/cppo/domain enriched) CBV adjunction model. We define a ($\mathcal{A}$-set/cppo/domain enriched) CBPV adjunction model $(\mathcal{C}', \mathcal{D}', \mathcal{O}')$ as follows.

- A val-object is an object A of the CBV model.
- A comp-object is a countable family $\{(A_i, B_i\}_{i \in I}$ of pairs of objects of the CBV model—to be thought of as $\prod_{i \in I}(A_i \to FB_i)$.

The homsets are given by

$$\begin{array}{rcl}
\mathcal{C}'_A B & = & \mathcal{C}_A B \\
\mathcal{D}'_A(\{(B_i, C_i)\}_{i \in I}, \{(D_j, E_j)\}_{j \in J}) & = & \prod_{j \in J} \sum_{i \in I} (\mathcal{C}_{A \times D_j} B_i \times \mathcal{D}_{A \times D_j}(C_i, E_j)) \\
\mathcal{O}'_A\{(D_j, E_j)\}_{j \in J} & = & \prod_{j \in J} \mathcal{O}_{A \times D_j} E_j
\end{array}$$

with the obvious identities, composition and reindexing. The connectives are given by

$$\begin{array}{rcl}
A \to \{(B_j, C_j)\}_{j \in J} & = & \{(A \times B_j, C_j)\}_{j \in J} \\
\prod_{i \in I}\{(B_{ij}, C_{ij})\}_{j \in J_i} & = & \{(B_{ij}, C_{ij})\}_{(i,j) \in \sum_{i \in I} J_i} \\
U\{(B_i, C_i)\}_{i \in I} & = & \{\prod^{\rightarrow}_{i \in I} B_i C_i\} \\
FA & = & \{(1, A)\}
\end{array}$$

Products and distributive coproducts of val-objects are as in the CBV model. □

Clearly the Kleisli part of this CBPV model is isomorphic to our original CBV adjunction model.

IV

CONCLUSIONS

Chapter 12

CONCLUSIONS, COMPARISONS AND FURTHER WORK

12.1 Summary of Achievements and Drawbacks

In Part I, we introduced the CBPV paradigm as a language of semantic primitives for higher order programming with computational effects (even just divergence). In Part II we saw a vast range of semantics that support CBPV's claim to be such a language, subsuming both CBV and CBN. That range includes models for printing, storage, divergence, erratic choice, errors and control effects; it also includes possible world models and interaction-based semantics such as jumping (using continuation semantics) and pointer game models. Again and again, we saw the advantages of using CBPV as a language of study. For example, in the interaction-based semantics, the explicit control flow in CBPV makes it closer to the detailed behaviour present in the model than CBN or CBV are.

However, there were some examples of CBV models that we were unable to decompose into CBPV in a natural way:

- the model for input based on Moggi's input monad [Moggi, 1991]
- the constrained model for finite erratic choice based on the strong monad $\mathcal{P}_{\mathsf{fin}}$ on **Set**, described in Sect. 5.5.3;
- the constrained model for store that incorporates parametricity in initializations, mentioned in Sect. 6.9.

But see Sect. 12.2 for further remarks on this.

In Part III, we saw that all of our semantics for CBPV are instances of adjunction models. It is therefore reasonable to say that what CBPV achieves is to decompose Moggi's strong monads into strong adjunctions.

In the PhD version of this monograph, that claim was problematic because of the absence of stack terms. However, it has now been validated through our incorporation of stacks into the equational theory.

12.2 Simplifying Algebra Semantics

Whilst we said that the CBV models for finite erratic choice and input do not have a natural CBPV decomposition, recent work of [Plotkin and Power, 2002] suggests a more optimistic conclusion. For many effects, including these two, the relevant strong monad on **Set** (and, sometimes, on other categories too) can be presented using *algebraic operations and equations*; and this leads to a simple characterization of algebras. Thus, for these effects, the algebra semantics for CBPV appears more natural in the light of this research than it did previously.

12.3 Advantages Of CBPV Over Monad And Linear Logic Decompositions

We contrast the success of the CBPV decompositions of CBV and CBN type constructors with 2 other well-known decompositions:

- Moggi's decomposition of $A \to_{\mathbf{CBV}} B$ as $A \to TB$ by contrast with our $U(A \to FB)$
- the linear logic decomposition of $\underline{A} \to_{\mathbf{CBN}} \underline{B}$ as $!\underline{A} \multimap \underline{B}$ by contrast with our $(U\underline{A}) \to \underline{B}$.

Both these decompositions/translations are given in [Benton and Wadler, 1996].

For discussion's sake, we will say that the types of Moggi's target language include

$$A ::= \ \mathtt{bool} \mid \mathtt{nat} \mid A \times A \mid A \to A \mid TA \mid \cdots$$

although Moggi does not explicitly include bool and nat. In categorical terms, Moggi's decomposition of $\to_{\mathbf{CBV}}$ is isomorphic to ours, because $U(A \to FB)$ must be the vertex of an exponent from A to TB. But much is lost in Moggi's translation, as we now explain.

- As we said in Sect. I.2.3, the monadic language does not have an operational semantics (at least not a simple one in the same way as CBV, CBN and CBPV) because it is a language of values.
- Although the monadic language does have a denotational semantics using cpos, in which $A \to B$ denotes the cpo of total functions from $[\![A]\!]$ to $[\![B]\!]$, this is not a predomain semantics (Def. 4.8). For example, $\mathtt{nat} \to \mathtt{nat}$ denotes an uncountable flat cpo, which is not a predomain.

- We cannot add general type recursion to the monadic language, because for example we could not interpret $\mu \mathtt{X}.(\mathtt{X} \to \mathtt{bool})$—there is no cpo X such that $X \cong X \to 2$. The usual technology for recursively defined cpos (Sect. 4.3.2–4.3.3) is not applicable, because $\to$ does not give a mixed variance, locally continuous functor on a bilimit-compact category.

- We cannot translate the monadic language into Jump-With-Argument (preserving semantics), so we do not have a jumping implementation for it. To put this another way, look at the continuation semantics for $A \to_{\mathbf{CBV}} B$ given by the two decompositions:

$$\begin{array}{rcl} A \to TB & \text{gives} & A \to \neg\neg B \\ U(A \to FB) & \text{gives} & \neg(A \times \neg B) \end{array}$$

  These are isomorphic. But the latter is meaningful in a jumping sense—it says that a CBV function is a point to which we jump taking both an argument and a return address for the result. By contrast, the $\to$ used in the former cannot be understood in jumping terms.

- For possible world semantics, we saw in Sect. 6.6.4 that $A \to_{\mathsf{CBV}} B$ has a simple and intuitive possible world semantics

$$[\![A \to_{\mathbf{CBV}} B]\!]w = \prod_{w' \geqslant w} (Sw' \to [\![A]\!]w' \to \sum_{w'' \geqslant w'} (Sw'' \times [\![B]\!]w'')) \tag{12.1}$$

  But if we use monadic decomposition of $\to_{\mathbf{CBV}}$, we must recall how to calculate an exponential in a functor category as an end

$$(A \to B)w = \int_{w' \geqslant w} (Aw' \to Bw')$$

  and we then obtain the following:

$$[\![A \to_{\mathbf{CBV}} B]\!]w = \int_{w' \geqslant w} (Aw' \to \prod_{w'' \geqslant w'} (Sw'' \to \sum_{w''' \geqslant w''} (Sw''' \times [\![B]\!]w''')))$$

  This *is* isomorphic to (12.1)—although the isomorphism is hardly obvious—but it is unintuitive and more complicated. Calculations using it will be more difficult.

The target language of the linear logic decomposition is a linear λ-calculus with types including

$$\underline{A} ::= \; !\underline{A} \mid \underline{A} \otimes \underline{A} \mid \underline{A} \multimap \underline{A} \mid \cdots$$

(We are underlining the types because they denote pointed cpos.)

Much is lost in the translation from CBN into this language. For example, O'Hearn's semantics of global store

$$[\![\underline{A} \to_{\mathbf{CBN}} \underline{B}]\!] = (S \to [\![\underline{A}]\!]) \to [\![\underline{B}]\!]$$

and the Streicher-Reus continuation semantics

$$[\![\underline{A} \to_{\mathbf{CBN}} \underline{B}]\!] = (\neg[\![\underline{A}]\!]) \times [\![\underline{B}]\!]$$

do not exhibit the linear decomposition. One reason for this is that, as we said in Sect. I.2.3, the linear λ-calculus assumes commutativity of effects.

For the same reason, the pointer game model for CBN (given by the CBPV model of Chap. 8) does not exhibit the linear decomposition of $\to_{\mathbf{CBN}}$. This is perhaps surprising, given the influence of linear logic on the development of game semantics.

12.4 Beyond Simple Types

In this book we have studied only simply typed languages, together with recursive and infinitary types. We have not looked at polymorphism or dependent typing.

12.4.1 Dependent Types

Dependent types and computational effects are a problematic combination, as Moggi found.

On the one hand, it is easy and even beneficial to add dependent types to CBPV. We can generalize $\times$ to a dependent sum connective, and generalize $\to$ to a dependent product connective. The one typing rule that requires care is the sequencing rule

$$\frac{\Gamma \vdash^{\mathsf{c}} M : FA \quad \Gamma, \mathtt{x} : A \vdash^{\mathsf{c}} N : \underline{B}}{\Gamma \vdash^{\mathsf{c}} M \ \mathtt{to}\ \mathtt{x}.\ N : \underline{B}}$$

In this rule, we must stipulate that $\underline{B}$ does not depend on $\mathtt{x}$, only on Γ. This restriction is indispensable if we want to retain our denotational models. The categorical semantics of this language remain to be studied, but we do not expect difficulties. And we can likewise extend JWA, generalizing $\times$ to a dependent sum connective.

But the problem: there is no obvious translation from λ-calculus with dependent sums and dependent products into this "dependently typed CBPV". In the simply typed case we have 2 translations (CBV and CBN); but the CBV translation of application and the CBN translation of pattern-matching both use sequencing, and when we try extend to the dependently typed setting, we violate the above restriction.

Thus it appears that this "dependently typed CBPV" may not be as expressive as a dependently typed language without effects.

12.4.2 Polymorphism

From an operational perspective, we speculate that we could add polymorphism to CBPV by adding the following type constructors:

$$\begin{array}{rl} A ::= & \cdots \mid X \mid \sum X.A \mid \sum \underline{X}.A \\ \underline{B} ::= & \cdots \mid \underline{X} \mid \prod X.\underline{B} \mid \prod \underline{X}.\underline{B} \end{array}$$

In the same way, we could add polymorphism to JWA by adding the following type constructors

$$A ::= \quad \cdots \mid X \mid \sum X.A$$

However, whether these extensions are operationally and denotationally well-behaved remains to be investigated.

12.5 Further Work

In addition to the investigation of dependent and polymorphic types mentioned above, there are a number of holes in various places in the book that need to be filled, and the problems involved are by no means trivial.

- Providing complex values and an equational theory in the setting of infinitely deep syntax.
- Development of models combining erratic choice and divergence (including the new may-testing model of Sect. 5.5.4).
- Development of parametric models for cell generation.
- Prove that the pointer game model of JWA does indeed describe the set of possible traces of a term according to the jumping operational semantics.
- In relating syntax to categorical semantics (Chap. 10), allow the object structure to vary.

Appendix A
Technical Treatment of CBV and CBN

A.1 The Jumbo λ-Calculus

A.1.1 Introduction

This chapter looks in detail at the relationship between CBV, CBN and CBPV. We will prove various technical results, including universality and full abstraction of transforms. The various languages, translations and results are listed in Sect. A.2.

Before we come to these, it is helpful to recollect what we were aiming to achieve in Chap. 1–2:

> We take a purely functional language, add a computational effect (e.g. printing) and explore various possible choices: about evaluation order, about observational equivalence etc. Then we seek a single language that includes all of these possible choices.

The wider the range of language possibilities that we explore and show to be included within CBPV, the stronger our claim that CBPV is a "subsuming paradigm".

In this chapter we carefully retread this path. In order that our exploration of possibilities be as thorough as possible, we want our starting point (the purely functional language) to have a wider range of type constructors than was provided in λ `bool+`.

For example, we want to look at product types. Rather than decide *a priori* whether these will be projection products or pattern-match products (in the sense of Sect. 1.3.2), we will provide both. We will then see how each is affected by the addition of computational effects, and eventually verify that each possibility is included within CBPV.

Similarly, we will provide not just unary but multi-ary functions. For although there are various isomorphisms that justify the decomposition of multi-ary function types into unary function types, it is not clear *a priori* that these will continue to be valid when effects are added.

The purely functional language that we use is called the *jumbo λ-calculus.* Because it is intended as an exploratory tool, as we have explained, it has extremely general *jumbo connectives*

tuple types include both sums and pattern-match products as special cases.

function types include both unary function types and projection products as special cases.

We work with infinitely wide languages (see Sect. 4.1 for a discussion). The reader wishing to consider only finitely wide languages should substitute "finite" for "countable" throughout this chapter.

A.1.2 Tuple Types

Tuple types are "sum of product" types in usual programming parlance. Something of tuple type is formed as a finite tuple consisting of a tag and *several* terms; the number of terms and their types depend on the tag. This is the most liberal tuple type formation possible within a simply typed language, because simple typing requires that the type of a term cannot depend on the type of another term, only on a tag.

Definition A.1 A *hyper-arity* is a countable family of natural numbers $\{r_i\}_{i \in I}$. □

Let $\{r_i\}_{i \in I}$ be a hyper-arity. We define the *tuple connective* for this hyper-arity as follows. Suppose that for each $i \in I$ we have a finite sequence of types $A_{i\,0}, \ldots, A_{i\,(r_i-1)}$. Then we form the tuple type $\sum_{i \in I}^{\times_{j \in \$ r_i}} A_{i\,j}$. This denotes the set

$\sum_{i\in I}([\![A_{i\,0}]\!] \times \cdots \times [\![A_{i\,(r_i-1)}]\!])$, or, isomorphically, the set of tuples $(i, a_0, \ldots, a_{r_i-1})$, where $a_j \in [\![A_{i\,j}]\!]$.

When each $r_i = 1$ this connective is the usual sum $\sum_{i\in I} A_i$. In particular, when $I = \{0, 1\}$ we obtain the binary sum $A + B$ and when I is empty we obtain the zero type 0.

When I is a singleton set $\{*\}$ and $r_* = 2$, we obtain a pattern-match binary product type $A \times B$. Similarly, when I is a singleton set $\{*\}$ and $r_* = 0$, we obtain the *pattern-match-unit* type 1.

In the effect-free setting, $\sum$,$\times$ and 1 are sufficient to give all tuple types because of this decomposition:

$$\sum_{i\in I}^{\times_{j\in \$r_i}} A_{i\,j} \cong \sum_{i\in I}(A_{i\,0} \times \cdots \times A_{i\,(r_i-1)}) \tag{A.1}$$

However, it is not apparent *a priori* whether (A.1) will remain valid when we add effects. This is the reason for providing general tuple types. (In fact, (A.1) is valid in CBV but not in CBN.)

Another special case of tuple types is when each $r_i = 0$. Such a tuple type is called a *ground type*. Its closed terms are of the form (i), for $i \in I$; sometimes we write this just as i. In particular when $I = \{\mathsf{true}, \mathsf{false}\}$ we call this type `bool` and when $I = \mathbb{N}$ we call this type `nat`. We write `true` for (true), `false` for (false) and `if` M `then` N `else` N' for `pm` M `as` $\{(\mathsf{true}).\ N, (\mathsf{false}).\ N'\}$.

A.1.3 Function Types

Jumbo λ-calculus function types are a generalization of the usual unary function types. Something of unary function type is applied to a single operand. By contrast, something of jumbo λ-calculus function type is applied to several operands. The first operand is a tag and the rest are terms. The number of terms and their types depend on the tag, and the type of the result also depends on the tag. This is the most liberal function type formation possible within a simply typed language, because simple typing requires that the type of a term cannot depend on the type of another term, only on a tag.

We use a novel notation: when we apply a function to several operands, we delimit these operands on the left with the symbol $\angle$. For example N applied to M_0 and M_1 is written $\angle M_0, M_1$‘N. The reason we do not write it (M_0, M_1)‘N is that this notation suggests that (M_0, M_1) is a subterm, which it is not.

Let $\{r_i\}_{i\in I}$ be a hyper-arity. We define the *function connective* for this hyper-arity as follows. Suppose that for each $i \in I$ we have a finite sequence of types $A_{i\,0}, \ldots, A_{i\,(r_i-1)}$ and another type B_i. Then we form the function type $\prod_{i\in I}^{\to_{j\in \$r_i}} A_{i\,j} B_i$. This denotes the set $\prod_{i\in I}([\![A_{i\,0}]\!] \to \cdots \to [\![A_{i\,(r_i-1)}]\!] \to [\![B_i]\!])$, or isomorphically the set of functions that takes the sequence of operands $\angle i, a_0, \ldots, a_{r_i-1}$, where $a_j \in [\![A_{i\,j}]\!]$, to an element of $[\![B_i]\!]$.

Where each $r_i = 1$ we obtain a projection product type written $\prod_{i\in I} A_i$. In particular, when $I = \{0, 1\}$ we obtain the binary projection-product which we write $A \sqcap B$, and when I is empty we obtain the projection-unit 1_{Π}.

When I is a singleton set, say $\{*\}$, and $r_* = 1$ we obtain the usual unary function type $A \to B$.

In the effect-free setting $\prod$ and $\to$ are sufficient to give all function types because of this decomposition:

$$\prod_{i\in I}^{\to_{j\in \$r_i}} A_{i\,j} B_i \cong \prod_{i\in I}(A_{i\,0} \to \cdots \to A_{i\,(r_i-1)} \to B_i) \tag{A.2}$$

However, it is not apparent *a priori* whether (A.2) will remain valid when we add effects. This is the reason for providing general function types. (In fact, (A.2) is valid in CBN but not in CBV.)

Types $A ::= \sum_{i \in I}^{\times j \in \$r_i} A_{ij} \mid \prod_{i \in I}^{\rightarrow j \in \$r_i} A_{ij} A_i$

where each set I is countable.

Terms

$$\frac{}{\Gamma, \mathtt{x} : A, \Gamma' \vdash \mathtt{x} : A} \qquad \frac{\Gamma \vdash M : A \quad \Gamma, \mathtt{x} : A \vdash N : B}{\Gamma \vdash \mathtt{let}\ M\ \mathtt{be}\ \mathtt{x}.\ N : B}$$

$$\frac{\Gamma \vdash M_0 : A_{\hat{\imath}\,0} \quad \cdots \quad \Gamma \vdash M_{r_{\hat{\imath}}-1} : A_{\hat{\imath}\,(r_{\hat{\imath}}-1)}}{\Gamma \vdash (\hat{\imath}, M_0, \ldots, M_{r_{\hat{\imath}}-1}) : \sum_{i \in I}^{\times j \in \$r_i} A_{ij}}$$

$$\frac{\Gamma \vdash M : \sum_{i \in I}^{\times j \in \$r_i} A_{ij} \quad \cdots \quad \Gamma, \mathtt{x}_0 : A_{i\,0}, \ldots, \mathtt{x}_{r_i-1} : A_{i\,(r_i-1)} \vdash N_i : B \quad \cdots\ {}_{i \in I}}{\Gamma \vdash \mathtt{pm}\ M\ \mathtt{as}\ \{\ldots, (i, \mathtt{x}_0, \ldots, \mathtt{x}_{r_i-1}).N_i, \ldots\} : B}$$

$$\frac{\cdots \quad \Gamma, \mathtt{x}_0 : A_{i\,0}, \ldots, \mathtt{x}_{r_i-1} : A_{i\,r_i-1} \vdash M_i : B_i \quad \cdots\ {}_{i \in I}}{\Gamma \vdash \lambda\{\ldots, \angle i, \mathtt{x}_0, \ldots, \mathtt{x}_{r_i-1}.M_i, \ldots\} : \prod_{i \in I}^{\rightarrow j \in \$r_i} A_{ij} B_i}$$

$$\frac{\Gamma \vdash M_0 : A_{\hat{\imath}\,0} \quad \cdots \quad \Gamma \vdash M_{r_{\hat{\imath}}-1} : A_{\hat{\imath}\,(r_{\hat{\imath}}-1)} \quad \Gamma \vdash N : \prod_{i \in I}^{\rightarrow j \in \$r_i} A_{ij} B_i}{\Gamma \vdash \angle\hat{\imath}, M_0, \ldots, M_{r_{\hat{\imath}-1}}`N : B_{\hat{\imath}}}$$

where $\hat{\imath}$ is any element of I.

β-laws

$$\begin{array}{rcl} \mathtt{let}\ M\ \mathtt{be}\ \mathtt{x}.\ N & = & N[M/\mathtt{x}] \\ \mathtt{pm}\ (\hat{\imath}, \overrightarrow{M_j})\ \mathtt{as}\ \{\ldots, (i, \overrightarrow{\mathtt{x}_j}).N_i, \ldots\} & = & N_{\hat{\imath}}[\overrightarrow{M_j/\mathtt{x}_j}] \\ \angle\hat{\imath}, \overrightarrow{M_j}`\lambda\{\ldots, \angle i, \overrightarrow{\mathtt{x}_j}.N_i, \ldots\} & = & N_{\hat{\imath}}[\overrightarrow{M_j/\mathtt{x}_j}] \end{array}$$

η-laws

$$\begin{array}{rcl} N[M/\mathtt{z}] & = & \mathtt{pm}\ M\ \mathtt{as}\ \{\ldots, (i, \overrightarrow{\mathtt{x}_j}).N[(i, \overrightarrow{\mathtt{x}_j})/\mathtt{z}], \ldots\} \\ M & = & \lambda\{\ldots, \angle i, \overrightarrow{\mathtt{x}_j}.(\angle i, \overrightarrow{\mathtt{x}_j}`M), \ldots\} \end{array}$$

Figure A.1. Syntax and Equations of Jumbo λ-calculus

A.2 Languages and Translations

In the rest of this chapter we will look in detail at various languages involving effects, and the translations between them, shown in Fig. A.2. Arrows of the form $\hookrightarrow$ indicate language extensions. For the three languages marked with an asterisk, we provide an equational theory.

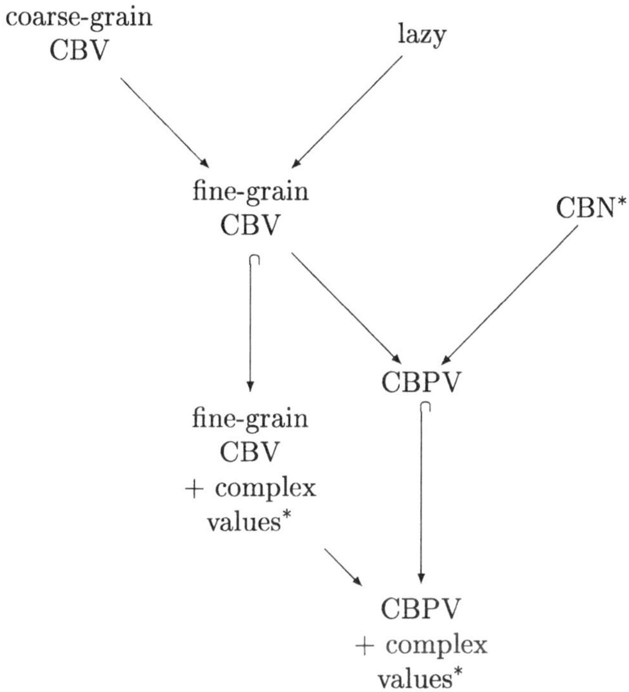

Figure A.2. Effectful Languages

Because we are considering so many languages, we have many variants of the same results. To reduce clutter, we omit many of these propositions, where they are straightforward. Furthermore, when treating CBV and CBN, we omit proofs when they are similar to the corresponding proofs for CBPV. We give a summary of the results here.

Each language has the following properties:

- the big-step semantics is total, because the only effect we are using in our languages is printing
- denotational semantics commutes with substitution and with weakening
- denotational semantics agrees with operational semantics (soundness)
- denotational equality implies observational equivalence.

Each equational theory has the following properties:

- provable equality commutes with substitution and weakening
- provable equality implies denotational equality
- provable equality implies observational equivalence.

Each translation has the following properties:

- the translation preserves denotational semantics
- the translation commutes with substitution and weakening up to provable equality

- the translation preserves provable equality (where the source language has an equational theory)
- the translation preserves and reflects operational semantics for ground terms
- the translation reflects observational equivalence.

We would like each translation to have the following properties:

- the translation commutes exactly with substitution and weakening
- the translation preserves and reflects operational semantics for all terms.

However, these properties usually fail, and to achieve them we have to extend the translation from a function between terms to a relation between terms.

The two translations into CBPV[1] have the following properties:

- every type in the target language is isomorphic to the translation of some type in the source language
- every term in the target language (of appropriate type and context) is provably equal to the translation of some term in the source language
- the translation reflects provable equality
- the translation preserves observational equivalence (full abstraction).

A.3 Call-By-Value

A.3.1 Coarse-Grain Call-By-Value

For CG-CBV we evaluate to the following *terminal* closed terms:

$$T ::= \quad (\hat{\imath}, T_0, \ldots, T_{r_i - 1}) \mid \lambda\{\ldots, \angle i, \overrightarrow{\mathtt{x}_j}.M_i, \ldots\}$$

The relation $M \Downarrow T$ is given inductively by the rules of Fig. A.3. We extend it to printing as in Sect. 2.4.1.

Proposition 132 Properties of big-step semantics:

1 If T is terminal, then $T \Downarrow T$.

2 For every closed term M, $M \Downarrow T$ for a unique T.

□

Denotational semantics for printing is given as in Sect. 1.6.2. The type constructors are interpreted as follows.

- $\sum_{i \in I}^{\times j \in \$ r_i} A_{i\,j}$ denotes $\sum_{i \in I}([\![A_{i\,0}]\!] \times \cdots \times [\![A_{i\,(r_i - 1)}]\!])$.
- $\prod_{i \in I}^{\rightarrow j \in \$ r_i} A_{i\,j} B_i$ denotes $\prod_{i \in I}([\![A_{i\,0}]\!] \rightarrow \cdots \rightarrow [\![A_{i\,(r_i - 1)}]\!] \rightarrow (\mathcal{A}^* \times [\![B_i]\!]))$.

[1] We have not investigated these properties for the other translations.

$$\frac{M \Downarrow T \quad N[T/\mathtt{x}] \Downarrow T'}{\mathtt{let}\ M\ \mathtt{be}\ \mathtt{x}.\ N \Downarrow T'}$$

$$\frac{M_0 \Downarrow T_0 \quad \cdots \quad M_{r_{\hat{\imath}}-1} \Downarrow T_{r_{\hat{\imath}}-1}}{(\hat{\imath}, M_0, \ldots, M_{r_{\hat{\imath}}-1}) \Downarrow (\hat{\imath}, T_0, \ldots, T_{r_{\hat{\imath}}-1})}$$

$$\frac{M \Downarrow (\hat{\imath}, \overrightarrow{T_j}) \quad N_{\hat{\imath}}[\overrightarrow{T_j/\mathtt{x}_j}] \Downarrow T}{\mathtt{pm}\ M\ \mathtt{as}\ \{\ldots, (i, \overrightarrow{\mathtt{x}_j}).N_i, \ldots\} \Downarrow T}$$

$$\frac{}{\lambda\{\ldots, \angle i, \overrightarrow{\mathtt{x}_j}.M_i, \ldots\} \Downarrow \lambda\{\ldots, \angle i, \overrightarrow{\mathtt{x}_j}.M_i, \ldots\}}$$

$$\frac{M_0 \Downarrow T_0 \quad \cdots \quad M_{r_{\hat{\imath}}-1} \Downarrow T_{r_{\hat{\imath}}-1} \quad N \Downarrow \lambda\{\ldots, \angle i, \overrightarrow{\mathtt{x}_j}.N_i, \ldots\} \quad N_{\hat{\imath}}[\overrightarrow{T_j/\mathtt{x}_j}] \Downarrow T}{\angle \hat{\imath}, M_0, \ldots, M_{r_{\hat{\imath}}-1}`N \Downarrow T}$$

Figure A.3. Big-Step Semantics for CG-CBV—No Effects

A.3.2 Fine-Grain Call-By-Value

The Language

The coarse-grain CBV language that we have seen suffers from two problems.

1 We had to make an arbitrary choice as to the order of evaluation of tuples and applications.

2 A value V has two denotations: $[\![V]\!]^{\mathsf{val}}$, its denotation as a value, and $[\![V]\!]^{\mathsf{ret}}$, its denotation as a returner.

How can we refine the language to avoid these problems, while leaving unchanged the types and their denotations?

In order to have a single denotation function $[\![-]\!]$, we make a syntactic distinction between values and returners, so that every term is either a value or a returner but not both. Thus we have two judgements

$$\Gamma \vdash^{\mathsf{v}} V : A \qquad \Gamma \vdash^{\mathsf{p}} M : A$$

which respectively say that V is a value of type A and that M is a returner of type A (i.e. M returns a value of type A).

The calculus that this leads to is called *fine-grain CBV*. The terms and big-step semantics are given in Fig. A.4. We do not evaluate values, so the operational semantics is defined only on returners.

It is convenient in FG-CBV to write TA as syntactic sugar for $\prod_{i \in \{*\}} A$. Thus in the printing semantics TA denotes $\mathcal{A}^* \times [\![A]\!]$, and in the Scott semantics TA denotes $[\![A]\!]_\perp$. We write $\mathtt{thunk}\ M$ for $\lambda *.M$ and $\mathtt{force}\ V$ for $*`V$. Because of the importance of TA, we present rules and equations for it explicitly, even though they are just special cases of the rules and equations for function types.

Note that in FG-CBV `let` is used only for binding identifiers. The sequencing of returners, which was written in CG-CBV as $\mathtt{let}\ M\ \mathtt{be}\ \mathtt{x}.\ N$ is in FG-CBV written

$$\frac{}{\Gamma, \mathtt{x} : A, \Gamma' \vdash^{\mathsf{v}} \mathtt{x} : A} \qquad \frac{\Gamma \vdash^{\mathsf{v}} V : A \quad \Gamma, \mathtt{x} : A \vdash^{\mathsf{p}} M : B}{\Gamma \vdash^{\mathsf{v}} \mathtt{let}\ V\ \mathtt{be}\ \mathtt{x}.\ M : B}$$

$$\frac{\Gamma \vdash^{\mathsf{v}} V : A}{\Gamma \vdash^{\mathsf{p}} \mathtt{return}\ V : A} \qquad \frac{\Gamma \vdash^{\mathsf{p}} M : A \quad \Gamma, \mathtt{x} : A \vdash^{\mathsf{p}} N : B}{\Gamma \vdash^{\mathsf{p}} M\ \mathtt{to}\ \mathtt{x}.\ N : B}$$

$$\frac{\overrightarrow{\Gamma \vdash^{\mathsf{v}} V_j : A_{\hat{\imath} j}}}{\Gamma \vdash^{\mathsf{v}} (\hat{\imath}, \overrightarrow{V_j}) : \sum_{i \in I}^{\times_{j \in \$ r_i}} A_{i j}}$$

$$\frac{\Gamma \vdash^{\mathsf{v}} V : \sum_{i \in I}^{\times_{j \in \$ r_i}} A_{i j} \quad \cdots \quad \Gamma, \overrightarrow{\mathtt{x}_j : A_{i j}} \vdash^{\mathsf{p}} M_i : B \quad \cdots_{\ i \in I}}{\Gamma \vdash^{\mathsf{p}} \mathtt{pm}\ V\ \mathtt{as}\ \{\ldots (i, \overrightarrow{\mathtt{x}}).M_i, \ldots\} : B}$$

$$\frac{\cdots \quad \Gamma, \overrightarrow{\mathtt{x}_j : A_{i j}} \vdash^{\mathsf{p}} M_i : B_i \quad \cdots_{\ i \in I}}{\Gamma \vdash^{\mathsf{v}} \lambda\{\ldots, \angle i, \overrightarrow{\mathtt{x}_j}.M_i, \ldots\} : \prod_{i \in I}^{\rightarrow_{j \in \$ r_i}} A_{i j} B_i}$$

$$\frac{\overrightarrow{\Gamma \vdash^{\mathsf{v}} V_j : A_{\hat{\imath} j}} \quad \Gamma \vdash^{\mathsf{v}} W : \prod_{i \in I}^{\rightarrow_{j \in \$ r_i}} A_{i j} B_i}{\Gamma \vdash^{\mathsf{p}} \angle \hat{\imath}, \overrightarrow{V_j}`W : B_{\hat{\imath}}}$$

$$\frac{\Gamma \vdash^{\mathsf{p}} M : A}{\Gamma \vdash^{\mathsf{v}} \mathtt{thunk}\ M : TA} \qquad \frac{\Gamma \vdash^{\mathsf{v}} V : TA}{\Gamma \vdash^{\mathsf{p}} \mathtt{force}\ V : A}$$

$$\frac{M[V/\mathtt{x}] \Downarrow W}{\mathtt{let}\ V\ \mathtt{be}\ \mathtt{x}.\ M \Downarrow W}$$

$$\frac{}{\mathtt{return}\ V \Downarrow V} \qquad \frac{M \Downarrow V \quad N[V/\mathtt{x}] \Downarrow W}{M\ \mathtt{to}\ \mathtt{x}.\ N \Downarrow W}$$

$$\frac{M_{\hat{\imath}}[\overrightarrow{V_j/\mathtt{x}_j}] \Downarrow W}{\mathtt{pm}\ (\hat{\imath}, \overrightarrow{V_j})\ \mathtt{as}\ \{\ldots, (i, \overrightarrow{\mathtt{x}_j}).M_i, \ldots\} \Downarrow W}$$

$$\frac{M_{\hat{\imath}}[\overrightarrow{V_j/\mathtt{x}_j}] \Downarrow W}{\angle \hat{\imath}, \overrightarrow{V_j}`\lambda\{\ldots, \angle i, \overrightarrow{\mathtt{x}_j}.M_i, \ldots\} \Downarrow W}$$

$$\frac{M \Downarrow W}{\mathtt{force}\ \mathtt{thunk}\ M \Downarrow W}$$

Figure A.4. Terms and Big-Step Semantics for FG-CBV

more suggestively as M `to` $\mathtt{x}$. N. Furthermore this is the *only* construct in FG-CBV that sequences returners, so both operational and denotational semantics for other terms (such as application) is much simpler than before. In particular, the arbitrariness present in CG-CBV (evaluation order for application and for tupling) is no longer present in FG-CBV.

FG-CBV is based on Moggi's "monadic metalanguage" [Moggi, 1991]. But unlike Moggi's calculus, FG-CBV distinguishes between a returner M of type A and its thunk, a value of type TA.

To add our example effect to the language, we add the rule

$$\frac{\Gamma \vdash^{\mathsf{p}} M : A}{\Gamma \vdash^{\mathsf{p}} \mathtt{print}\ c.\ M : A}$$

and adapt and then extend the big-step semantics exactly as in Sect. 2.4.1. We then obtain:

Proposition 133 For every closed returner M, there is a unique m, V such that $M \Downarrow m, V$. □

Observational Equivalence

Definition A.2 Given two returners $\Gamma \vdash^{\mathsf{p}} M, M' : B$, we say that

1 $M \simeq_{\mathsf{ground}} M'$ when for all ground returner contexts $\mathcal{C}[\cdot]$, $\mathcal{C}[M] \Downarrow m, i$ iff $\mathcal{C}[M'] \Downarrow m, i$

2 $M \simeq_{\mathsf{anytype}} M'$ when for all returner contexts $\mathcal{C}[\cdot]$ of any type, $\mathcal{C}[M] \Downarrow m, T$ for some T iff $\mathcal{C}[M'] \Downarrow m, T$ for some T

Similarly for values. □

Proposition 134 The two relations $\simeq_{\mathsf{ground}}$ and $\simeq_{\mathsf{anytype}}$ are the same. □

Denotational Semantics for Printing

The semantics of types is the same as CG-CBV.

If $\Gamma \vdash^{\mathsf{v}} V : A$ then V denotes a function from $[\![\Gamma]\!]$ to $[\![A]\!]$, whereas if $\Gamma \vdash^{\mathsf{p}} M : A$ then M denotes a function from $[\![\Gamma]\!]$ to $\mathcal{A}^* \times [\![A]\!]$. We omit the semantics of terms.

Complex Values and Equational Theory

For all purposes except operational semantics, we extend the FG-CBV calculus with the following rules for *complex values*:

$$\frac{\Gamma \vdash^{\mathsf{v}} V : A \quad \Gamma, \mathtt{x} : A \vdash^{\mathsf{v}} W : A}{\Gamma \vdash^{\mathsf{v}} \mathtt{let}\ V\ \mathtt{be}\ \mathtt{x}.\ W : A}$$

$$\frac{\Gamma \vdash^{\mathsf{v}} V : \sum_{i \in I}^{\times_{j \in \$ r_i}} A_{ij} \quad \cdots \quad \Gamma, \overrightarrow{\mathtt{x}_j : A_{ij}} \vdash^{\mathsf{v}} W_i : B \quad \cdots_{\, i \in I}}{\Gamma \vdash^{\mathsf{v}} \mathtt{pm}\ V\ \mathtt{as}\ \{\ldots, (i, \overrightarrow{\mathtt{x}_j}).W_i, \ldots\} : B}$$

We then form the equational theory shown in Fig. A.5. M,N and P range over returners, while V and W range over values.

β-laws

$$\begin{array}{rcl}
\texttt{let } V \texttt{ be x. } M & = & M[V/\texttt{x}] \\
\texttt{let } V \texttt{ be x. } W & = & W[V/\texttt{x}] \\
(\texttt{return } V) \texttt{ to x. } M & = & M[V/\texttt{x}] \\
\texttt{pm } (\hat{\imath}, \overrightarrow{V_j}) \texttt{ as } \{\ldots, (i, \overrightarrow{\texttt{x}_j}).M_i, \ldots\} & = & M_{\hat{\imath}}[\overrightarrow{V_j/\texttt{x}_j}] \\
\texttt{pm } (\hat{\imath}, \overrightarrow{V_j}) \texttt{ as } \{\ldots, (i, \overrightarrow{\texttt{x}_j}).W_i, \ldots\} & = & W_{\hat{\imath}}[\overrightarrow{V_j/\texttt{x}_j}] \\
\angle \hat{\imath}, \overrightarrow{V_j}\text{`}\lambda\{\ldots, \angle i, \overrightarrow{\texttt{x}_j}.M_i, \ldots\} & = & M_{\hat{\imath}}[\overrightarrow{V_j/\texttt{x}_j}] \\
\texttt{force thunk } M & = & M
\end{array}$$

η-laws

$$\begin{array}{rcl}
M & = & M \texttt{ to x. return x} \\
M[V/\texttt{z}] & = & \texttt{pm } V \texttt{ as } \{\ldots, (i, \overrightarrow{\texttt{x}_j}).M[(i, \overrightarrow{\texttt{x}_j})/\texttt{z}], \ldots\} \\
W[V/\texttt{z}] & = & \texttt{pm } V \texttt{ as } \{\ldots, (i, \overrightarrow{\texttt{x}_j}).W[(i, \overrightarrow{\texttt{x}_j})/\texttt{z}], \ldots\} \\
V & = & \lambda\{\ldots, \angle i, \overrightarrow{\texttt{x}_j}.(\angle i, \overrightarrow{\texttt{x}_j}\text{`}V), \ldots\} \\
V & = & \texttt{thunk force } V
\end{array}$$

sequencing laws

$$(M \texttt{ to x. } N) \texttt{ to y. } P \quad = \quad M \texttt{ to x. } (N \texttt{ to y. } P)$$

`print` laws

$$(\texttt{print } c.\ M) \texttt{ to x. } N \quad = \quad \texttt{print } c.\ (M \texttt{ to x. } N)$$

Figure A.5. CBV equations, using conventions of Sect. I.4.2

The equation for `print` is of course specific to our example effect, but there are directly analogous equations for many other effects. For example, if we were considering divergence, we would have an equation

$$\texttt{diverge to x. } M = \texttt{diverge}$$

We call this theory (without the `print` equation) the *CBV equational theory.*

Proposition 135 1 There is an effective procedure that given a returner $\Gamma \vdash^{\mathsf{p}} M : A$, possibly containing complex values, returns a returner $\Gamma \vdash^{\mathsf{p}} \tilde{M} : A$ without complex values, such that $M = \tilde{M}$ is provable.

2 There is an effective procedure that, given a closed value $\vdash^{\mathsf{v}} V : A$, possibly containing complex values, returns a closed value $\vdash^{\mathsf{v}} \tilde{V} : A$ without complex values, such that $V = \tilde{V}$ is provable.

□

This is proved like Prop. 25.

A.3.3 From CG-CBV To FG-CBV

The translation from CG-CBV to FG-CBV is given in Fig. A.6. It is in two parts: $-^{\mathsf{cg}}$ is defined on returners and $-^{\mathsf{cgval}}$ on values. The translation $-^{\mathsf{cg}}$ reflects our choice of evaluation order for CG-CBV.

$\Gamma \vdash M : A$	$\Gamma \vdash^{\mathsf{p}} M^{\mathsf{cg}} : A$
$\mathtt{x}$	$\mathtt{return}\ \mathtt{x}$
$\mathtt{let}\ M\ \mathtt{be}\ \mathtt{x}.\ N$	$M^{\mathsf{cg}}\ \mathtt{to}\ \mathtt{x}.\ N^{\mathsf{cg}}$
$(\hat{\imath}, M_0, \ldots, M_{r_{\hat{\imath}}-1})$	$M_0^{\mathsf{cg}}\ \mathtt{to}\ \mathtt{x}_0.\ \ldots M^{\mathsf{cg}}_{r_{\hat{\imath}}-1}\ \mathtt{to}\ \mathtt{x}_{r_{\hat{\imath}}-1}.\ \mathtt{return}\ (\hat{\imath}, \overrightarrow{\mathtt{x}_j})$
$\mathtt{pm}\ M\ \mathtt{as}\ \{\ldots, (i, \overrightarrow{\mathtt{x}_j}).N_i, \ldots\}$	$M^{\mathsf{cg}}\ \mathtt{to}\ \mathtt{z}.\ \mathtt{pm}\ \mathtt{z}\ \mathtt{as}\ \{\ldots, (i, \overrightarrow{\mathtt{x}_j}).N_i^{\mathsf{cg}}, \ldots\}$
$\lambda\{\ldots, \angle i, \overrightarrow{\mathtt{x}_j}.M_i, \ldots\}$	$\mathtt{return}\ \lambda\{\ldots, \angle i, \overrightarrow{\mathtt{x}_j}.M_i^{\mathsf{cg}}, \ldots\}$
$\angle\hat{\imath}, M_0, \ldots, M_{r_{\hat{\imath}}-1}`N$	$M_0^{\mathsf{cg}}\ \mathtt{to}\ \mathtt{x}_0.\ \ldots M^{\mathsf{cg}}_{r_{\hat{\imath}}-1}\ \mathtt{to}\ \mathtt{x}_{r_{\hat{\imath}}-1}.\ N^{\mathsf{cg}}\ \mathtt{to}\ f.\ \angle\hat{\imath}, \overrightarrow{\mathtt{x}_j}`f$
$\mathtt{print}\ c.\ M$	$\mathtt{print}\ c.\ M^{\mathsf{cg}}$

$\Gamma \vdash V : A$	$\Gamma \vdash^{\mathsf{v}} V^{\mathsf{cgval}} : A$
$\mathtt{x}$	$\mathtt{x}$
$(\hat{\imath}, V_0, \ldots, V_{r_{\hat{\imath}}-1})$	$(\hat{\imath}, V_0^{\mathsf{cgval}}, \ldots, V_{r_{\hat{\imath}}-1}^{\mathsf{cgval}})$
$\lambda\{\ldots, \angle i, \overrightarrow{\mathtt{x}_j}.M_i, \ldots\}$	$\lambda\{\ldots, \angle i, \overrightarrow{\mathtt{x}_j}.M_i^{\mathsf{cg}}, \ldots\}$

Figure A.6. Translation from CG-CBV to FG-CBV

We now address the technical properties of the translation. It neither commutes with substitution nor preserves operational semantics.

substitution It is not the case that $(M[V/\mathtt{x}])^{\mathsf{cg}}$ and $M^{\mathsf{cg}}[V^{\mathsf{cgval}}/\mathtt{x}]$ are the same term (although they will are provably equal in the CBV equational theory of Fig. A.5). To see this, let M be $\mathtt{x}$ and V be $(0,(0))$, where 0 is a tag. Then

$$\begin{array}{rcl} (M[V/\mathtt{x}])^{\mathsf{cg}} & = & \mathtt{return}\ (0)\ \mathtt{to}\ \mathtt{x}.\ \mathtt{return}\ (0, \mathtt{x}) \\ M^{\mathsf{cg}}[V^{\mathsf{cgval}}/\mathtt{x}] & = & \mathtt{return}\ (0,(0)) \end{array}$$

operational semantics It is not the case that $M \Downarrow m, V$ implies $M^{\mathsf{cg}} \Downarrow m, V^{\mathsf{cgval}}$. For example, let M be $\mathtt{let}\ (0,(0))\ \mathtt{be}\ \mathtt{x}.\ \lambda * .\mathtt{x}$.

The issue is that CG-CBV cannot recognize that what is substituted is not a certain kind of returner (to be trivially reevaluated when required) but something genuinely different—a value.

To salvage what we can, we proceed as follows. First, we write the two translations as relations $\mapsto^{\mathsf{cg}}$ and $\mapsto^{\mathsf{cgval}}$ from CG-CBV to FG-CBV terms. We can present Fig. A.6 by rules such as these:

$$\frac{M \mapsto^{\mathsf{cg}} M' \quad N \mapsto^{\mathsf{cg}} N'}{\mathtt{let}\ M\ \mathtt{be}\ \mathtt{x}.\ N \mapsto^{\mathsf{cg}} M'\ \mathtt{to}\ \mathtt{x}.\ N'}$$

$$\frac{\cdots\ P_i \mapsto^{\mathsf{cg}} P_i' \ \cdots_{\ i \in I}}{\lambda\{\ldots, \angle i, \overrightarrow{\mathtt{x}_j}.P_i, \ldots\} \mapsto^{\mathsf{cgval}} \lambda\{\ldots, \angle i, \overrightarrow{\mathtt{x}_j}.P_i', \ldots\}}$$

To these rules we add the following

$$\frac{M \mapsto^{\mathsf{cg}} \mathtt{return}\ V_0\ \mathtt{to}\ \mathtt{x}_0.\ \ldots \mathtt{return}\ V_{r_{\hat{\imath}}-1}\ \mathtt{to}\ \mathtt{x}_{r_{\hat{\imath}}-1}.\ \mathtt{return}\ (\hat{\imath}, \overrightarrow{\mathtt{x}_j})}{M \mapsto^{\mathsf{cg}} \mathtt{return}\ (\hat{\imath}, \overrightarrow{V_j})}$$

(Strictly, we should write weakenings to indicate that $\mathtt{x}_j$ is not in the context of $V_{j'}$.) We have thus defined non-functional relations $\mapsto^{\mathsf{cg}}$ and $\mapsto^{\mathsf{cgval}}$, and we will show that they commute with substitution and preserve and reflect operational semantics.

Proposition 136 For any returner M, we have $M \mapsto^{\mathsf{cg}} M^{\mathsf{cg}}$, and if $M \mapsto^{\mathsf{cg}} N$ then $N = M^{\mathsf{cg}}$ is provable in the CBV equational theory of Fig. A.5; similarly for values. □

Proposition 137 1 If $V \mapsto^{\mathsf{cgval}} V'$ then $V \mapsto^{\mathsf{cg}} \mathtt{return}\ V'$.

2 If $M \mapsto^{\mathsf{cg}} M'$ and $V \mapsto^{\mathsf{cgval}} V'$ then $M[V/\mathtt{x}] \mapsto^{\mathsf{cg}} M'[V'/\mathtt{x}]$.

3 If $W \mapsto^{\mathsf{cg}} W'$ and $V \mapsto^{\mathsf{cgval}} V'$ then $W[V/\mathtt{x}] \mapsto^{\mathsf{cg}} W'[V'/\mathtt{x}]$.

□

Proposition 138 1 If $M \Downarrow V$ and $M \mapsto^{\mathsf{cg}} M'$, then, for some V', $M' \Downarrow V'$ and $V \mapsto^{\mathsf{cgval}} V'$.

2 If $M \mapsto^{\mathsf{cg}} M'$ and $M' \Downarrow V'$, then, for some V, $M \Downarrow V$ and $V \mapsto^{\mathsf{cgval}} V'$.

□

To prove this, we introduce the following.

Definition A.3 1 In CG-CBV, the following returners are *safe*:

$$\begin{array}{rcl} S & ::= & \mathtt{x} \mid \mathtt{let}\ S\ \mathtt{be}\ \mathtt{x}.\ S \mid (\hat{\imath}, \overrightarrow{S_j}) \\ & & \mid \mathtt{pm}\ S\ \mathtt{as}\ \{\ldots, (i, \overrightarrow{\mathtt{x}_j}).S_i, \ldots\} \mid \lambda\{\ldots, \angle i, \overrightarrow{\mathtt{x}_j}.M_i, \ldots\} \end{array}$$

2 In FG-CBV the following returners are *safe*:

$$\begin{array}{rcl} S & ::= & \mathtt{return}\ V \mid \mathtt{let}\ V\ \mathtt{be}\ \mathtt{x}.\ S \mid S\ \mathtt{to}\ \mathtt{x}.\ S \\ & & \mid \mathtt{pm}\ V\ \mathtt{as}\ \{\ldots, (i, \overrightarrow{\mathtt{x}_j}).S_i, \ldots\} \end{array}$$

□

In summary, a returner is safe iff, inside it, every application occurs in the scope of a λ.

Lemma 139 Suppose $A_0, \ldots, A_{n-1} \vdash M \mapsto^{\mathsf{cg}} M' : B$. Then

1 M is safe iff M' is safe.

2 Suppose M is safe and $U_0 \mapsto^{\mathsf{cgval}} U'_0, \ldots, U_{n-1} \mapsto^{\mathsf{cgval}} U'_{n-1}$ (U_i and U'_i may not be safe).

- If $M[\overrightarrow{U_i/\mathtt{x}_i}] \Downarrow V$, then, for some V', $M'[\overrightarrow{U'_i/\mathtt{x}_i}] \Downarrow V'$ and $V \mapsto^{\mathsf{cgval}} V'$.
- If $M'[\overrightarrow{U'_i/\mathtt{x}_i}] \Downarrow V'$, then, for some V, $M[\overrightarrow{U_i/\mathtt{x}_i}] \Downarrow V$ and $V \mapsto^{\mathsf{cgval}} V'$.

□

We prove this by induction on $M \mapsto^{\mathsf{cg}} M'$. Finally we prove Prop. 138 by induction on $M \Downarrow V$ (for (1)) and on $M' \Downarrow V'$ (for (2)).

$$\frac{N[M/\mathtt{x}] \Downarrow T}{\mathtt{let}\ M\ \mathtt{be}\ \mathtt{x}.\ N \Downarrow T}$$

$$\frac{}{(\hat{\imath}, \overrightarrow{M_j}) \Downarrow (\hat{\imath}, \overrightarrow{M_j})} \qquad \frac{M \Downarrow (\hat{\imath}, \overrightarrow{N_j}) \quad P_{\hat{\imath}}[\overrightarrow{N_j/\mathtt{x}_j}] \Downarrow T}{\mathtt{pm}\ M\ \mathtt{as}\ \{\ldots, (i, \overrightarrow{\mathtt{x}_j}).P_i, \ldots\} \Downarrow T}$$

$$\frac{}{\lambda\{\ldots, \angle i, \overrightarrow{\mathtt{x}_j}.M_i, \ldots\} \Downarrow \lambda\{\ldots, \angle i, \overrightarrow{\mathtt{x}_j}.M_i, \ldots\}}$$

$$\frac{N \Downarrow \lambda\{\ldots, \angle i, \overrightarrow{\mathtt{x}_j}.N_i, \ldots\} \quad N_{\hat{\imath}}[\overrightarrow{M_j/\mathtt{x}_j}] \Downarrow T}{\angle \hat{\imath}, \overrightarrow{M_j}`N \Downarrow T}$$

Figure A.7. Big-Step Semantics for CBN—No Effects

A.4 Call-By-Name

For CBN we evaluate to the following *terminal* closed terms:

$$T ::= \quad (\hat{\imath}, \overrightarrow{M_j}) \mid \quad \mid \lambda\{\ldots, \angle i, \overrightarrow{\mathtt{x}_j}.M_i, \ldots\}$$

The relation $M \Downarrow T$ is defined in Fig. A.7.

Proposition 140 For each closed term M, we have $M \Downarrow T$ for a unique terminal term T. □

To add our example effect, we adapt and extend Fig. A.7 exactly as in Sect. 2.4.1. We then have

Proposition 141 For every closed term M, there is a unique m, T such that $M \Downarrow m, T$. □

The printing semantics follows Sect. 1.7.4. The interpretation of type constructors is given by

- If $[\![A_{ij}]\!] = (X_{ij}, *)$, then $[\![\sum_{i \in I}^{\times_{j \in \$ r_i}} A_{ij}]\!]$ is the free $\mathcal{A}$-set on $\sum_{i \in I}(X_{i0} \times \cdots \times X_{i(r_i-1)})$.
- If $[\![A_{ij}]\!] = (X_{ij}, *)$ and $[\![B_i]\!] = (Y_i, *)$ then $[\![\prod_{i \in I}^{\rightarrow_{j \in \$ r_i}} A_{ij} B_i]\!]$ is the $\mathcal{A}$-set $\prod_{i \in I}(X_{i0} \to \cdots \to X_{i(r_i-1)} \to (Y_i, *))$

We define an equational theory for CBN , whose axioms are the equations in Fig. A.8.

The equations for `print` are of course specific to our example effect, but there would be directly analogous equations for many other effects. For example, if we were considering divergence, we would have equations:

$$\begin{array}{lcl} \mathtt{diverge} & = & \mathtt{pm}\ \mathtt{diverge}\ \mathtt{as}\ \{\ldots, (i, \overrightarrow{\mathtt{x}_j}).N_i, \ldots\} \\ \mathtt{diverge} & = & \lambda\{\ldots, \angle i, \overrightarrow{\mathtt{x}_j}.\mathtt{diverge}, \ldots\} \end{array}$$

We call this theory (without the `print` equations) the *CBN equational theory.*

β-laws

$$\begin{array}{rcl}
\texttt{let } M \texttt{ be x. } N & = & N[M/\mathtt{x}] \\
\texttt{pm } (\hat{\imath}, \overrightarrow{M_j}) \texttt{ as } \{\ldots, (i, \overrightarrow{\mathtt{x}_j}).N_i, \ldots\} & = & N_{\hat{\imath}}[\overrightarrow{M_j/\mathtt{x}_j}] \\
\angle\hat{\imath}, \overrightarrow{M_j}`\lambda\{\ldots, \angle i, \overrightarrow{\mathtt{x}_j}.N_i, \ldots\} & = & N_{\hat{\imath}}[\overrightarrow{M_j/\mathtt{x}_j}]
\end{array}$$

η-laws

$$M = \lambda\{\ldots, \angle i, \overrightarrow{\mathtt{x}_j}.(\angle i, \overrightarrow{\mathtt{x}_j}`M), \ldots\}$$

pattern-matching laws

$$\begin{array}{l}
M = \texttt{pm } M \texttt{ as } \{\ldots, (i, \overrightarrow{\mathtt{x}_j}).(i, \overrightarrow{\mathtt{x}_j}), \ldots\} \\
\texttt{pm } M \texttt{ as } \{\ldots, (i, \overrightarrow{\mathtt{x}_j}).(\texttt{pm } N_i \texttt{ as } \{\ldots, (k, \overrightarrow{\mathtt{y}_l}.P_k, \ldots\}), \ldots\} \\
\quad = \texttt{pm } (\texttt{pm } M \texttt{ as } \{\ldots, (i, \overrightarrow{\mathtt{x}_j}).N_i, \ldots\}) \texttt{ as } \{\ldots, (k, \overrightarrow{\mathtt{y}_l}).P_k, \ldots\} \\
\texttt{pm } M \texttt{ as } \{\ldots, (i, \overrightarrow{\mathtt{x}_j}).\lambda\{\ldots, \angle k, \overrightarrow{\mathtt{y}_l}.N_{ik}, \ldots\}, \ldots\} \\
\quad = \lambda\{\ldots, \angle k, \overrightarrow{\mathtt{y}_l}.(\texttt{pm } M \texttt{ as } \{\ldots, (i, \overrightarrow{\mathtt{x}_j}).N_{ik}, \ldots\}), \ldots\}
\end{array}$$

print laws

$$\begin{array}{rcl}
\texttt{print } c.\ (\texttt{pm } M \texttt{ as } \{\ldots, (i, \overrightarrow{\mathtt{x}_j}).N_i, \ldots\}) & = & \texttt{pm } (\texttt{print } c.\ M) \texttt{ as } \{\ldots, (i, \overrightarrow{\mathtt{x}_j}).N_i, \ldots\} \\
\texttt{print } c.\ \lambda\{\ldots, \angle i, \overrightarrow{\mathtt{x}_j}.M_i, \ldots\} & = & \lambda\{\ldots, \angle i, \overrightarrow{\mathtt{x}_j}.(\texttt{print } c.\ M_i), \ldots\}
\end{array}$$

Figure A.8. CBN equations, using conventions of Sect. I.4.2

A.5 The Lazy Paradigm

Recall from Sect. 1.7.3 that the lazy paradigm is defined to have the same operational semantics as CBN, but to use $\simeq_{\mathsf{anytype}}$ rather than $\simeq_{\mathsf{ground}}$ as its observational equivalence. Thus its models must not satisfy, for example, the η-law for function types.

We shall say little about the lazy paradigm, because it is subsumed within FG-CBV. First, as in CG-CBV, we introduce a value/returner terminology.

Definition A.4

A *value* is a lazy term whose substitution instances are all terminal. Unlike in CBV an identifier is not a value because any term can be substituted for it, so values are just the following:

$$V ::= (\hat{\imath}, \overrightarrow{M_j}) \mid \mid \lambda\{\ldots, \angle i, \overrightarrow{\mathtt{x}_j}.M_i, \ldots\}$$

A *returner* is a lazy term (so-called because a closed returner *returns* a closed value). □

We translate the lazy language into FG-CBV in Fig. A.9. The translation is essentially that given in [Hatcliff and Danvy, 1997]. Like the translation from CG-CBV to FG-CBV, the translation on terms is defined in two parts: $-^{\mathsf{lazy}}$ is defined on returners, and $-^{\mathsf{lazyval}}$ is defined on values.

The translation is motivated as follows.

- Identifiers in the lazy language are bound to unevaluated terms, so we regard them (from a CBV perspective) as bound to thunks.
- Similarly, tuple-components and operands in the lazy language are unevaluated terms, so we regard them as thunks.

- Consequently, identifiers, tuple-components and operands all have type of the form TA—a thunk type.

The translation from lazy to FG-CBV is very similar to the translation from CBN to CBPV.

C	C^{lazy}
$\sum_{i \in I}^{\times j \in \$ r_i} A_{i\,j}$	$\sum_{i \in I}^{\times j \in \$ r_i} TA_{i\,j}^{\mathsf{lazy}}$
$\prod_{i \in I}^{\rightarrow j \in \$ r_i} A_{i\,j} B_i$	$\prod_{i \in I}^{\rightarrow j \in \$ r_i} TA_{i\,j}^{\mathsf{lazy}} B_i^{\mathsf{lazy}}$

$A_0, \ldots, A_{n-1} \vdash M : B$	$TA_0^{\mathsf{lazy}}, \ldots, TA_{n-1}^{\mathsf{lazy}} \vdash^{\mathsf{p}} M^{\mathsf{lazy}} : B^{\mathsf{lazy}}$
$\mathtt{x}$	$\mathtt{force\ x}$
$\mathtt{let}\ M\ \mathtt{be\ x.}\ N$	$\mathtt{let\ thunk}\ M^{\mathsf{lazy}}\ \mathtt{be\ x.}\ N^{\mathsf{lazy}}$
$(\hat{\imath}, M_0, \ldots, M_{r_{\hat{\imath}}-1})$	$\mathtt{return}\ (\hat{\imath}, \mathtt{thunk}\ M_0^{\mathsf{lazy}}, \ldots, \mathtt{thunk}\ M_{r_{\hat{\imath}}-1}^{\mathsf{lazy}})$
$\mathtt{pm}\ M\ \mathtt{as}\ \{\ldots, (i, \overrightarrow{\mathtt{x}_j}).N_j, \ldots\}$	$M^{\mathsf{lazy}}\ \mathtt{to\ z.\ pm\ z\ as}\ \{\ldots, (i, \overrightarrow{\mathtt{x}_j}).N_j^{\mathsf{lazy}}, \ldots\}$
$\lambda\{\ldots, \angle i, \overrightarrow{\mathtt{x}_j}.M_i, \ldots\}$	$\mathtt{return}\ \lambda\{\ldots, \angle i, \overrightarrow{\mathtt{x}_j}.M_i^{\mathsf{lazy}}, \ldots\}$
$\angle\hat{\imath}, M_0, \ldots, M_{r_{\hat{\imath}}-1}`N$	$N^{\mathsf{lazy}}\ \mathtt{to\ f.}\ \angle\hat{\imath}, (\mathtt{thunk}\ M_0), \ldots, (\mathtt{thunk}\ M_{r_{\hat{\imath}}-1})`f$
$\mathtt{print}\ c.\ M$	$\mathtt{print}\ c.\ M^{\mathsf{lazy}}$

$A_0, \ldots, A_{n-1} \vdash V : B$	$TA_0^{\mathsf{lazy}}, \ldots, TA_{n-1}^{\mathsf{lazy}} \vdash^{\mathsf{v}} V^{\mathsf{lazyval}} : B^{\mathsf{lazy}}$
$(\hat{\imath}, M_0, \ldots, M_{r_{\hat{\imath}}-1})$	$(\hat{\imath}, \mathtt{thunk}\ M_0^{\mathsf{lazy}}, \ldots, \mathtt{thunk}\ M_{r_{\hat{\imath}}-1}^{\mathsf{lazy}})$
$\lambda\{\ldots, \angle i, \overrightarrow{\mathtt{x}_j}.M_i, \ldots\}$	$\lambda\{\ldots, \angle i, \overrightarrow{\mathtt{x}_j}.M_i^{\mathsf{lazy}}, \ldots\}$

Figure A.9. Translation from lazy to FG-CBV: types, returners, values

This translation does not commute with substitution or preserve operational semantics precisely—only up to the prefix `force thunk`. (Recall that $\mathtt{force\ thunk}\ M = M$ is provable in the CBV equational theory.)

substitution It is not the case that $(M[N/\mathtt{x}])^{\mathsf{lazy}}$ and $M^{\mathsf{lazy}}[\mathtt{thunk}\ N^{\mathsf{lazy}}/\mathtt{x}]$ are the same term (although they will be provably equal in the CBV equational theory).. To see this, let M be $\mathtt{x}$. Then $(M[N/\mathtt{x}])^{\mathsf{lazy}} = N^{\mathsf{lazy}}$ but $M^{\mathsf{lazy}}[\mathtt{thunk}\ N^{\mathsf{lazy}}/\mathtt{x}] = \mathtt{force\ thunk}\ N^{\mathsf{lazy}}$.

operational semantics It is not the case that $M \Downarrow m, V$ implies $M^{\mathsf{lazy}} \Downarrow m, V^{\mathsf{lazyval}}$. For example, let M be $\mathtt{let}\ (0)\ \mathtt{be\ x.}\ (0, \mathtt{x})$.

To make the relationship precise, we proceed as follows. First, we write the two translations as relations $\mapsto^{\mathsf{lazy}}$ and $\mapsto^{\mathsf{lazyval}}$ from lazy to FG-CBV terms. We can present Fig. A.9 by rules such as these:

$$\frac{M \mapsto^{\mathsf{lazy}} M' \quad N \mapsto^{\mathsf{lazy}} N'}{\mathtt{let}\ M\ \mathtt{be\ x.}\ N \mapsto^{\mathsf{lazy}} \mathtt{let}\ (\mathtt{thunk}\ M')\ \mathtt{be\ x.}\ N'}$$

$$\frac{\cdots \quad P_i \mapsto^{\mathsf{lazy}} P_i' \quad \cdots_{\ i \in I}}{\lambda\{\ldots, \angle i, \overrightarrow{\mathtt{x}_j}.P_i, \ldots\} \mapsto^{\mathsf{lazyval}} \lambda\{\ldots, \angle i, \overrightarrow{\mathtt{x}_j}.P_i', \ldots\}}$$

To these rules we add the following:

$$\frac{M \mapsto^{\mathsf{lazy}} M'}{M \mapsto^{\mathsf{lazy}} \mathtt{force\ thunk}\ M'}$$

Proposition 142 For any returner M, we have $M \mapsto^{\mathsf{lazy}} M^{\mathsf{lazy}}$, and if $M \mapsto^{\mathsf{lazy}} M'$ then $M' = M^{\mathsf{lazy}}$ is provable in the CBV equational theory; similarly for values. □

Proposition 143 1 If $V \mapsto^{\mathsf{lazyval}} V'$ then $V \mapsto^{\mathsf{lazy}} \mathtt{return}\ V'$.

2 If $M \mapsto^{\mathsf{lazy}} M'$ and $N \mapsto^{\mathsf{lazyval}} N'$ then $M[N/\mathtt{x}] \mapsto^{\mathsf{lazy}} M'[\mathtt{thunk}\ N'/\mathtt{x}]$.

3 If $W \mapsto^{\mathsf{lazyval}} W'$ and $N \mapsto^{\mathsf{lazyval}} N'$ then $W[N/\mathtt{x}] \mapsto^{\mathsf{lazyval}} W'[\mathtt{thunk}\ N'/\mathtt{x}]$.

□

Proposition 144 1 If $M \Downarrow V$ and $M \mapsto^{\mathsf{lazy}} M'$, then, for some V', $M' \Downarrow V'$ and $V \mapsto^{\mathsf{lazyval}} V'$.

2 If $M \mapsto^{\mathsf{lazy}} M'$ and $M' \Downarrow V'$, then, for some V, $M \Downarrow V$ and $V \mapsto^{\mathsf{lazyval}} V'$.

□

We prove these by induction primarily on $\Downarrow$ and secondarily on $\mapsto^{\mathsf{lazy}}$.

A.6 Subsuming FG-CBV and CBN

The translations into CBPV are easily obtained by considering denotational semantics.

A.6.1 From FG-CBV to CBPV

The translation from FG-CBV is given in Fig. A.10.

Proposition 145 For a returner M, $(M[V/\mathtt{x}])^{\mathsf{v}}$ and $M^{\mathsf{v}}[V^{\mathsf{v}}/\mathtt{x}]$ are the same term; and similarly for values. □

Proposition 146 For returners $\Gamma \vdash^{\mathsf{p}} M, N : A$, if $M = N$ is provable in the CBV theory then $M^{\mathsf{v}} = N^{\mathsf{v}}$ is provable in the CBPV theory; and similarly for values. □

Proposition 147 The translation $-^{\mathsf{v}}$ preserves and reflects operational semantics:

1 if $M \Downarrow V$ then $M^{\mathsf{v}} \Downarrow \mathtt{return}\ V^{\mathsf{v}}$

2 if $M^{\mathsf{v}} \Downarrow \mathtt{return}\ V'$ then $M \Downarrow V$ for some V such that $V^{\mathsf{v}} = V'$.

□

A.6.2 From CBPV Back To FG-CBV

Our aim is to prove the following.

Proposition 148 1 Any CBPV value type A is isomorphic to B^{v} for some CBV type B.

2 For any CBPV returner ${A_0}^{\mathsf{v}}, \ldots, {A_{n-1}}^{\mathsf{v}} \vdash^{\mathsf{c}} N : FB^{\mathsf{v}}$ there is an FG-CBV returner $A_0, \ldots, A_{n-1} \vdash^{\mathsf{p}} M : B$ such that $M^{\mathsf{v}} = N$ is provable in CBPV; and similarly for values.

3 For any FG-CBV returners $\Gamma \vdash^{\mathsf{p}} M, M' : B$, if $M^{\mathsf{v}} = M'^{\mathsf{v}}$ is provable in CBPV then $M = M'$ is provable in FG-CBV; and similarly for values.

(2)–(3) can be extended to terms with holes i.e. contexts. □

C	C^{v} (a value type)
$\sum_{i\in I}^{\times_{j\in\$r_i}} A_{ij}$	$\sum_{i\in I}(A_{i\,0}{}^{\mathsf{v}} \times \cdots \times A_{i\,(r_i-1)}{}^{\mathsf{v}})$
$\prod_{i\in I}^{\rightarrow_{j\in\$r_i}} A_{ij} B_i$	$U\prod_{i\in I}(A_{i\,0}{}^{\mathsf{v}} \to \cdots \to A_{i\,(r_i-1)}{}^{\mathsf{v}} \to FB_i{}^{\mathsf{v}})$
TA	UFA^{v}

$A_0,\ldots,A_{n-1} \vdash^{\mathsf{v}} V : B$	$A_0{}^{\mathsf{v}},\ldots,A_{n-1}{}^{\mathsf{v}} \vdash^{\mathsf{v}} V^{\mathsf{v}} : B^{\mathsf{v}}$
$\mathtt{x}$	$\mathtt{x}$
$(\hat{\imath}, V_0,\ldots,V_{r_{\hat{\imath}}-1})$	$(\hat{\imath}, (V_0{}^{\mathsf{v}},\ldots,V_{r_{\hat{\imath}}-1}{}^{\mathsf{v}}))$
$\lambda\{\ldots,\angle i,\mathtt{x}_0,\ldots,\mathtt{x}_{r_i-1}.M_i,\ldots\}$	$\mathtt{thunk}\ \lambda\{\ldots,i.\lambda\mathtt{x}_0.\ldots\lambda\mathtt{x}_{r_i-1}.M_i{}^{\mathsf{v}},\ldots\}$
$\mathtt{thunk}\ M$	$\mathtt{thunk}\ M^{\mathsf{v}}$

$A_0,\ldots,A_{n-1} \vdash^{\mathsf{p}} M : C$	$A_0{}^{\mathsf{v}},\ldots,A_{n-1}{}^{\mathsf{v}} \vdash^{\mathsf{c}} M^{\mathsf{v}} : FC^{\mathsf{v}}$
$\mathtt{let}\ V\ \mathtt{be\ x.}\ M$	$\mathtt{let}\ V^{\mathsf{v}}\ \mathtt{be\ x.}\ M^{\mathsf{v}}$
$\mathtt{return}\ V$	$\mathtt{return}\ V^{\mathsf{v}}$
$M\ \mathtt{to\ x.}\ N$	$M^{\mathsf{v}}\ \mathtt{to\ x.}\ N^{\mathsf{v}}$
$\mathtt{pm}\ V\ \mathtt{as}\ \{\ldots,(i,\overrightarrow{\mathtt{x}_j}).M_i,\ldots\}$	$\mathtt{pm}\ V^{\mathsf{v}}\ \mathtt{as}\ \{\ldots,(i,(\overrightarrow{\mathtt{x}_j})).M_i{}^{\mathsf{v}},\ldots\}$
$\angle\hat{\imath}, V_0,\ldots,V_{r_{\hat{\imath}}-1}`W$	$V_{r_{\hat{\imath}}-1}{}^{\mathsf{v}}`\ldots`V_0{}^{\mathsf{v}}`\hat{\imath}`\mathtt{force}\ W^{\mathsf{v}}$
$\mathtt{force}\ V$	$\mathtt{force}\ V^{\mathsf{v}}$
$\mathtt{print}\ c.\ M$	$\mathtt{print}\ c.\ M^{\mathsf{v}}$

When we add complex values to the source language FG-CBV, we must add them also to the target language CBPV, and we then extend the translation as follows:

$A_0,\ldots,A_{n-1} \vdash^{\mathsf{v}} V : B$	$A_0{}^{\mathsf{v}},\ldots,A_{n-1}{}^{\mathsf{v}} \vdash^{\mathsf{v}} V^{\mathsf{v}} : B^{\mathsf{v}}$
$\mathtt{let}\ V\ \mathtt{be\ x.}\ W$	$\mathtt{let}\ V^{\mathsf{v}}\ \mathtt{be\ x.}\ W^{\mathsf{v}}$
$\mathtt{pm}\ V\ \mathtt{as}\ \{\ldots,(i,\overrightarrow{\mathtt{x}_j}).W_i,\ldots\}$	$\mathtt{pm}\ V^{\mathsf{v}}\ \mathtt{as}\ \{\ldots,(i,(\overrightarrow{\mathtt{x}_j})).W_i{}^{\mathsf{v}},\ldots\}$

Figure A.10. Translation of FG-CBV types, values and returners

Corollary 149 (Full Abstraction) For any FG-CBV returners $A_0,\ldots,A_{n-1} \vdash^{\mathsf{p}} M, M' : B$, if $M \simeq M'$ then $M^{\mathsf{v}} \simeq M'^{\mathsf{v}}$; and similarly for values. (The converse is trivial.) □

Proof Suppose $\mathcal{C}[M^{\mathsf{v}}] \Downarrow m, \mathtt{return}\ i$, for some CBPV ground context $\mathcal{C}$ of type $F\sum_{i\in I}1$. Construct a FG-CBV context $\mathcal{C}'$ such that $\mathcal{C}'^{\mathsf{v}} = \mathcal{C}$ is provable in CBPV. Then we reason as follows.

1 Since in CBPV provable equality implies observational equivalence, $(\mathcal{C}'[M])^{\mathsf{v}} \Downarrow m, \mathtt{return}\ i$ using the fact that $(\mathcal{C}'[M])^{\mathsf{v}}$ is precisely $\mathcal{C}'^{\mathsf{v}}[M^{\mathsf{v}}]$.

2 By Prop. 147(2), $\mathcal{C}'[M] \Downarrow m, i$.

3 Since $M \simeq M'$, we have $\mathcal{C}'[M'] \Downarrow m, i$.

4 By Prop. 147(1), $(\mathcal{C}'[M'])^{\mathsf{v}} \Downarrow m, i$.

5 Since in CBPV provable equality implies observational equivalence, $\mathcal{C}[M'^{\mathsf{v}}] \Downarrow m, \mathtt{return}\ i$.

The proof for values is similar. □

To prove Prop. 148, we give a translation $-^{\mathsf{v}^{-1}}$ from CBPV to FG-CBV. (We shall see that, up to isomorphism, it is inverse to $-^{\mathsf{v}}$.) At first glance, it is not apparent how to make such a translation. For while it will translate a value type into a CBV type, what will it translate a computation type into? The answer is: a family of pairs of CBV types $\{(B_i, C_i)\}_{i \in I}$. To see the principle of the translation, we notice that in CBPV (following Prop. 27) any computation type must be isomorphic to a type of the form $\prod_{i \in I}(B_i \rightarrow FC_i)$. This motivates the following.

Definition A.5 ▪ A *CBV pseudo-computation-type* is a family $\{(B_i, C_i)\}_{i \in I}$ of pairs of CBV types.

▪ Let $A_0, \ldots,, A_{n-1}$ be a sequence of CBV types and let $\{(B_i, C_i)\}_{i \in I}$ be a CBV pseudo-computation-type. We write

$$A_0, \ldots, A_{n-1} \vdash^{\mathsf{c}}_{\mathsf{CBV}} \{M_i\}_{i \in I} : \{(B_i, C_i)\}_{i \in I}$$

to mean that, for each $i \in I$, M_i is an FG-CBV returner $A_0, \ldots, A_{n-1}, \mathtt{n} : B_i \vdash M_i : C_i$. We say that $\{M_i\}_{i \in I}$ is an *FG-CBV pseudo-computation* of type $\{(B_i, C_i)\}_{i \in I}$ on the context $A_0, \ldots,, A_{n-1}$. Given two such pseudo-computations $\{M_i\}_{i \in I}$ and $\{N_i\}_{i \in I}$, we say that they are *provably equal in CBV* when for each $i \in I$ the equation $M_i = N_i$ is provable in CBV.

□

Before presenting the reverse translation, we extend the forward translation as follows:

▪ Given a CBV pseudo-computation-type $\{(A_i, B_i)\}_{i \in I}$, we write $\{(A_i, B_i)\}_{i \in I}{}^{\mathsf{v}}$ for $\prod_{i \in I}(A_i{}^{\mathsf{v}} \rightarrow FB_i{}^{\mathsf{v}})$.

▪ Given an FG-CBV pseudo-computation

$$A_0, \ldots, A_{n-1} \vdash^{\mathsf{c}}_{\mathsf{CBV}} \{M_i\}_{i \in I} : \{(B_i, C_i)\}_{i \in I}$$

we write $\{M_i\}_{i \in I}{}^{\mathsf{v}}$ for

$$A_0{}^{\mathsf{v}}, \ldots, A_{n-1}{}^{\mathsf{v}} \vdash^{\mathsf{c}} \lambda\{\ldots, i.\lambda \mathtt{n} M_i{}^{\mathsf{v}}, \ldots\} : \{(A_i, B_i)\}_{i \in I}{}^{\mathsf{v}}$$

It is clear that the extended translation preserves provable equality.

The reverse translation, presented in Fig. A.11 is organized as follows:

▪ a value type A is translated into a CBV type $A^{\mathsf{v}^{-1}}$

▪ a computation type $\underline{B}$ is translated into a CBV pseudo-computation-type $\underline{B}^{\mathsf{v}^{-1}}$

▪ a CBPV value $A_0, \ldots, A_{n-1} \vdash^{\mathsf{v}} V : B$ is translated into a CBV value $A_0^{\mathsf{v}^{-1}}, \ldots, A_{n-1}^{\mathsf{v}^{-1}} \vdash^{\mathsf{v}} V^{\mathsf{v}^{-1}} : B^{\mathsf{v}^{-1}}$

▪ a computation $A_0, \ldots, A_{n-1} \vdash^{\mathsf{c}} M : \underline{B}$ is translated into an FG-CBV pseudo-computation $A_0^{\mathsf{v}^{-1}}, \ldots, A_{n-1}^{\mathsf{v}^{-1}} \vdash^{\mathsf{c}}_{\mathsf{CBV}} M^{\mathsf{v}^{-1}} : \underline{B}^{\mathsf{v}^{-1}}$.

It is easy to show that the reverse translation commutes with substitution (of values), and thence that it preserves provable equality.

Using this translation (and a result that it agrees with operational semantics in a certain sense) we can obtain from the printing denotational semantics for FG-CBV an alternative semantics for CBPV. Thus a computation type will denote a family of pairs of sets and a computation will denote a family of functions. More generally we can obtain a CBPV semantics from any FG-CBV semantics. However, the construction is artificial and we are not aware of any natural model for CBPV that arises in this way.

A	$A^{\mathsf{v}^{-1}}$	where
$U\underline{B}$	$\prod^{\to}_{i\in I} A_i B_i$	$\underline{B}^{\mathsf{v}^{-1}} = \{(A_i, B_i)\}_{i\in I}$
$\sum_{k\in K} A_k$	$\sum_{k\in K} A_k^{\mathsf{v}^{-1}}$	
$A \times A'$	$A^{\mathsf{v}^{-1}} \times A'^{\mathsf{v}^{-1}}$	

$\underline{B}$	$\underline{B}^{\mathsf{v}^{-1}}$	where
FA	$\{(1, A^{\mathsf{v}^{-1}})\}_{i\in\{*\}}$	
$\prod_{k\in K}\underline{B}_k$	$\{(A_{kl}, B_{kl})\}_{k\in K, l\in L_k}$	$\underline{B}_k^{\mathsf{v}^{-1}} = \{(A_{kl}, B_{kl})\}_{l\in L_k}$
$A \to \underline{B}$	$\{(A^{\mathsf{v}^{-1}} \times A_i, B_i)\}_{i\in I}$	$\underline{B}^{\mathsf{v}^{-1}} = \{(A_i, B_i)\}_{i\in I}$

V	$V^{\mathsf{v}^{-1}}$
$\mathtt{x}$	$\mathtt{x}$
$\mathtt{let}\ V\ \mathtt{be\ x.}\ W$	$\mathtt{let}\ V^{\mathsf{v}^{-1}}\ \mathtt{be\ x.}\ W^{\mathsf{v}^{-1}}$
$(\hat{k}, V)$	$(\hat{k}, V^{\mathsf{v}^{-1}})$
$\mathtt{pm}\ V\ \mathtt{as}\ \{\ldots, (k, \mathtt{x}).W_k, \ldots\}$	$\mathtt{pm}\ V^{\mathsf{v}^{-1}}\ \mathtt{as}\ \{\ldots, (k, \mathtt{x}).W_k^{\mathsf{v}^{-1}}, \ldots\}$
(V, V')	$(V^{\mathsf{v}^{-1}}, V'^{\mathsf{v}^{-1}})$
$\mathtt{pm}\ V\ \mathtt{as}\ (\mathtt{x}, \mathtt{y}).W$	$\mathtt{pm}\ V^{\mathsf{v}^{-1}}\ \mathtt{as}\ (\mathtt{x}, \mathtt{y}).W^{\mathsf{v}^{-1}}$
$\mathtt{thunk}\ M$	$\lambda\{\ldots, \angle i, \mathtt{n}.M_i^{\mathsf{v}^{-1}}, \ldots\}$

M	$M_i^{\mathsf{v}^{-1}}$	where
$\mathtt{let}\ V\ \mathtt{be\ x.}\ M$	$\mathtt{let}\ V^{\mathsf{v}^{-1}}\ \mathtt{be\ x.}\ M_i^{\mathsf{v}^{-1}}$	
$\mathtt{return}\ V$	$\mathtt{return}\ V^{\mathsf{v}^{-1}}$	
$M\ \mathtt{to\ x.}\ N$	$M_*^{\mathsf{v}^{-1}}[()/\mathtt{n}]\ \mathtt{to\ x.}\ N_i^{\mathsf{v}^{-1}}$	
$\mathtt{force}\ V$	$\angle i, \mathtt{n}`\mathtt{force}\ V^{\mathsf{v}^{-1}}$	
$\mathtt{pm}\ V\ \mathtt{as}\ \{\ldots, (k, \mathtt{x}).M_k, \ldots\}$	$\mathtt{pm}\ V^{\mathsf{v}^{-1}}\ \mathtt{as}\ \{\ldots, (k, \mathtt{x}).(M_k)_i^{\mathsf{v}^{-1}}, \ldots\}$	
$\mathtt{pm}\ V\ \mathtt{as}\ (\mathtt{x}, \mathtt{y}).M$	$\mathtt{pm}\ V^{\mathsf{v}^{-1}}\ \mathtt{as}\ (\mathtt{x}, \mathtt{y}).M_i^{\mathsf{v}^{-1}}$	
$\lambda\{\ldots, k.M_k, \ldots\}$	$(M_{\hat{k}})_{\hat{l}}^{\mathsf{v}^{-1}}$	$\hat{\imath} = (\hat{k}, \hat{l})$
$\hat{k}`M$	$M_{\hat{k}\hat{\imath}}^{\mathsf{v}^{-1}}$	
$\lambda\mathtt{x}.M$	$\mathtt{pm\ n\ as}\ (\mathtt{x}, \mathtt{n}).\ M_i^{\mathsf{v}^{-1}}$	
$V`M$	$M_i^{\mathsf{v}^{-1}}[(V^{\mathsf{v}^{-1}}, \mathtt{n})/\mathtt{n}]$	
$\mathtt{print}\ c.\ M$	$\mathtt{print}\ c.\ M_i^{\mathsf{v}^{-1}}$	

Figure A.11. The Reverse Translation $-^{\mathsf{v}^{-1}}$

To show that the two translations are inverse up to isomorphism, we first construct the isomorphisms.

1 For each CBV type A we define a function α_A from CBV values $\Gamma \vdash^{\mathsf{v}} V : A$ to CBV values $\Gamma \vdash^{\mathsf{v}} W : A^{\mathsf{v}\mathsf{v}^{-1}}$, and a function α_A^{-1} in the opposite direction.

2 For each CBPV value type A we define a function β_A from CBPV values $\Gamma \vdash^{\mathsf{v}} V : A$ to values $\Gamma \vdash^{\mathsf{v}} W : A^{\mathsf{v}^{-1}\mathsf{v}}$, and a function β_A^{-1} in the opposite direction.

3 For each CBPV computation type $\underline{B}$ we define a function $\beta_{\underline{B}}$ from CBPV computations $\Gamma \vdash^{\mathsf{c}} M : \underline{B}$ to CBPV computations $\Gamma \vdash^{\mathsf{c}} N : \underline{B}^{\mathsf{v}^{-1}\mathsf{v}}$ and a function $\beta_{\underline{B}}^{-1}$ in the opposite direction.

(1) is defined by induction on A. (2) and (3) are defined by mutual induction on A and $\underline{B}$. We omit the definitions, which are straightforward.

Lemma 150 (properties of α and β)
The following equations are provable in the CBV equational theory.

$$\begin{aligned}\alpha_A(W[V/\mathtt{x}]) &= (\alpha_A W)[V/\mathtt{x}]\\ \alpha_A\alpha_A^{-1}V &= V\\ \alpha_A^{-1}\alpha_A V &= V\end{aligned}$$

The following equations are provable in the CBPV equational theory.

$$\begin{aligned}\beta_A(W[V/\mathtt{x}]) &= (\beta_A W)[V/\mathtt{x}]\\ \beta_{\underline{B}}(M[V/\mathtt{x}]) &= (\beta_{\underline{B}}M)[W/\mathtt{x}]\\ \beta_A\beta_A^{-1}V &= V\\ \beta_A^{-1}\beta_A V &= V\\ \beta_{\underline{B}}\beta_{\underline{B}}^{-1}M &= M\\ \beta_{\underline{B}}^{-1}\beta_{\underline{B}}M &= M\\ \beta_{\underline{B}}(M\ \mathtt{to}\ \mathtt{x}.\ N) &= M\ \mathtt{to}\ \mathtt{x}.\ \beta_{\underline{B}}N\\ \beta_{\underline{B}}(\mathtt{print}\ c.\ N) &= \mathtt{print}\ c.\ \beta_{\underline{B}}N\\ (\alpha_A V)^{\mathsf{v}} &= \beta_{A^{\mathsf{v}}}(V^{\mathsf{v}})\end{aligned}$$

□

These are each proved by induction over types. Using β_A, we have now proved Prop. 148(1).

The translations are inverse up to these isomorphisms, in the following sense:

Lemma 151 For an FG-CBV value $A_0, \ldots, A_{n-1} \vdash^{\mathsf{v}} V : B$, we can prove in CBV

$$V^{\mathsf{vv}^{-1}}[\overrightarrow{\alpha_{A_i}\mathtt{x}_i/\mathtt{x}_i}] = \alpha_B V$$

For an FG-CBV returner $A_0, \ldots, A_{n-1} \vdash^{\mathsf{p}} M : B$, we can prove in CBV

$$M_*^{\mathsf{vv}^{-1}}[\overrightarrow{\alpha_{A_i}\mathtt{x}_i/\mathtt{x}_i}, ()/\mathtt{n}] = M\ \mathtt{to}\ \mathtt{x}.\ \mathtt{return}\ \alpha_B\mathtt{x}$$

For a CBPV value $A_0, \ldots, A_{n-1} \vdash^{\mathsf{v}} V : B$, we can prove in CBPV

$$V^{\mathsf{v}^{-1}\mathsf{v}}[\overrightarrow{\beta_{A_i}\mathtt{x}_i/\mathtt{x}_i}] = \beta_B V$$

For a CBPV computation $A_0, \ldots, A_{n-1} \vdash^{\mathsf{c}} M : \underline{B}$, we can prove in CBPV

$$M^{\mathsf{v}^{-1}\mathsf{v}}[\overrightarrow{\beta_{A_i}\mathtt{x}_i/\mathtt{x}_i}] = \beta_{\underline{B}} M$$

These results can be extended to terms with holes i.e. contexts. □

This is proved by induction over terms using Lemma 150.

We are now in a position to prove Prop. 148(2)–(3). Fix a CBV context $A_0, \ldots, A_{n-1}$ and CBV type B.

Definition A.6 1 For any CBPV value ${A_0}^{\mathsf{v}}, \ldots, {A_{n-1}}^{\mathsf{v}} \vdash^{\mathsf{v}} W : B^{\mathsf{v}}$ define the FG-CBV value θW to be

$$A_0, \ldots, A_{n-1} \vdash^{\mathsf{v}} \alpha_B^{-1} W^{\mathsf{v}^{-1}}[\overrightarrow{\alpha A_i \mathtt{x}_i/\mathtt{x}_i}] : B$$

2 For any CBPV returner ${A_0}^{\mathsf{v}}, \ldots, {A_{n-1}}^{\mathsf{v}} \vdash^{\mathsf{c}} N : FB^{\mathsf{v}}$ define the FG-CBV returner θN to be

$$A_0, \ldots, A_{n-1} \vdash^{\mathsf{p}} N_*^{\mathsf{v}^{-1}}[\overrightarrow{\alpha A_i \mathtt{x}_i/\mathtt{x}_i}, ()/\mathtt{n}] \; \mathtt{to} \; \mathtt{y}. \; \mathtt{return} \; \alpha_B^{-1}\mathtt{y} : B$$

□

Lemma 152 1 For any CBPV returner ${A_0}^{\mathsf{v}}, \ldots, {A_{n-1}}^{\mathsf{v}} \vdash^{\mathsf{c}} N : FB^{\mathsf{v}}$, the equation $(\theta N)^{\mathsf{v}} = N$ is provable in CBPV.

2 For any FG-CBV returner $A_0, \ldots, A_{n-1} \vdash^{\mathsf{p}} M : B$, the equation $\theta(M^{\mathsf{v}}) = M$ is provable in CBV.

Similarly for values. This can all be extended to terms with holes i.e. contexts. □

Proof

(1) By Lemma 151, $N = \beta_{FB^{\mathsf{v}}}^{-1} N^{\mathsf{v}^{-1}\mathsf{v}}[\overrightarrow{\beta_{A_i}\mathtt{x}_i/\mathtt{x}_i}]$ is provable, and we can see that $\beta_{FB^{\mathsf{v}}}^{-1} N^{\mathsf{v}^{-1}\mathsf{v}}[\overrightarrow{\beta_{A_i}\mathtt{x}_i/\mathtt{x}_i}] = (\theta N)^{\mathsf{v}}$ by expanding both sides and using Lemma 150.

(2) This follows directly from Lemma 151.

The proof for values is similar. □

Prop. 148(2)–(3) follows immediately from Lemma 152.

A.6.3 From CBN to CBPV

C	C^{n} (a computation type)
$\sum_{i \in I}^{\times_{j \in \$ r_i}} A_{ij}$	$F\sum_{i \in I}(UA_{i0}^{\mathsf{n}} \times \cdots \times UA_{i(r_i-1)}^{\mathsf{n}})$
$\prod_{i \in I}^{\rightarrow_{j \in \$ r_i}} A_{ij} B_i$	$\prod_{i \in I}(UA_{i0}^{\mathsf{n}} \rightarrow \cdots \rightarrow UA_{i(r_i-1)}^{\mathsf{n}} \rightarrow B_i^{\mathsf{n}})$

$A_0, \ldots, A_{n-1} \vdash M : C$	$UA_0^{\mathsf{n}}, \ldots, UA_{n-1}^{\mathsf{n}} \vdash^{\mathsf{c}} M^{\mathsf{n}} : C^{\mathsf{n}}$
$\mathtt{x}$	$\mathtt{force}\ \mathtt{x}$
$\mathtt{let}\ M\ \mathtt{be}\ \mathtt{x}.\ N$	$\mathtt{let}\ \mathtt{thunk}\ M^{\mathsf{n}}\ \mathtt{be}\ \mathtt{x}.\ N^{\mathsf{n}}$
$(\hat{\imath}, M_0, \ldots, M_{r_{\hat{\imath}}-1})$	$\mathtt{return}\ (\hat{\imath}, (\mathtt{thunk}\ M_0^{\mathsf{n}}, \ldots, \mathtt{thunk}\ M_{r_{\hat{\imath}}-1}^{\mathsf{n}}))$
$\mathtt{pm}\ M\ \mathtt{as}\ \{\ldots, (i, \overrightarrow{\mathtt{x}_j}).N_i, \ldots\}$	$M\ \mathtt{to}\ \mathtt{z}.\ \mathtt{pm}\ \mathtt{z}\ \mathtt{as}\ \{\ldots, (i, (\overrightarrow{\mathtt{x}_j})).M_i, \ldots\}$
$\lambda\{\ldots, \angle i, \mathtt{x}_0, \ldots, \mathtt{x}_{r_i-1}.M_i, \ldots\}$	$\lambda\{\ldots, i.\lambda\mathtt{x}_0. \ldots \lambda\mathtt{x}_{r_i-1}.M_i^{\mathsf{n}}, \ldots\}$
$\angle\hat{\imath}, M_0, \ldots, M_{r_{\hat{\imath}}-1}$‘$N$	$(\mathtt{thunk}\ M_{r_{\hat{\imath}}-1}^{\mathsf{n}})$‘$\cdots(\mathtt{thunk}\ M_0^{\mathsf{n}})$‘$\hat{\imath}$‘$N$

Figure A.12. Translation of CBN types and terms

Lemma 153 Given CBN terms $\Gamma \vdash N : A$ and $\Gamma, \mathtt{x} : A \vdash M : B$, the equation $M[N/\mathtt{x}]^{\mathsf{n}} = M^{\mathsf{n}}[\mathtt{thunk}\ N^{\mathsf{n}}/\mathtt{x}]$ is provable in CBPV. □

Proposition 154 If $M = N$ is provable in the CBN equational theory then $M^{\mathsf{n}} = N^{\mathsf{n}}$ is provable in the CBPV equational theory. □

The technical treatment of this translation is also very similar to the treatment of the translation in Sect. A.5.

This translation does not commute with substitution or preserve operational semantics precisely—only up to the prefix **force thunk**. (Recall that in the CBV equational theory, we have **force thunk** $M = M$.)

substitution It is not the case that $(M[N/\mathtt{x}])^{\mathsf{n}}$ and $M^{\mathsf{n}}[\mathtt{thunk}\ N^{\mathsf{n}}/\mathtt{x}]$ are the same term (although they will be provably equal in the CBPV equational theory). To see this, let M be $\mathtt{x}$. Then $(M[N/\mathtt{x}])^{\mathsf{n}} = N^{\mathsf{n}}$ but $M^{\mathsf{n}}[\mathtt{thunk}\ N^{\mathsf{n}}/\mathtt{x}] = \mathtt{force\ thunk}\ N^{\mathsf{n}}$.

operational semantics It is not the case that $M \Downarrow m, T$ implies $M^{\mathsf{n}} \Downarrow m, T^{\mathsf{n}}$. For example, let M be $\mathtt{let}\ (0)\ \mathtt{be\ x.}\ (0, \mathtt{x})$.

To make the relationship precise, we proceed as follows. First, we write the translation as a relation $\mapsto^{\mathsf{n}}$ from CBN terms to CBPV computations. We can present Fig. A.12 by rules such as these:

$$\frac{M \mapsto^{\mathsf{n}} M' \quad N \mapsto^{\mathsf{n}} N'}{\mathtt{let}\ M\ \mathtt{be\ x.}\ N \mapsto^{\mathsf{n}} \mathtt{let\ thunk}\ M'\ \mathtt{be\ x.}\ N'}$$

To these rules we add the following:

$$\frac{M \mapsto^{\mathsf{n}} M'}{M \mapsto^{\mathsf{n}} \mathtt{force\ thunk}\ M'}$$

Lemma 155 For any term M, we have $M \mapsto^{\mathsf{n}} M^{\mathsf{n}}$, and if $M \mapsto^{\mathsf{n}} M'$ then $M' = M^{\mathsf{n}}$ is provable in the CBPV equational theory. □

Lemma 156 If $M \mapsto^{\mathsf{n}} M'$ and $N \mapsto^{\mathsf{n}} N'$ then $M[N/\mathtt{x}] \mapsto^{\mathsf{n}} M'[\mathtt{thunk}\ N'/\mathtt{x}]$, and similarly for multiple substitution. □

Proposition 157 1 If $M \Downarrow N$ and $M \mapsto^{\mathsf{n}} M'$, then, for some N', $M' \Downarrow N'$ and $N \mapsto^{\mathsf{n}} N'$.

2 If $M \mapsto^{\mathsf{n}} M'$ and $M' \Downarrow N'$, then, for some N, $M \Downarrow N$ and $N \mapsto^{\mathsf{n}} N'$.

□

We prove these by induction primarily on $\Downarrow$ and secondarily on $\mapsto^{\mathsf{n}}$.

A.6.4 From CBPV Back To CBN

Our aim is to prove the following.

Proposition 158 1 Every CBPV computation type $\underline{B}$ is isomorphic to B^{n} for some CBN type B.

2 For any CBPV computation $UA_0^{\mathsf{n}}, \ldots, UA_{n-1}^{\mathsf{n}} \vdash^{\mathsf{c}} N : B^{\mathsf{n}}$ there is a CBN term $A_0, \ldots, A_{n-1} \vdash M : B$ such that $M^{\mathsf{n}} = N$ is provable in CBPV.

3 For any CBN terms $\Gamma \vdash M, M' : B$, if $M^{\mathsf{n}} = M'^{\mathsf{n}}$ is provable in CBPV then $M = M'$ is provable in CBN.

(2)–(3) can be extended to terms with holes i.e. contexts. □

Corollary 159 (Full Abstraction) For any CBN terms $A_0, \ldots, A_{n-1} \vdash M, M' : B$, if $M \simeq M'$ then $M^{\mathsf{n}} \simeq M'^{\mathsf{n}}$. (The converse is trivial.) □

The proof is similar to that of Cor. 149.

To prove Prop. 158, we first give a translation $-^{\mathsf{n}^{-1}}$ from CBPV to CBN. (We shall see that, up to isomorphism, it is inverse to $-^{\mathsf{n}}$.)

At first glance, it is not apparent how to make such a translation. For while it will translate a computation type into a CBN type, what will it translate a value type into? The answer is: a family of CBN types $\{A_i\}_{i \in I}$. This approach is based on [Abramsky and McCusker, 1998].

To see the principle of the translation, we notice that in CBPV (following Prop. 27) any value type must be isomorphic to a type of the form $\sum_{i \in I} U \underline{A}_i$. This motivates the following.

Definition A.7 ■ A *CBN pseudo-value-type* is a family $\{A_i\}_{i \in I}$ of CBN types.

■ Let $\{A_{0i}\}_{i \in I_0}, \ldots, \{A_{(n-1)i}\}_{i \in I_{n-1}}$ be a sequence of CBN pseudo-value-types and let $\{B_j\}_{j \in J}$ be another CBN pseudo-value-type. We write

$$\{A_{0i}\}_{i \in I_0}, \ldots, \{A_{(n-1)i}\}_{i \in I_{n-1}} \vdash^{\mathsf{v}}_{\mathsf{CBN}} \{(V \bullet \overrightarrow{i_j}, V_{\overrightarrow{i_j}})\}_{\overrightarrow{i_j \in I_j}} : \{B_j\}_{j \in J}$$

to mean that, for each $i_0 \in I_0, \ldots, i_{n-1} \in I_{n-1}$, we have $V \bullet \overrightarrow{i_j} \in J$ and $V_{\overrightarrow{i_j}}$ is a CBN term $A_{0i_0}, \ldots, A_{(n-1)i_{n-1}} \vdash V_{\overrightarrow{i_j}} : B_{V \bullet \overrightarrow{i_j}}$. We say that $\{(V \bullet \overrightarrow{i_j}, V_{\overrightarrow{i_j}})\}_{\overrightarrow{i_j \in I_j}}$ is a *CBN pseudo-value* of type $\{B_j\}_{j \in J}$ on the context $\{A_{0i}\}_{i \in I_0}, \ldots, \{A_{(n-1)i}\}_{i \in I_{n-1}}$. Given two such pseudo-values $\{(V \bullet \overrightarrow{i_j}, V_{\overrightarrow{i_j}})\}_{\overrightarrow{i_j \in I_j}}$ and $\{(W \bullet \overrightarrow{i_j}, W_{\overrightarrow{i_j}})\}_{\overrightarrow{i_j \in I_j}}$, we say that they are *provably equal in CBN* when for each $\overrightarrow{i_j \in I_j}$ we have $V \bullet \overrightarrow{i_j} = W \bullet \overrightarrow{i_j}$ and the equation $V_{\overrightarrow{i_j}} = W_{\overrightarrow{i_j}}$ is provable in CBN.

■ Let $\{A_{0i}\}_{i \in I_0}, \ldots, \{A_{(n-1)i}\}_{i \in I_{n-1}}$ be a sequence of CBN pseudo-value-types and let B be a CBN type. We write

$$\{A_{0i}\}_{i \in I_0}, \ldots, \{A_{(n-1)i}\}_{i \in I_{n-1}} \vdash^{\mathsf{c}}_{\mathsf{CBN}} \{M_{\overrightarrow{i_j}}\}_{\overrightarrow{i_j \in I_j}} : B$$

to mean that, for each $i_0 \in I_0, \ldots, i_{n-1} \in I_{n-1}$, we have that $M_{\overrightarrow{i_j}}$ is a CBN term $A_{0i_0}, \ldots, A_{(n-1)i_{n-1}} \vdash M_{\overrightarrow{i_j}} : B$. We say that $\{M_{\overrightarrow{i_j}}\}_{\overrightarrow{i_j \in I_j}}$ is a *CBN pseudo-computation* of type B on the context $\{A_{0i}\}_{i \in I_0}, \ldots, \{A_{(n-1)i}\}_{i \in I_{n-1}}$. Given two such pseudo-computations $\{M_{\overrightarrow{i_j}}\}_{\overrightarrow{i_j \in I_j}}$ and $\{N_{\overrightarrow{i_j}}\}_{\overrightarrow{i_j \in I_j}}$, we say that they are *provably equal in CBN* when for all $\overrightarrow{i_j \in I_j}$ the equation $M_{\overrightarrow{i_j}} = N_{\overrightarrow{i_j}}$ is provable in CBN.

□

Before presenting the reverse translation, we extend the forward translation as follows:

■ Given a CBN pseudo-value-type $\{A_i\}_{i \in I}$ we write $\{A_i\}^{\mathsf{n}}_{i \in I}$ for $\sum_{i \in I} U A_i^{\mathsf{n}}$.

■ Given a CBN pseudo-value

$$\{A_{0i}\}_{i \in I_0}, \ldots, \{A_{(n-1)i}\}_{i \in I_{n-1}} \vdash^{\mathsf{v}}_{\mathsf{CBN}} \{(V \bullet \overrightarrow{i_j}, V_{\overrightarrow{i_j}})\}_{\overrightarrow{i_j \in I_j}} : \{B_j\}_{j \in J}$$

we write $\{(V \bullet \vec{i_j}, V_{\vec{i_j}})\}^{\mathsf{n}}_{\vec{i_j \in I_j}}$ to mean the CBPV value

$$\{A_{0i}\}^{\mathsf{n}}_{i \in I_0}, \ldots, \{A_{(n-1)i}\}^{\mathsf{n}}_{i \in I_{n-1}} \vdash^{\mathsf{v}} \\ \mathtt{pm}\ (\vec{\mathtt{x}_j})\ \mathtt{as}\ \{\ldots, (\vec{i_j}, (\vec{\mathtt{x}_j})).(V \bullet \vec{i_j}, \mathtt{thunk}\ V^{\mathsf{n}}_{\vec{i_j}}), \ldots\} : \{B_j\}^{\mathsf{n}}_{j \in J}$$

- Given a CBN pseudo-computation

$$\{A_{0i}\}_{i \in I_0}, \ldots, \{A_{(n-1)i}\}_{i \in I_{n-1}} \vdash^{\mathsf{c}}_{\mathsf{CBN}} \{M_{\vec{i_j}}\}_{\vec{i_j \in I_j}} : B$$

we write $\{M_{\vec{i_j}}\}^{\mathsf{n}}_{\vec{i_j \in I_j}}$ to mean the CBPV computation

$$\{A_{0i}\}^{\mathsf{n}}_{i \in I_0}, \ldots, \{A_{(n-1)i}\}^{\mathsf{n}}_{i \in I_{n-1}} \vdash^{\mathsf{c}} \mathtt{pm}\ (\vec{\mathtt{x}_j})\ \mathtt{as}\ \{\ldots, (\vec{i_j}, (\vec{\mathtt{x}_j})).M^{\mathsf{n}}_{\vec{i_j}}, \ldots\} : B^{\mathsf{n}}$$

It is clear that this extended translation preserves provable equality.

The reverse translation, presented in Fig. A.13 is organized as follows:

- a value type A is translated into a CBN pseudo-value-type $A^{\mathsf{n}^{-1}}$
- a computation type $\underline{B}$ is translated into a CBN type $\underline{B}^{\mathsf{n}^{-1}}$
- a CBPV value $A_0, \ldots, A_{n-1} \vdash^{\mathsf{v}} B$ is translated into a CBN pseudo-value $A_0^{\mathsf{n}^{-1}}, \ldots, A_{n-1}^{\mathsf{n}^{-1}} \vdash^{\mathsf{v}}_{\mathsf{CBN}} V^{\mathsf{n}^{-1}} : B^{\mathsf{n}^{-1}}$
- a CBPV computation $A_0, \ldots, A_{n-1} \vdash^{\mathsf{c}} M : \underline{B}$ is translated into a CBN pseudo-computation $A_0^{\mathsf{n}^{-1}}, \ldots, A_{n-1}^{\mathsf{n}^{-1}} \vdash^{\mathsf{c}}_{\mathsf{CBN}} M^{\mathsf{n}^{-1}} : \underline{B}^{\mathsf{n}^{-1}}$.

It is easy to show that the reverse translation commutes with substitution (of values) and thence that it preserves provable equality.

Using this translation (and a result that it agrees with operational semantics in a certain sense) we can obtain from the printing denotational semantics for CBN an alternative semantics for CBPV. Thus a computation type will denote a family of sets and a computation will denote a family of functions. More generally we can obtain a CBPV semantics from any CBN semantics. Several important models, such as the Scott model and the game model, can be seen as arising in this way [Abramsky and McCusker, 1998].

To show that the two translations are inverse up to isomorphism, we first construct the isomorphisms.

1. For each CBN type A we define a function α_A from CBN terms $\Gamma \vdash M : A$ to CBN terms $\Gamma \vdash N : A^{\mathsf{nn}^{-1}}$, and a function α_A^{-1} in the opposite direction.
2. For each CBPV value type A we define a function β_A from CBPV values $\Gamma \vdash^{\mathsf{v}} V : A$ to values $\Gamma \vdash^{\mathsf{v}} W : A^{\mathsf{n}^{-1}\mathsf{n}}$, and a function β_A^{-1} in the opposite direction.
3. For each CBPV computation type $\underline{B}$ we define a function $\beta_{\underline{B}}$ from CBPV computations $\Gamma \vdash^{\mathsf{c}} M : \underline{B}$ to CBPV computations $\Gamma \vdash^{\mathsf{c}} N : \underline{B}^{\mathsf{n}^{-1}\mathsf{n}}$ and a function $\beta_{\underline{B}}^{-1}$ in the opposite direction.

(1) is defined by induction on A. (2) and (3) are defined by mutual induction on A and $\underline{B}$. We omit the definitions, which are straightforward.

Lemma 160 (properties of α and β)

A	$A^{\mathsf{n}^{-1}}$	where
$U\underline{B}$	$\{\underline{B}^{\mathsf{n}^{-1}}\}$	
$\sum_{k\in K} A_k$	$\{A_{kl}\}_{k\in K, l\in L_k}$	$A_k^{\mathsf{n}^{-1}} = \{A_{kl}\}_{l\in L_k}$
$A \times A'$	$\{A_k \sqcap A'_l\}_{k\in K, l\in L}$	$A^{\mathsf{n}^{-1}} = \{A_k\}_{k\in K}$ and $A'^{\mathsf{n}^{-1}} = \{A_l\}_{l\in L}$

$\underline{B}$	$\underline{B}^{\mathsf{n}^{-1}}$	where
FA	$\sum_{i\in I} A_i$	$A^{\mathsf{n}^{-1}} = \{A_i\}_{i\in I}$
$\prod_{k\in K} \underline{B}_k$	$\prod_{k\in K} \underline{B}_k^{\mathsf{n}^{-1}}$	
$A \to \underline{B}$	$\prod_{i\in I}(A_i \to \underline{B}^{\mathsf{n}^{-1}})$	$A^{\mathsf{n}^{-1}} = \{A_i\}_{i\in I}$

V	$V^{\mathsf{n}^{-1}} \bullet \overrightarrow{\hat{\imath}_j}$	$V^{\mathsf{n}^{-1}}_{\overrightarrow{\hat{\imath}_j}}$	where
$\mathtt{x}_r$	i_r	$\mathtt{x}_r$	
$\mathtt{let}\ V\ \mathtt{be}\ \mathtt{x}.\ W$	$W^{\mathsf{n}^{-1}} \bullet \overrightarrow{\hat{\imath}_j}\hat{k}$	$\mathtt{let}\ V^{\mathsf{n}^{-1}}_{\overrightarrow{\hat{\imath}_j}}\ \mathtt{be}\ \mathtt{x}.\ W^{\mathsf{n}^{-1}}_{\overrightarrow{\hat{\imath}_j}\hat{k}}$	$V^{\mathsf{n}^{-1}} \bullet \overrightarrow{\hat{\imath}_j} = \hat{k}$
(V, V')	$(V^{\mathsf{n}^{-1}} \bullet \overrightarrow{\hat{\imath}_j}, V'^{\mathsf{n}^{-1}} \bullet \overrightarrow{\hat{\imath}_j})$	$\lambda\{0.V^{\mathsf{n}^{-1}}_{\overrightarrow{\hat{\imath}_j}}, 1.V'^{\mathsf{n}^{-1}}_{\overrightarrow{\hat{\imath}_j}}\}$	
$\mathtt{pm}\ V\ \mathtt{as}\ (\mathtt{x},\mathtt{y}).W$	$W^{\mathsf{n}^{-1}} \bullet \overrightarrow{\hat{\imath}_j}\hat{k}\hat{l}$	$\mathtt{let}\ 0`V^{\mathsf{n}^{-1}}_{\overrightarrow{\hat{\imath}_j}}\ \mathtt{be}\ \mathtt{x}, 1`V^{\mathsf{n}^{-1}}_{\overrightarrow{\hat{\imath}_j}}\ \mathtt{be}\ \mathtt{y}.\ W^{\mathsf{n}^{-1}}_{\overrightarrow{\hat{\imath}_j}\hat{k}\hat{l}}$	$V^{\mathsf{n}^{-1}} \bullet \overrightarrow{\hat{\imath}_j} = (\hat{k}, \hat{l})$
$(\hat{k}, V)$	$(\hat{k}, V^{\mathsf{n}^{-1}} \bullet \overrightarrow{\hat{\imath}_j})$	$V^{\mathsf{n}^{-1}}_{\overrightarrow{\hat{\imath}_j}}$	
$\mathtt{pm}\ V\ \mathtt{as}$ $\{\ldots,(k,\mathtt{x}).W_k,\ldots\}$	$(W_{\hat{k}})^{\mathsf{n}^{-1}} \bullet \overrightarrow{\hat{\imath}_j}\hat{l}$	$\mathtt{let}\ V^{\mathsf{n}^{-1}}_{\overrightarrow{\hat{\imath}_j}}\ \mathtt{be}\ \mathtt{x}.\ (W_{\hat{k}})^{\mathsf{n}^{-1}}_{\overrightarrow{\hat{\imath}_j}\hat{l}}$	$V^{\mathsf{n}^{-1}} \bullet \overrightarrow{\hat{\imath}_j} = (\hat{k}, \hat{l})$
$\mathtt{thunk}\ M$	$*$	$M^{\mathsf{n}^{-1}}_{\overrightarrow{\hat{\imath}_j}}$	

M	$M^{\mathsf{n}^{-1}}_{\overrightarrow{\hat{\imath}_j}}$	where
$\mathtt{let}\ V\ \mathtt{be}\ \mathtt{x}.\ M$	$\mathtt{let}\ V^{\mathsf{n}^{-1}}_{\overrightarrow{\hat{\imath}_j}}\ \mathtt{be}\ \mathtt{x}.\ M^{\mathsf{n}^{-1}}_{\overrightarrow{\hat{\imath}_j}\hat{k}}$	$V^{\mathsf{n}^{-1}} \bullet \overrightarrow{\hat{\imath}_j} = \hat{k}$
$\mathtt{pm}\ V\ \mathtt{as}\ (\mathtt{x},\mathtt{y}).M$	$\mathtt{let}\ 0`V^{\mathsf{n}^{-1}}_{\overrightarrow{\hat{\imath}_j}}\ \mathtt{be}\ \mathtt{x}, 1`V^{\mathsf{n}^{-1}}_{\overrightarrow{\hat{\imath}_j}}\ \mathtt{be}\ \mathtt{y}.\ M^{\mathsf{n}^{-1}}_{\overrightarrow{\hat{\imath}_j}\hat{k}\hat{l}}$	$V^{\mathsf{n}^{-1}} \bullet \overrightarrow{\hat{\imath}_j} = (\hat{k}, \hat{l})$
$\mathtt{pm}\ V\ \mathtt{as}$ $\{\ldots,(k,\mathtt{x}).M_k,\ldots\}$	$\mathtt{let}\ V^{\mathsf{n}^{-1}}_{\overrightarrow{\hat{\imath}_j}}\ \mathtt{be}\ \mathtt{x}.\ (M_{\hat{k}})^{\mathsf{n}^{-1}}_{\overrightarrow{\hat{\imath}_j}\hat{l}}$	$V^{\mathsf{n}^{-1}} \bullet \overrightarrow{\hat{\imath}_j} = (\hat{k}, \hat{l})$
$\mathtt{force}\ V$	$V^{\mathsf{n}^{-1}}_{\overrightarrow{\hat{\imath}_j}}$	
$\mathtt{return}\ V$	$(V^{\mathsf{n}^{-1}} \bullet \overrightarrow{\hat{\imath}_j}, V^{\mathsf{n}^{-1}}_{\overrightarrow{\hat{\imath}_j}})$	
$M\ \mathtt{to}\ \mathtt{x}.\ N$	$\mathtt{pm}\ M^{\mathsf{n}^{-1}}_{\overrightarrow{\hat{\imath}_j}}\ \mathtt{as}\ \{\ldots,(i,\mathtt{x}).N^{\mathsf{n}^{-1}}_{\overrightarrow{\hat{\imath}_j}i},\ldots\}$	
$\lambda\mathtt{x}.M$	$\lambda\{\ldots,i.\lambda\mathtt{x}.M^{\mathsf{n}^{-1}}_{\overrightarrow{\hat{\imath}_j}i},\ldots\}$	
$V`M$	$V^{\mathsf{n}^{-1}}_{\overrightarrow{\hat{\imath}_j}}`V^{\mathsf{n}^{-1}} \bullet \overrightarrow{\hat{\imath}_j}`M^{\mathsf{n}^{-1}}_{\overrightarrow{\hat{\imath}_j}}$	
$\lambda\{\ldots,k.M_k,\ldots\}$	$\lambda\{\ldots,k.(M_k)^{\mathsf{n}^{-1}}_{\overrightarrow{\hat{\imath}_j}},\ldots\}$	
$\hat{k}`M$	$\hat{k}`M^{\mathsf{n}^{-1}}_{\overrightarrow{\hat{\imath}_j}}$	
$\mathtt{print}\ c.\ M$	$\mathtt{print}\ c.\ M^{\mathsf{n}^{-1}}_{\overrightarrow{\hat{\imath}_j}}$	

Figure A.13. The Reverse Translation $-^{\mathsf{n}^{-1}}$

The following equations are provable in the CBN equational theory.

$$\begin{aligned}
\alpha_A(N[M/\mathtt{x}]) &= (\alpha_A N)[M/\mathtt{x}] \\
\alpha_A \alpha_A^{-1} M &= M \\
\alpha_A^{-1} \alpha_A M &= M \\
\alpha_B(\mathtt{pm}\ M\ \mathtt{as}\ \{\ldots,(i,\overrightarrow{\mathtt{x}_j}).N_i,\ldots\}) &= \mathtt{pm}\ M\ \mathtt{as}\ \{\ldots,(i,\overrightarrow{\mathtt{x}_j}).\alpha_B N_i,\ldots\}
\end{aligned}$$

The following equations are provable in the CBPV equational theory.

$$\begin{aligned}
\beta_A(W[V/\mathtt{x}]) &= (\beta_A W)[V/\mathtt{x}] \\
\beta_{\underline{B}}(M[V/\mathtt{x}]) &= (\beta_{\underline{B}} M)[W/\mathtt{x}] \\
\beta_A \beta_A^{-1} V &= V \\
\beta_A^{-1} \beta_A V &= V \\
\beta_{\underline{B}} \beta_{\underline{B}}^{-1} M &= M \\
\beta_{\underline{B}}^{-1} \beta_{\underline{B}} M &= M \\
\beta_{\underline{B}}(M\ \mathtt{to}\ \mathtt{x}.\ N) &= M\ \mathtt{to}\ \mathtt{x}.\ \beta_{\underline{B}} N \\
\beta_{\underline{B}}(\mathtt{print}\ c.\ N) &= \mathtt{print}\ c.\ \beta_{\underline{B}} N \\
(\alpha_A M)^{\mathsf{n}} &= \beta_{A^{\mathsf{n}}}(M^{\mathsf{n}})
\end{aligned}$$

□

These are each proved by induction over types. Using $\beta_{\underline{B}}$, we have now proved Prop. 158(1).

The translations are inverse up to these isomorphisms, in the following sense:

Lemma 161 For a CBN term $A_0, \ldots, A_{n-1} \vdash M : B$, we can prove in CBV

$$M^{\mathsf{nn}^{-1}}_{*\ldots*}[\overrightarrow{\alpha_{A_i}\mathtt{x}_i/\mathtt{x}_i}] = \alpha_B M$$

For a CBPV value $A_0, \ldots, A_{n-1} \vdash^{\mathsf{v}} V : B$, we can prove in CBPV

$$V^{\mathsf{n}^{-1}\mathsf{n}}[\overrightarrow{\beta_{A_i}\mathtt{x}_i/\mathtt{x}_i}] = \beta_B V$$

For a CBPV computation $A_0, \ldots, A_{n-1} \vdash^{\mathsf{c}} M : \underline{B}$, we can prove in CBPV

$$M^{\mathsf{n}^{-1}\mathsf{n}}[\overrightarrow{\beta_{A_i}\mathtt{x}_i/\mathtt{x}_i}] = \beta_{\underline{B}} M$$

These results can be extended to terms with holes i.e. contexts. □

This is proved by induction over terms using Lemma 160.

We are now in a position to prove Prop. 158(2)–(3). Fix a CBN context $A_0, \ldots, A_{n-1}$ and CBN type B.

Definition A.8 For any CBPV computation $UA_0^{\mathsf{n}}, \ldots, UA_{n-1}^{\mathsf{n}} \vdash^{\mathsf{c}} N : B^{\mathsf{n}}$ define the CBN term ϕN to be

$$A_0, \ldots, A_{n-1} \vdash \alpha_B N^{\mathsf{n}^{-1}}_{*\ldots*}[\overrightarrow{\alpha_{A_i}\mathtt{x}_i/\mathtt{x}_i}] : B$$

□

Lemma 162 1 For any CBPV computation $UA_0^{\mathsf{n}}, \ldots, UA_{n-1}^{\mathsf{n}} \vdash^{\mathsf{c}} N : B^{\mathsf{n}}$, the equation $(\phi N)^{\mathsf{n}} = N$ is provable in CBPV.

2 For any CBN term $A_0, \ldots, A_{n-1} \vdash M : B$, the equation $\phi(M^{\mathsf{n}}) = M$ is provable in CBN.

□

Proof

(1) By Lemma 161, $N = \beta_B^{-1} N^{\mathsf{n}^{-1}\mathsf{n}}[\overrightarrow{\beta_{UA_i}\mathtt{x}_i/\mathtt{x}_i}]$ is provable, and we can see that $\beta_B^{-1} N^{\mathsf{n}^{-1}\mathsf{n}}[\overrightarrow{\beta_{UA_i}\mathtt{x}_i/\mathtt{x}_i}] = (\phi N)^{\mathsf{n}}$ by expanding both sides and using Lemma 160.

(2) This follows directly from Lemma 161.

This can all be extended to terms with holes i.e. contexts. □

Prop. 158(2)–(3) follows immediately from Lemma 162.

Appendix B
Models In The Style Of Power-Robinson

B.1 Introduction

In this appendix we give categorical semantics for CBPV without the stack judgement $\vdash^{\mathsf{k}}$. Our purpose is to prove

Proposition 163 The equational theory for CBPV+stacks (Fig. 3.6) conservatively extends the equational theory for CBPV without stacks (Fig. 3.3). This remains true when we include the printing laws in each. □

The categorical semantics presented here does not appear to be a useful way of organizing CBPV models in practice, because all the models we have seen model stacks in a natural way, and we would want to describe this explicitly.

As usual we first define a judgement model (a "staggered Freyd category") and then treat the various connectives. We see, by means of a full reflection, that each model of CBPV of this kind arises from a CBPV adjunction model. We see how to generate a categorical model from a theory, and use this to prove the conservation result.

B.2 Actions of Monoidal Categories

In this section, we review some well-known material.

Definition B.9 1 A *monoidal category* consists of

- a category $\mathcal{C}$;
- an object 1;
- a functor $\otimes : \mathcal{C} \times \mathcal{C} \longrightarrow \mathcal{C}$;
- natural isomorphisms

$$\begin{array}{rcl} A \otimes (B \otimes C) & \cong & (A \otimes B) \otimes C \\ A & \cong & 1 \otimes A \end{array}$$

such that the two structural isomorphisms from $1 \otimes 1$ to 1 are equal and the diagrams

$$\begin{array}{ccccc} A \otimes (B \otimes (C \otimes D)) & \xrightarrow{\cong} & (A \otimes B) \otimes (C \otimes D) & \xrightarrow{\cong} & ((A \otimes B) \otimes C) \otimes D \\ \downarrow{\cong} & & & \nearrow{\cong} & \\ A \otimes ((B \otimes C) \otimes D) & \xrightarrow{\cong} & (A \otimes (B \otimes C)) \otimes D & & \end{array}$$

$$\begin{array}{ccc} & & A \otimes B \\ & \swarrow{\cong} & \downarrow{\cong} \\ (A \otimes 1) \otimes B & \xrightarrow{\cong} & A \otimes (1 \otimes B) \end{array}$$

commute.

2 Let $(\mathcal{C}, 1, \otimes)$ be a monoidal category and $\mathcal{D}$ a category. A *left $\mathcal{C}$-action* on $\mathcal{D}$ consists of a functor $\oslash$ from $\mathcal{C} \times \mathcal{D}$ to $\mathcal{D}$, together with natural isomorphisms

$$\begin{array}{rcl} A \oslash (B \oslash Z) & \cong & (A \otimes B) \oslash Z \\ Z & \cong & 1 \oslash Z \end{array}$$

such that the diagrams

$$\begin{array}{ccccc} A \oslash (B \oslash (C \oslash Z)) & \xrightarrow{\cong} & (A \otimes B) \oslash (C \oslash Z) & \xrightarrow{\cong} & ((A \otimes B) \otimes C) \oslash Z \\ \downarrow \cong & & & \nearrow \cong & \\ A \oslash ((B \otimes C) \oslash Z) & \xrightarrow{\cong} & (A \otimes (B \otimes C)) \oslash Z & & \end{array}$$

$$\begin{array}{ccc} & & A \oslash Z \\ & \swarrow \cong & \downarrow \cong \\ (A \otimes 1) \oslash Z & \xrightarrow{\cong} & A \oslash (1 \oslash Z) \end{array}$$

commute. A *right* $\mathcal{C}$*-action* on $\mathcal{D}$ is defined similarly.

□

It is clear that

- every cartesian category is monoidal;
- every monoidal category has a canonical left action and right action on itself.

Proposition 164 (coherence) 1 Given a monoidal category $\mathcal{C}$, every diagram built from structural isomorphisms commutes.

2 Given a monoidal category $\mathcal{C}$ and a left $\mathcal{C}$-action on a category $\mathcal{D}$, every diagram built from structural isomorphisms commutes.

□

A precise statement and a proof of Prop. 164(1) can be found in [Mac Lane, 1971]. Prop. 164(2) is stated and proved similarly.

B.3 Freyd Categories

Despite its name, the following has no connection with Freyd and is not a category (although, as observed in [Power, 2002], it can be seen as a $[\cdot \to \cdot, \mathbf{Set}]$-enriched category). We use a different, but equivalent, definition to that which appears in [Power and Thielecke, 1999].

Definition B.10 Let $\mathcal{C}$ be a cartesian category and $\mathcal{K}$ a category such that $\mathsf{ob}\,\mathcal{K} = \mathsf{ob}\,\mathcal{C}$—we write a morphism in $\mathcal{K}$ as $A \xrightarrow{\quad f \quad} B$. A *Freyd category* from $\mathcal{C}$ to $\mathcal{K}$ consists of

- an identity-on-objects functor ι from $\mathcal{C}$ to $\mathcal{K}$;
- a left $\mathcal{C}$-action on $\mathcal{K}$, extending (along ι) the canonical left $\mathcal{C}$-action on $\mathcal{C}$.

We call the left action $\times$, because it is given on objects by $\times$. □

Suppose we are given a Freyd category from $\mathcal{C}$ to $\mathcal{K}$. We can obtain a *right* $\mathcal{C}$-action on $\mathcal{K}$, extending (along ι) the canonical right $\mathcal{C}$-action on $\mathcal{C}$. This action too we call $\times$, because it is given on objects by $\times$. In summary, we write $f \times g$ in 3 situations:

- if f and g are both $\mathcal{C}$-morphisms, then $f \times g$ is a $\mathcal{C}$-morphism;
- if f is a $\mathcal{C}$-morphism and g is a $\mathcal{K}$-morphism, then $f \times g$ is a $\mathcal{K}$-morphism;
- if f is a $\mathcal{K}$-morphism and g is a $\mathcal{C}$-morphism, then $f \times g$ is a $\mathcal{K}$-morphism.

By contrast, if f and g are both $\mathcal{K}$-morphisms, then $f \times g$ is not defined. So it is not the case (in general) that $\mathcal{K}$ forms a monoidal category under $\times$.

In the sense of Power and Robinson [Power and Robinson, 1997], $\mathcal{K}$ is a *symmetric premonoidal category* and ι is a *strict symmetric premonoidal functor*. This provides the definition of Freyd category in [Power and Thielecke, 1999].

A Freyd category models the $1, \times$-fragment of fine-grain CBV (the language defined in Sect. A.3.2). This means that it models computations of F type only.

B.4 Judgement Model

Whereas computations of F type form a category, computations of arbitrary type do not, and we therefore require a more general structure.

Definition B.11 A *staggered category* $\mathcal{E}$ consists of

- a class of *source objects* $\mathsf{sourceob}\ \mathcal{E}$;
- a class of *target objects* $\mathsf{targetob}\ \mathcal{E}$, whose elements we underline;
- a *source-to-target function* $F : \mathsf{sourceob}\ \mathcal{E} \rightarrow \mathsf{targetob}\ \mathcal{E}$;
- for each $A \in \mathsf{sourceob}\ \mathcal{E}$ and $\underline{B} \in \mathsf{targetob}\ \mathcal{E}$ a set $\mathcal{E}(A, \underline{B})$ of *morphisms from A to $\underline{B}$*;
- for each $A \in \mathsf{sourceob}\ \mathcal{E}$, an identity morphism $A \xrightarrow{\mathsf{id}_A} FA$;
- for each $A \xrightarrow{f} FB$ and $B \xrightarrow{g} \underline{C}$, a composite morphism $A \xrightarrow{f;g} \underline{C}$, which we sometimes write as

$$A \xrightarrow{f} B \xrightarrow{g} \underline{C}$$

satisfying identity and associativity laws

$$\begin{aligned} \mathsf{id}; f &= f \\ f; \mathsf{id} &= f \\ (f; g); h &= f; (g; h) \end{aligned}$$

□

Definition B.12 Let $\mathcal{E}$ be a staggered category. We write $\mathcal{E}_F$ for the ordinary category given by

$$\begin{aligned} \mathsf{ob}\ \mathcal{E}_F &= \mathsf{sourceob}\ \mathcal{E} \\ \mathcal{E}_F(A, B) &= \mathcal{E}(A, FB) \end{aligned}$$

with the evident composition. □

With this construction $\mathcal{E}_F$ in mind, we often write $A \xrightarrow{f} B$ to say that f is a $\mathcal{E}$-morphism from A to FB. This agrees with the notation for composition in Def. B.11.

Now we give a judgement model for CBPV (without stacks):

Definition B.13 A *staggered Freyd category* consists of

- a cartesian category $\mathcal{C}$, whose objects we call *val-objects*
- a staggered category $\mathcal{E}$ such that $\mathsf{sourceob}\ \mathcal{E} = \mathsf{ob}\ \mathcal{C}$—we call the target objects of $\mathcal{E}$ *comp-objects*
- a Freyd category $(\iota, \times)$ from $\mathcal{C}$ to $\mathcal{E}_F$.

□

It is clear that this models the $1, \times, F$ fragment of CBPV (without stacks).

- A value type (and hence a context) denotes a val-object.
- A value $\Gamma \vdash^{\mathsf{v}} V : A$ denotes a $\mathcal{C}$-morphism from $[\![\Gamma]\!]$ to $[\![A]\!]$.
- A computation type $\underline{B}$ denotes a comp-object.
- A computation $\Gamma \vdash^{\mathsf{c}} M : \underline{B}$ denotes a $\mathcal{E}$-morphism from $[\![\Gamma]\!]$ to $[\![\underline{B}]\!]$.

Note that a Freyd category can be described as a staggered Freyd category in which $\mathsf{sourceob}\ \mathcal{E} = \mathsf{targetob}\ \mathcal{E}$ and F is identity (on objects).

B.5 Enrichment

As in Sect. 9.5.5, we define enrichment for judgement models.

Definition B.14 An *$\mathcal{A}$-set enriched staggered Freyd category* is a staggered Freyd category where each homset of $\mathcal{E}$ is equipped with an $\mathcal{A}$-set structure, and the equations

$$\begin{array}{rcl} (c*f);g & = & c*(f;g) \\ (\iota f);(c*g) & = & c*((\iota f);g) \\ A\times(c*g) & = & c*(A\times g) \end{array}$$

are satisfied. □

In a similar way we can define cppo enriched and domain enriched staggered Freyd categories.

B.6 Connectives

Definition B.15 Let $(\mathcal{C},\mathcal{E},\iota,\times)$ be a staggered Freyd category.

1 Write $\mathcal{N}$ for the family of left $\mathcal{C}$-modules $\mathcal{E}(\iota-,\underline{Y})$ indexed by target objects $\underline{Y}$. A *distributive coproduct* for a family of $\mathcal{C}$-objects $\{A_i\}_{i\in I}$ is a distributive coproduct for $\{A_i\}_{i\in I}$ in $(\mathcal{C},\mathcal{N})$, in the sense of Def. 9.29 and Def. 11.16.

2 A *product* for a family of comp-objects $\{\underline{B}_i\}_{i\in I}$ is a representing object for

$$\prod_{i\in I}\mathcal{E}(-,\underline{B}_i) : \mathcal{E}_F^{\mathsf{op}} \to \mathbf{Set}$$

3 An *exponential* from a val-object A to a comp-object $\underline{B}$ is a representing object for

$$\mathcal{E}(-\times A,\underline{B}) : \mathcal{E}_F^{\mathsf{op}} \to \mathbf{Set}$$

item A *right adjunctive* for a comp-object $\underline{B}$ in a staggered Freyd category $(\mathcal{C},\mathcal{E},\iota,\times)$ is a representing object for

$$\mathcal{E}(\iota-,\underline{B}) : \mathcal{C}^{\mathsf{op}} \to \mathbf{Set}$$

□

Proposition 165 Let $(\mathcal{C},\mathcal{E},\iota,\times)$ be a staggered Freyd category with all right adjunctives. Then a distributive coproduct for $\{A_i\}_{i\in I}$ in $\mathcal{C}$ is automatically a distributive coproduct in $(\mathcal{C},\mathcal{E},\iota,\times)$. (The converse is trivial.) □

Proof Similar to Prop. 100. □

We can also define composite connectives in the manner of Sect. 9.8.1.

Definition B.16 Let $(\mathcal{C},\mathcal{E},\iota,\times)$ be a staggered Freyd category.

1 An *exponential-product* for a family of val-object/comp-object pairs $(A_i,\underline{B}_i)$ is a representing object for

$$\prod_{i\in I}\mathcal{E}(-\times A_i,\underline{B}_i) : \mathcal{E}_F^{\mathsf{op}} \to \mathbf{Set}$$

2 A *right product-adjunctive* for a family of comp-objects $\{\underline{B}_i\}_{i \in I}$ is a representing object for

$$\prod_{i \in I} \mathcal{E}(\iota -, \underline{B}_i) : \mathcal{C}^{\mathsf{op}} \to \mathbf{Set}$$

3 A *right exponential-adjunctive* from a val-object A to a comp-object $\underline{B}$ is a representing object for

$$\mathcal{E}(\iota - \times A, \underline{B}) : \mathcal{C}^{\mathsf{op}} \to \mathbf{Set}$$

4 An *right exponential-product-adjunctive* for a family of val-object/comp-object pairs $(A_i, \underline{B}_i)$ is a representing object for

$$\prod_{i \in I} \mathcal{E}(\iota - \times A_i, \underline{B}_i) : \mathcal{C}^{\mathsf{op}} \to \mathbf{Set}$$

□

B.7 Modelling CBPV

We now come to our main definition.

Definition B.17 A ($\mathcal{A}$-set/cppo/domain enriched) *CBPV staggered Freyd category* is a ($\mathcal{A}$-set/cppo/domain enriched) staggered Freyd category with all countable distributive coproducts, countable products, exponentials and right adjunctives. □

It is easy to see how to interpret CBPV in such a structure, and we obtain a direct model/categorical model equivalence as in Chap. 10. For a CBPV object structure τ, we define a *τ-sequent* to be either $A_0, \ldots, A_{n-1} \vdash^{\mathsf{v}} B$ or $A_0, \ldots, A_{n-1} \vdash^{\mathsf{c}} \underline{B}$, where $A_0, \ldots, A_{n-1}$ and B and $\underline{B}$ are τ-objects. A *τ-signature* s provides a set $s(Q)$ for each τ-sequent Q.

Given a τ-signature s, we define the set of terms built from s. These are inductively defined by the rules of JWA+complex values, together with the rules

$$\frac{\Gamma \vdash^{\mathsf{v}} V_0 : A_0 \quad \cdots \quad \Gamma \vdash^{\mathsf{v}} V_{r-1} : A_{r-1}}{\Gamma \vdash^{\mathsf{v}} f(V_0, \ldots, V_{r-1}) : B} \; f \in s\,(A_0, \ldots, A_{r-1} \vdash^{\mathsf{v}} B)$$

$$\frac{\Gamma \vdash^{\mathsf{v}} V_0 : A_0 \quad \cdots \quad \Gamma \vdash^{\mathsf{v}} V_{r-1} : A_{r-1}}{\Gamma \vdash^{\mathsf{c}} f(V_0, \ldots, V_{r-1}) : \underline{B}} \; f \in s\,(A_0, \ldots, A_{r-1} \vdash^{\mathsf{c}} \underline{B})$$

We take the least congruence $\sim_s$ on these terms containing the equational laws of JWA, which as in Prop. 108 and Prop. 110 is a hypercongruence (closed under substitution), and quotienting by this congruence obtain a new signature called $\mathcal{T}s$. This gives a monad $\mathcal{T}$ on the category of τ-signatures.

Proposition 166 Let τ be a CBPV object structure. The "categorical semantics" functor from CBPV staggered Freyd categories with object structure τ to the category of algebras for $\mathcal{T}$ on Sig_τ is an equivalence. □

This is proved in the same way as Prop. 109 and Prop. 111.

B.8 The Full Reflection

For a fixed CBPV type structure τ, write

- **Adj**$_\tau$ for the category of ($\mathcal{A}$-set/cppo/domain enriched) CBPV adjunction models on τ
- **ValProd**$_\tau$ for the category of ($\mathcal{A}$-set/cppo/domain enriched) CBPV staggered Freyd categories on τ

where, in each case, the morphisms are identity-on-objects maps preserving all structure. We will construct a full reflection (i.e. adjunction whose counit is an isomorphism—in our case, it will be the identity)

$$\mathbf{ValProd}_\tau \underset{\mathcal{F}}{\overset{\mathcal{U}}{\rightleftarrows}} \mathbf{Adj}_\tau \quad (\top)$$

The easy direction is the following.

Definition B.18 Suppose we have a ($\mathcal{A}$-set/cppo/domain enriched) CBPV judgement model $\mathcal{A} = (\mathcal{C}, \mathcal{D}, \mathcal{O})$ with all left adjunctives. We obtain a ($\mathcal{A}$-set/cppo/domain enriched) staggered Freyd category $\mathcal{FA} = (\mathcal{C}, \mathcal{E}, \iota, \times)$ as follows.

- The val-objects and comp-objects are as in the CBPV judgement model.
- The homsets of $\mathcal{E}$ are given by

$$\mathcal{E}(A, \underline{B}) = \mathcal{O}_A\underline{B}$$

- The identity on A is given by return.
- The composite of $\xrightarrow[A]{f} FB$ with $\xrightarrow[B]{g} \underline{C}$ is given by $f; (\mathsf{q}^F\pi'^*g)$.
- If $B \xrightarrow{f} C$ in $\mathcal{C}$, then ιf is given by $f^*\mathsf{return}$.
- If $\xrightarrow[B]{g} FC$ then $A \times g$ is given by $(\pi'^*g); (\pi^*\mathsf{q}^F\mathsf{return})$.

Furthermore, if we are given a distributive coproduct, product, exponential, or right adjunctive in $\mathcal{A}$, we obtain the corresponding structure in $\mathcal{FA}$ in the evident way. □

The harder direction is the following.

Definition B.19 Let $\mathcal{M} = (\mathcal{C}, \mathcal{E}, \iota, \times)$ be a ($\mathcal{A}$-set/cppo/domain enriched) staggered Freyd category. We obtain a ($\mathcal{A}$-set/cppo/domain enriched) CBPV judgement model $\mathcal{UM} = (\mathcal{C}, \mathcal{D}, \mathcal{O})$ with all left adjunctives as follows.

- The objects of $\mathcal{D}$ are the comp-objects.
- A $\mathcal{D}$-morphism over A from $\underline{B}$ to $\underline{C}$ is a natural transformation from $\mathcal{E}(-, \underline{B})$ to $\mathcal{E}(A \times -, \underline{C})$.
- The identity over A on $\underline{B}$ takes $X \xrightarrow{g} \underline{B}$ to $(\iota\pi'); f$.
- The composite of

$$\underline{B} \xrightarrow[A]{h} \underline{C} \xrightarrow[A]{h'} \underline{D}$$

  takes $X \xrightarrow{g} \underline{B}$ to $(\iota(\pi, (\pi, \pi'))); h'(h(g))$

- The reindexing of $\underline{B} \xrightarrow[A]{h} \underline{C}$ along $A' \xrightarrow{k} A$ takes $X \xrightarrow{g} \underline{B}$ to $(\iota(k \times X)); h(g)$.

- The $\mathcal{O}$ homsets are given by

$$\mathcal{O}_A\underline{B} = \mathcal{E}(A, \underline{B})$$

- The composite of

$$\xrightarrow[A]{g} \underline{B} \xrightarrow[A]{h} \underline{C}$$

  is $(\iota(\mathsf{id}, \mathsf{id})); h(g)$.

- The reindexing of $\xrightarrow[A]{g} \underline{B}$ along $A' \xrightarrow{k} A$ is $(\iota k); g$.
- The left adjunctive of A has vertex FA and return $= \iota\mathsf{id}$. The required inverse of

$$\begin{array}{rcl} \mathcal{D}_X(FA, \underline{Y}) & \longrightarrow & \mathcal{O}_{X \times A}\underline{Y} \\ h & \longmapsto & (\pi'^{*}\mathsf{return}); (\pi^{*}h) \end{array} \qquad \text{for all } X, \underline{Y}$$

  takes $X \times A \xrightarrow{g} \underline{Y}$ to the natural transformation that at W takes $W \xrightarrow{f} FA$ to $(X \times f); g$.

Furthermore, if we are given a distributive coproduct, product, exponential, or right adjunctive in $\mathcal{M}$, we obtain the corresponding structure in $\mathcal{UM}$ in the evident way. □

Proposition 167 Let $\mathcal{M}$ be a CBPV staggered Freyd category. Then $\mathcal{FUM} = \mathcal{M}$. □

This is straightforward to prove. Thus we can set the counit of our full reflection to be the identity, while the unit is defined as follows.

Definition B.20 Let $\mathcal{A} = (\mathcal{C}, \mathcal{D}, \mathcal{O})$ be a CBPV adjunction model. We define an identity-on-objects adjunction model morphism $\mathcal{A} \xrightarrow{\mathsf{unit}\,\mathcal{A}} \mathcal{UFA}$ that

- is identity on $\mathcal{C}$-morphisms and $\mathcal{O}$-morphisms
- takes a $\mathcal{D}$-morphism $\underline{B} \xrightarrow[A]{h} \underline{C}$ to the natural transformation which at X takes $\xrightarrow[X]{g} \underline{B}$ to $(\pi'^{*}g); (\pi^{*}h)$.

□

Finally, we have to verify the triangle laws for an adjunction. Because the counit is identity, these are quite simple.

- $\mathcal{F}\mathsf{unit}_{\mathcal{A}} = \mathsf{id}_{\mathcal{FA}}$. This simply says that $\mathsf{unit}_{\mathcal{A}}$ is identity on $\mathcal{C}$-morphisms, which is true by definition.
- $\mathsf{unit}_{\mathcal{UM}} = \mathsf{id}_{\mathcal{UM}}$. This is easily verified.

B.9 Theories

Each of our equivalences between direct models and categorical models can be formulated as an equivalence between *theories* and categorical models. A theory is a presentation of a direct model, similar to a presentation of a group by generators and relations. We need this formulation for our conservativity result.

We present this material for $\times$-calculus, but it adapts straightforwardly to CBPV with stacks, to JWA, and, most importantly for our present purposes, to CBPV without stacks.

Recall that the *kernel* of a function $A \xrightarrow{f} B$ is the equivalence relation

$$\ker f = \{(x, y) \in A \times A | fx = fy\}$$

Definition B.21 Let τ be a $\times$ object structure.

1 A *theory* on τ for $\times$-calculus is a τ-signature s together with a hypercongruence $\sim$ on terms generated from r.

2 An *interpretation* of a τ-theory $(s, \sim)$ in a direct τ-model (r, θ) is a signature morphism $s \xrightarrow{i} r$ such that, writing $[\![-]\!]_i$ for

$$\mathcal{T}s \xrightarrow{\mathcal{T}i} \mathcal{T}r' \xrightarrow{\theta} r$$

we have

$$\sim \subset \ker [\![-]\!]_i$$

i.e. terms related by $\sim$ have the same denotation.

3 A *model* of $(s, \sim)$ is a direct τ-model (r, θ) together with an interpretation i of $(s, \sim)$ in (r, θ).

4 A *model morphism* from a τ-model (r, θ, i) to a τ-model (r', θ', i'), of $(s, \sim)$ is a τ signature morphism $r \xrightarrow{f} r'$ such that $i; f = i'$.

□

Proposition 168 Let τ be a $\times$ object structure, and let $(s, \sim)$ be a theory on τ. There the category of models for $(s, \sim)$ has an initial object (r, i), and this satisfies

$$\sim = \ker[\![-]\!]i$$

i.e. terms are related by $\sim$ iff they have the same denotation. □

This is entirely analogous to the construction of a group from generators and relations. It is sometimes called the *classifying model* of the theory.

Proof We define r to consist of equivalence classes under $\sim$ of terms under $\sim$, and obtain a quotienting map $\mathcal{T}s \xrightarrow{q} r$. We define θ by induction in the obvious way, and show by induction that $(\mathcal{T}q); \theta = \mu_s; r$. We define i to be $\eta_r; q$. The required properties are straightforward. □

We now reformulate Def. B.21 and Prop. 168 using categorical models rather than direct models.

Definition B.22 Let τ be a $\times$ object structure.

1 A *categorical model* of a τ-theory $(s, \sim)$ on τ is a τ-cartesian category $\mathcal{C}$ on τ together with an interpretation i of $(s, \sim)$ in $\mathcal{IC}$.

2 A *model morphism* from a τ categorical model $(\mathcal{C}, i)$ to another τ categorical model $(\mathcal{C}', i')$ is an identity-on-objects structure-preserving functor $i \xrightarrow{F} i'$ such that $i; \mathcal{I}F = i'$

□

Unpacking, we see that (1) is the usual definition of categorical model, whereas (2) is more restrictive than usual in its requirement that the functor be identity-on-objects. This is in keeping with our fixing of object structure throughout.

Proposition 169 Let τ be a $\times$ object structure, and let $(s, \sim)$ be a theory on τ. There the category of categorical models for $(s, \sim)$ has an initial object $(\mathcal{C}, i)$, and this satisfies

$$\sim \; = \; \ker[\![-]\!]i$$

i.e. terms are related by $\sim$ iff they have the same denotation. □

This is sometimes called the *classifying category* of the theory.

B.10 Conservativity

Proposition 170 Let $\mathcal{A}$ be a ($\mathcal{A}$-set/cppo/domain enriched) CBPV adjunction model. If P is a term of CBPV without stacks (with printing/recursion), then the denotation of P in $\mathcal{A}$ is equal to the denotation of P in $\mathcal{FA}$. □

Proposition 171 There is a ($\mathcal{A}$-set enriched) CBPV staggered Freyd category $\mathcal{M}$ such that parallel (i.e. same context and type) terms P and Q (with printing) have the same denotation if they are provably equal in the equational theory for CBPV without stacks (with printing). □

Proof This follows from Prop. 166 the same way that Prop. 169 follows from Prop. 109. □

Finally, we can achieve the goal of the chapter: to prove Prop. 163.

Proof Let $\mathcal{M}$ be defined as in Prop. 171. Let P, Q be parallel terms in CBPV without stacks (with printing). Suppose P and Q are provably equal in CBPV+stacks (with printing). Then they have the same denotation in every adjunction model, in particular in $\mathcal{UM}$, and so (by Prop. 170) the same denotation in $\mathcal{FUM}$, which is $\mathcal{M}$ (Prop. 167). By the definition of $\mathcal{M}$, they must be provably equal in CBPV without stacks (with printing). □

References

Abramsky, S. (1990). The lazy lambda-calculus. In *Research topics in Functional Programming*, pages 65–117. Addison Wesley. (p 20)

Abramsky, S. (1992). Games and full completeness for multiplicative linear logic (extended abstract). In Shyamasundar, R., editor, *Proceedings of Foundations of Software Technology and Theoretical Computer Science*, volume 652 of *LNCS*, pages 291–301, Berlin, Germany. Springer. (p 169)

Abramsky, S., Honda, K., and McCusker, G. (1998). A fully abstract game semantics for general references. In *Proc., 13th Annual IEEE Symposium on Logic in Computer Science.* (pp 125, 169, 171, 172, 186, 193)

Abramsky, S., Jagadeesan, R., and Malacaria, P. (1994). Full abstraction for PCF (extended abstract). In Hagiya, M. and Mitchell, J. C., editors, *Theoretical Aspects of Computer Software. International Symposium TACS'94*, volume 789 of *LNCS*, pages 1–15, Sendai, Japan. Springer-Verlag. (pp 70, 169)

Abramsky, S. and McCusker, G. (1997). Linearity, sharing and state: a fully abstract game semantics for Idealized Algol with active expressions. In O'Hearn, P. W. and Tennent, R. D., editors, *Algol-like languages.* Birkhaüser. (p 169)

Abramsky, S. and McCusker, G. (1998). Call-by-value games. In Nielsen, M. and Thomas, W., editors, *Computer Science Logic: 11th International Workshop Proceedings*, LNCS. Springer-Verlag. (pp xxix, 84, 175, 187, 246, 288, 289, 321, 322)

Aczel, P. (1988). *Non-Well-Founded Sets.* Center for the Study of Language and Information, Stanford University. CSLI Lecture Notes, Volume 14. (p 80)

Altenkirch, T. and Reus, B. (1999). Monadic presentations of lambda terms using generalized inductive types. In Flum, J. and Rodríguez-Artalejo, M., editors, *Proc. of 13th Int. Workshop on Computer Science Logic, CSL'99, Madrid, Spain, 20–25 Sept. 1999*, volume 1683 of *Lecture Notes in Computer Science*, pages 453–468. Springer-Verlag, Berlin. (pp 81, 217)

Appel, A. W. (1991). *Compiling with Continuations.* CUP. (p 141)

Appel, A. W. and Jim, T. (1989). Continuation-passing, closure-passing style. In *Conference Record of the 16th Annual ACM Symposium on Principles of Programming Languages*, pages 293–302. ACM, ACM. (p 141)

Benton, N. and Wadler, P. (1996). Linear logic, monads and the lambda calculus. In *Proceedings, 11th Annual IEEE Symposium on Logic in Computer Science*, pages 420–431, New Brunswick. IEEE Computer Society Press. (pp xxxi, 83, 294)

Bierman, G. M. (1998). A computational interpretation of the $\lambda\mu$-calculus. In Brim, L., Gruska, J., and Zlatuska, J., editors, *Mathematical Foundations of Computer Science, Proc. 23rd Int. Symp.*, volume 1450 of *LNCS*, pages 336–345, Brno, Czech Republic. Springer-Verlag, Berlin. (p 32)

Borceux, F. (1994). *Handbook of Categorical Algebra 1.* Cambridge University Press. (p 208)

Carboni, A., Kelly, G. M., Verity, D., and Wood, R. J. (1998). A 2-categorical approach to change of base and geometric morphisms ii. *Theory and Applications of Categories*, 4(5):82–136. (pp 208, 262, 269, 270)

Carboni, A., Lack, S., and Walters, R. F. C. (1993). Introduction to extensive and distributive categories. *Journal of Pure and Applied Algebra*, 84:145–158. (p 16)

Cockett, J. R. B. (1993). Introduction to distributive categories. *Mathematical Structures in Computer Science*, 3(3):277–307. (p 16)

Curien, P.-L. (1998). Abstract Böhm trees. *Mathematical Structures in Computer Science*, 8(6):559–591. (p 174)

Curien, P.-L. and Herbelin, H. (2000). The duality of computation. In *Proc., ACM Sigplan International Conference on Functional Programming (ICFP-00)*, volume 35.9 of *ACM Sigplan Notices*, pages 233–243, N.Y. ACM Press. (p 288)

Danos, V., Herbelin, H., and Regnier, L. (1996). Game semantics and abstract machines. In *Proceedings of the Eleventh Annual IEEE Symposium On Logic In Computer Science (LICS'96)*, pages 394–405, New York. IEEE Computer Society Press. (pp 173, 174)

Danos, V. and Regnier, L. (1999). Reversible, irreversible and optimal λ-machines. *Theoretical Computer Science*, 227(1–2):79–97. (p 173)

Danvy, O. (1992). Back to direct style. In *ESOP'92, Proc. European Symposium on Programming, Rennes*, volume 582 of *LNCS*, pages 130–150. Springer Verlag. (pp 141, 142)

Duba, B., Harper, R., and MacQueen, D. (1991). Typing first-class continuations in ML. In ACM-SIGACT, A.-S., editor, *Conference Record of the 18th Annual ACM Symposium on Principles of Programming Languages (POPL '91)*, pages 163–173, Orlando, FL, USA. ACM Press. (pp 96, 150)

Felleisen, M. and Friedman, D. P. (1986). Control operators, the SECD-machine, and the λ-calculus. In Wirsing, M., editor, *Formal Description of Programming Concepts*, pages 193–217. North-Holland. (pp xxxviii, 27, 32)

Filinski, A. (1996). *Controlling Effects.* PhD thesis, School of Computer Science, Carnegie Mellon University, Pittsburgh, Pennsylvania. (pp xxii, xxxi, xxxii)

Fiore, M., Plotkin, G. D., and D.Turi (1999). Abstract syntax and variable binding. In Longo, G., editor, *Proceedings of the 14th Annual Symposium on Logic in Computer Science (LICS'99)*, pages 193–202, Trento, Italy. IEEE Computer Society Press. (pp 81, 217)

Freyd, P. J. (1991). Algebraically complete categories. In Carboni, A. et al., editors, *Proc. 1990 Como Category Theory Conference*, pages 95–104, Berlin. Springer-Verlag. Lecture Notes in Mathematics Vol. 1488. (pp 72, 74)

Friedman, H. (1978). Classically and intuitionistically provably recursive functions. In Muller, S., editor, *Higher Set Theory*, volume 699 of *Lecture Notes in Mathematics*, pages 21–28. Springer-Verlag. (pp 142, 158)

Ghica, D. R. (1997). Semantics of dynamic variables in algol-like languages. Master's thesis, Queens' University, Kingston,Ontario. (pp xxxix, 125, 136)

Girard, J.-Y. (1987). Linear logic. *Theoretical Computer Science*, 50:1–102. (p xxxi)

Girard, J.-Y., Lafont, Y., and Taylor, P. (1988). *Proofs and Types.* Cambridge Tracts in Theoretical Computer Science 7. Cambridge University Press. (pp 7, 158)

Goguen, J., Thatcher, J., and Wagner, E. (1979). An initial algebra approach to the specification, correctness, and implementation of abstract data types. In Yeh, R., editor, *Current Trends in Programming Methodology IV*, pages 80–149. Prentice-Hall. (p 83)

Griffin, T. G. (1990). The formulae-as-types notion of control. In *Conf. Record 17th Annual ACM Symp. on Principles of Programming Languages, POPL'90, San Francisco, CA, USA, 17–19 Jan 1990*, pages 47–57. ACM Press, New York. (pp 142, 158)

Gunter, C. A. (1995). *Semantics of Programming Languages.* MIT Press. (p xxix)

Harmer, R. and McCusker, G. (1999). A fully abstract game semantics for finite nondeterminism. In *LICS: IEEE Symposium on Logic in Computer Science.* (p 169)

Hatcliff, J. and Danvy, O. (1997). Thunks and the λ-calculus. *Journal of Functional Programming*, 7(3):303–319. (pp xxx, xxxi, 20, 151, 312)

Hennessy, M. C. B. (1980). The semantics of call-by-value and call-by-name in a nondeterministic environment. *SIAM Journal on Computing*, 9(1):67–84. (p xxix)

Hennessy, M. C. B. and Ashcroft, E. A. (1980). A mathematical semantics for a nondeterministic typed lambda -calculus. *Theoretical Computer Science*, 11(3):227–245. (p 20)

Hofmann, M. (1995). Sound and complete axiomatisations of call-by-value control operators. *Mathematical Structures in Computer Science*, 5(4):461–482. (p 238)

Hofmann, M. and Streicher, T. (1997). Continuation models are universal for $\lambda\mu$-calculus. In *Proc., Twelfth Annual IEEE Symposium on Logic in Computer Science*, pages 387–395, Warsaw, Poland. IEEE Computer Society Press. (pp 103, 151, 158)

Honda, K. and Yoshida, N. (1997). Game theoretic analysis of call by value computation. In Degano, P., Gorrieri, R., and Marchetti-Spaccamela, A., editors, *Automata, Languages and Programming, 24th International Colloquium*, volume 1256 of *LNCS*, pages 225–236, Bologna, Italy. Springer-Verlag. (pp xxix, 175)

Huth, M., Jung, A., and Keimel, K. (2000). Linear types and approximation. *Math. Struct. in Comp. Science*, 10:719–745. (p 110)

Hyland, J. M. E. and Ong, C.-H. L. (2000). On full abstraction for PCF: I, II, and III. *Information and Computation*, 163(2):285–408. (pp xxix, 47, 66, 169, 174, 175, 176, 201)

Ingerman, P. Z. (1961). Thunks: A way of compiling procedure statements with some comments on procedure declarations. *Communications of the ACM*, 4(1):55–58. (p xxxiv)

Jacobs, B. (1999). *Categorial Logic and Type Theory.* Elsevier, Amsterdam. (p 268)

Jeffrey, A. (1999). A fully abstract semantics for a higher-order functional language with nondeterministic computation. *TCS: Theoretical Computer Science*, 228. (p 251)

Jung, A. (1990). Colimits in DCPO. 3-page manuscript, available by fax. (p 136)

Krivine, J.-L. (1985). Un interpréteur de λ-calcul. Unpublished. (p 32)

Krivine, J.-L. (2001). personal communication. (p 54)

Lafont, Y., Reus, B., and Streicher, T. (1993). Continuation semantics or expressing implication by negation. Technical Report 9321, Ludwig-Maximilians-Universität, München. (pp 142, 158)

Laird, J. (1997). Full abstraction for functional languages with control. In *Proceedings, Twelfth Annual IEEE Symposium on Logic in Computer Science*, pages 58–67, Warsaw, Poland. IEEE Computer Society Press. (pp 20, 169)

Laird, J. (1998). *A Semantic Analysis of Control.* PhD thesis, University of Edinburgh. (p 103)

Laird, J. (2003). A categorical semantics of higher-order store. In *Proc., 9th Conference on Category Theory and Computer Science, Ottawa, 2002*, volume 69 of *Electronic Notes in Theoretical Computer Science.* (p 171)

Lambek, J. and Scott, P. (1986). *Introduction to Higher Order Categorical Logic.* Cambridge University Press, Cambridge. (p 235)

Landin, P. J. (1964). The mechanical evaluation of expressions. *The Computer Journal*, 6(4):308–320. (p 32)

Landin, P. J. (1966). The next 700 programming languages. *Communications of the ACM*, 9(3):157–164. Originally presented at the Proceedings of the ACM Programming Language and Pragmatics Conference, August 8–12, 1965. (p xxii)

Laurent, O. (1999). Polarized proof-nets: proof-nets for LC (extended abstract). In Girard, J.-Y., editor, *Typed Lambda Calculi and Applications '99*, volume 1581 of *Lecture Notes in Computer Science*, pages 213–227. Springer. (p xxxi)

Laurent, O. (2002a). *Etude de la polarisation en logique.* PhD thesis, Université Aix-Marseille II. (p 175)

Laurent, O. (2002b). Polarized games. In *Logic in Computer Science*, pages 265–274, Los Alamitos, CA, USA. IEEE Computer Society. (p 174)

Lawvere, F. W. (1963). *Functional Semantics of Algebraic Theories.* PhD thesis, Columbia University. (p 216)

Levy, P. B. (1996). λ-calculus and cartesian closed categories. Essay for Part III of the Mathematical Tripos, Cambridge University, manuscript. (p 251)

Levy, P. B. (2001). *Call-by-push-value.* PhD thesis, Queen Mary, University of London. (p xxi)

Levy, P. B. (2002). Possible world semantics for general storage in call-by-value. In Bradfield, J., editor, *Proc., 16th Annual Conference in Computer Science Logic, Edinburgh, 2002*, volume 2471 of *LNCS*, pages 232–246. Springer. (pp xxxix, 126)

Levy, P. B. (2003). Adjunction models for call-by-push-value with stacks. In *Proc., 9th Conference on Category Theory and Computer Science, Ottawa, 2002*, volume 69 of *Electronic Notes in Theoretical Computer Science.* (p xxii)

Levy, P. B., Thielecke, H., and Power, A. J. (2003). Modelling environments in call-by-value programming languages. *Information and Computation*, 185:182–210. (p 287)

Longley, J. and Plotkin, G. (1997). Logical full abstraction and PCF. In Ginzburg, J., editor, *Tbilisi Symposium on Language, Logic and Computation.* SiLLI/CSLI. Also available as LFCS Report ECS-LFCS-97-353. (p 169)

Mac Lane, S. (1971). *Categories for the Working Mathematician*, volume 5 of *Graduate Texts in Mathematics.* Springer-Verlag, New York. (pp 24, 236, 237, 264, 284, 329)

Marz, M. (2000). *A Fully Abstract Model for Sequential Computation.* PhD thesis, Technische Universität Darmstadt. published by Logos-Verlag, Berlin. (p xxxi)

Marz, M., Rohr, A., and Streicher, T. (1999). Full abstraction and universality via realisability. In *LICS: IEEE Symposium on Logic in Computer Science.* (p xxxi)

McCusker, G. (1996). *Games and Full Abstraction for a Functional Metalanguage with Recursive Types.* PhD thesis, University of London. (pp 80, 174, 176, 179)

McCusker, G. (1997). Games and definability for **FPC**. *The Bulletin of Symbolic Logic*, 3(3):347–362. (p 169)

Moggi, E. (1988). Computational lambda-calculus and monads. LFCS Report ECS-LFCS-88-66, University of Edinburgh. (pp xxxi, 18)

Moggi, E. (1991). Notions of computation and monads. *Information and Computation*, 93:55–92. (pp xxii, xxviii, xxxi, xxxii, 14, 18, 20, 89, 90, 114, 216, 293, 307)

Moggi, E. (90). An abstract view of programming languages. Technical Report ECS-LFCS-90-113, Dept. of Computer Science, Edinburgh Univ. (pp xxix, 125, 136)

Murthy, C. (1990). *Extracting Constructive Content from Classical Proofs*. PhD thesis, Cornell University, Department of Computer Science. (p 158)

Nickau, H. (1996). *Hereditarily Sequential Functionals: A Game-Theoretic Approach to Sequentiality*. Shaker-Verlag. Dissertation, Universität Gesamthochschule Siegen. (pp xxix, 169)

Odersky, M. (1994). A functional theory of local names. In ACM, editor, *Proceedings of 21st Annual ACM SIGACT-SIGPLAN Symposium on Principles of Programming Languages (POPL)*, pages 48–59, New York, NY, USA. ACM Press. (p 137)

O'Hearn, P. W. (1993). Opaque types in algol-like languages. manuscript. (pp 90, 116)

O'Hearn, P. W. and Tennent, R. D. (1995). Parametricity and local variables. *Journal of the ACM*, 42(3):658–709. (pp 136, 137)

Oles, F. J. (1982). *A Category-Theoretic Approach to the Semantics of Programming Languages*. Ph. D. dissertation, Syracuse University. (pp xxix, 125, 136)

Ong, C. H. L. (1988). *The Lazy Lambda Calculus: An Investigation into the Foundations of Functional Programming*. PhD thesis, Imperial College of Science and Technology. (p 20)

Ong, C.-H. L. (1996). A semantics view of classical proofs: type-theoretic, categorical, denotational characterizations. In *Proc. 11th IEEE Annual Symposium on Logic in Computer Science*, pages 230–241. IEEE Computer Society Press. (p 160)

Parigot, M. (1992). $\lambda\mu$-calculus: An algorithmic interpretation of classical natural deduction. In Voronkov, A., editor, *Proceedings of the International Conference on Logic Programming and Automated Reasoning (LPAR'92)*, volume 624 of *LNAI*, pages 190–201, St. Petersburg, Russia. Springer Verlag. (p 160)

Park, D. (1968). Some semantics for data structures. In Michie, D., editor, *Machine Intelligence 3*, pages 351–371. American Elsevier, New York. (p xxxiii)

Pitts, A. M. (1996). Relational properties of domains. *Information and Computation*, 127:66–90. (pp 69, 74, 76, 140)

Pitts, A. M. and Stark, I. D. B. (1993). Observable properties of higher order functions that dynamically create local names, or: What's new? In *Mathematical Foundations of Computer Science, Proc. 18th Int. Symp., Gdańsk, 1993*, volume 711 of *LNCS*, pages 122–141. Springer-Verlag, Berlin. (p 125)

Plotkin, G. D. (1976). Call-by-name, call-by-value and the λ-calculus. *Theoretical Computer Science*, 1(1):125–159. (pp xxxi, 20, 150, 151)

Plotkin, G. D. (1977). LCF as a programming language. *Theoretical Computer Science*, 5. (pp xxviii, 20, 47)

Plotkin, G. D. (1983). Domains. 1992 TeXed edition of course notes prepared by Yugo Kashiwagi and Hidetaka Kondoh from notes by Tatsuya Hagino. (p 84)

Plotkin, G. D. (1985). Lectures on predomains and partial functions. Course notes, Center for the Study of Language and Information, Stanford. (pp xxviii, 13)

Plotkin, G. D. and Power, A. J. (2002). Notions of computation determine monads. In *Proc., Foundations of Software Science and Computation Structures 2002, Grenoble, France*, volume 2303. LNCS. (pp 260, 294)

Power (2002). Premonoidal categories as categories with algebraic structure. *Theoretical Computer Science*, 278(1–2):302–331. (p 329)

Power, A. J. and Robinson, E. P. (1997). Premonoidal categories and notions of computation. *Math. Struct. in Comp. Sci.*, 7(5):453–468. (p 329)

Power, A. J. and Thielecke, H. (1999). Closed Freyd- and kappa-categories. In *Proc. ICALP '99*, volume 1644 of *LNCS*, pages 625–634. Springer-Verlag, Berlin. (p 329)

Reynolds, J. C. (1981). The essence of Algol. In de Bakker, J. W. and van Vliet, J. C., editors, *Algorithmic Languages*, pages 345–372, Amsterdam. North-Holland. (pp 46, 125)

Sabry, A. and Felleisen, M. (1993). Reasoning about programs in continuation-passing style. *Lisp and Symbolic Computation*, 6(3/4):289–360. (pp 141, 142, 161)

Sangiorgi, D. (1999). Interpreting functions as pi-calculus processes: a tutorial. RR 3470, INRIA Sophia-Antipolis, France. (pp xxx, xxxi)

Sazonov, V. Y. (1976). Expressibility of functions in D. Scott's LCF language. *Algebra i Logika*, 15(3):308–330. Translation. (p 47)

Scott, D. S. (1982). Domains for denotational semantics. In Nielson, M. and Schmidt, E. M., editors, *Automata, Languages and Programming: Proceedings 1982*. Springer-Verlag, Berlin. LNCS **140**. (p 80)

Selinger, P. (2001). Control categories and duality: On the categorical semantics of the lambda-mu calculus. *Mathematical Structures in Computer Science*, 11(2):207–260. (pp 261, 288)

Simpson, A. K. (1992). Recursive types in Kleisli categories. Unpublished manuscript. (p 24)

Smyth, M. and Plotkin, G. D. (1982). The category-theoretic solution of recursive domain equations. *SIAM J. Computing*, 11. (pp 69, 70, 86)

Stark, I. (1996). A fully abstract domain model for the π-calculus. In *Proceedings of the Eleventh Annual IEEE Symposium on Logic in Computer Science*, pages 36–42. IEEE Computer Society Press. (p 70)

Stark, I. D. B. (1994). *Names and Higher-Order Functions*. PhD thesis, University of Cambridge. (pp xxix, 125, 136)

Steele, G. L. (1978). RABBIT: A compiler for SCHEME. Technical Report 474, Artificial Intelligence Laboratory, Massachusetts Institute of Technology, Cambridge, MA. MSc Dissertation. (pp xxxix, 141, 142)

Stoltenberg-Hansen, V., Lindstroem, I., and Griffor, E. R. (1994). *Mathematical Theory of Domains*. Cambridge University Press, Cambridge, 1 edition. (pp 84, 85)

Streicher, T. and Reus, B. (1998). Classical logic, continuation semantics and abstract machines. *Journal of Functional Programming*, 8(6):543–572. (pp xxix, 32, 90, 103, 116, 288)

Tait, W. W. (1967). Intensional interpretation of functionals of finite type I. *Journal of Symbolic Logic*, 32(2):198–212. (pp 30, 67, 134, 153)

Tennent, R. D. and Ghica, D. R. (2000). Abstract models of storage. *Higher-Order and Symbolic Computation*, 13(1–2):119–129. (p xxxiii)

Thielecke, H. (1997a). *Categorical Structure of Continuation Passing Style*. PhD thesis, University of Edinburgh. (pp 103, 141, 142, 149, 151)

Thielecke, H. (1997b). Continuation semantics and self-adjointness. In *Proceedings MFPS XIII*, Electronic Notes in Theoretical Computer Science. Elsevier. (pp 226, 261)

Wadler, P. (2003). Call-by-value is dual to call-by-name. Proceedings, 8th ACM SIGPLAN International Conference on Functional Programming, Uppsala, Sweden, to appear. (p 288)

Wadsworth, C. P. (1976). The relation between computational and denotational properties for Scott's D_∞-models of the lambda-calculus. *SIAM Journal on Computing*, 5(3):488–521. (p 20)

Index

$*$ (concatenation of strings), 8
\$, xxxii
$\sum_{i \in I}^{\times_{j \in \$ r_i}}$, 300
$\prod_{i \in I}^{\rightarrow_{j \in \$ r_i}}$, 301
$\angle$ (delimits several arguments), 301
$\backslash$ (composition), 191
β-law, 6, 53, 59, 163
$\blacktriangleleft$ (where we are now), 145
$\bot$ as a term, 78
$\bot$ as a value type, 79
$\bullet$ (dismantling a stack), 36
$\bullet$ (glueing right module), 269, 270
$\mathcal{C}_A B$, 210
$\mathcal{N}_A$, 210
$+\!\!+$ (concatenation of stacks), 36
:: (stack formation), 33
$\cdot$:: (stack formation), 98
$\uplus$ (disjoint union of arenas), 177
$\emptyset$ (empty arena), 177
η-expansion, 188
η-law, 6, 26, 51, 53, 59, 151, 163, 312
$\triangleleft_{\mathsf{fin}}$, 78, 79, 179
γ (point in JWA), 143
$\hat{\imath}$, 28
λ_{c}-calculus, xxxi, 18
$\bullet\!\!\rightarrow$, 72
let, 4
$A \xrightarrow{\quad g \quad}$, 208
$\xrightarrow{\quad g \quad} B$, 208
μ (recursion), 67
$\nearrow$ (jump in JWA), 143
$\neg$ (type of jump-points in JWA), 142, 143
$\simeq_{\mathsf{anytype}}$, 17, 18
$\simeq_{\mathsf{ground}}$, 17, 18
$\$_C$, 138
ϕ (specifies type of cell contents), 122
π-calculus, xxx, xxxi
$\mapsto^{\mathsf{n}}$, 320
$\mapsto^{\mathsf{cg}}$, 309
$\mapsto^{\mathsf{cgval}}$, 309
$\mapsto^{\mathsf{lazy}}$, 313
$\mapsto^{\mathsf{lazyval}}$, 313
$\upharpoonright$ (arena operation), 177, 196
$\upharpoonright$ (thread), 191
$\simeq$, *see* observational equivalence
$\times$-calculus, 216, 249
$\underline{\bot}$ as a computation type, 79
$\vdash$ in an arena, 177
$\vdash^{\mathsf{c}}$, 29
$\vdash^{\mathsf{k}}$, 33
$\vdash^{\mathsf{n}}$, 143
$\vdash^{\mathsf{p}}$, 305
$\vdash^{\mathsf{v}}$, 29, 143, 305
* (composition), 210
$^{\mathsf{x}}M$ (weakening), xxxiii, 98
‘ (application backwards), xxxii
; (sequencing of commands), 46
; (composition in diagrammatic order), xxxii
=: (assignment backwards), 180
""(empty string), 8

$\mathcal{A}$ (set of printable characters), 8
$\mathcal{A}^*$, 8
$\mathcal{A}$-set, 3, 21
  free, 21
$\mathcal{A}$Set, 132, 225
$\mathcal{A}$-set enriched, *see* enriched
abstract material, 207
acceptable infinitely deep term, 77, 78
action
  left, 328, 329
  right, 329
adequacy
  cell generation + divergence, 134
  erratic choice, 108
  infinitely deep terms, 79
  infinitely deep types, 79

pointer games, 194, 195
recursion, 67
recursive types, 76
thunk storage, 140
adjoint
left, 235
right, 235
adjunction, 235, 278–286
admissible subcategory, 72
admissible subset, 67
algebra for monad, 213, 236, 283
denoted by CBN type, 24–25
denoted by computation type, 41
direct model as, 253
exponential, 25
free, 24
homomorphism denoted by stack, 41
product, 25
via algebraic operations, 294
algebraic operation, 260, 294
Ans in continuation semantics, 101
<u>Ans</u> in continuation semantics, 105
answer-move, 195
answer-move pointing to answer-move, 197
arena, 174
definition of, 176
Q/A-labelled, 195
stackfree-in-stack, 199, 248
stackfree-in-value, 199, 248
universal, 174
ArenaIso, 177, 190
arity of operation symbol, 251
assignment, 91, 118

Böhm tree, 77
bad cells, 172
Beck-Chevalley condition, 281
behaviours
computation type denotes set of, 92, 107, 126
big-step semantics
for CBN, 19
for CBPV, 31
with cell generation, 121
with erratic choice, 107
with global store, 92
with printing, 38
for CBPV errors, 112
for CBV, 10
unsuitable for control effects, 95
bilimit, 70
bimodule, 209
bracketable game, 200
bracketed move, 200
bracketing condition, 171, 202, 248

$\mathcal{C}_A B$, 210

$^c p$, 269, 270
C-machine, 151–154
with printing, 153
calculus
λbool+, 4
$\times\sum\prod\rightarrow$, 61, 143
$\times$
direct model, 253–254
semantics of types, 250
jumbo λ, *see* jumbo λ-calculus
call-by-name, xxix, 9, 18–25, 311
equational theory, 311
call-by-need, xxix, 20
call-by-push (another name for CBN), 54
call-by-push-value, 27–336
reason for name, 54
call-by-value, xxix, 9–18
coarse-grain, 17, 304, 308–310
equational theory, 307
fine-grain, 18, 305–310
carrier
of $\mathcal{A}$-set, 21
of algebra, 24
CartCat$_\tau$, 251
Cat, 216
category
bicartesian closed, 17, 26, 63
bilimit-compact, 70, 85, 111, 140
cartesian, 15, 250
cartesian closed, 23, 26, 83
countably distributive, 237
distributive, 16, 63
enriched, 70
Kleisli for a monad, 15
monoidal, 328
sub-bilimit-compact, 72–73, 85, 111
value, 63
CBN, *see* call-by-name
CBN adjunction model, 287, 289
CBPV, *see* call-by-push-value
CBPV adjunction model, 234
CBPV judgement model, 222
CBPV pre-families model, 247
CBV, *see* call-by-value
CBV adjunction model, 287, 290
cell, xxxiii, 91, 117
cell generation, 117–140, 171, 192
cell type, 122
changestk, 95
CK-machine, 32–37, 52–54, 95–102, 154
with printing, 38
classical logic, 157–160
classifying category, 336
classifying model, 335
co-Kleisli
strong adjunction, 286
co-Kleisli adjunction, 284

co-Kleisli part, 285
of a strong adjunction, 286, 287
cocone, 70
code jumpabout, 145
coerce, 46
coherence, 254
of monoidal categories and actions, 329
coinduction, 81
comm, 136
command, 46, 136
as prefix, 8, 118
nonreturning, *see* nonreturning command, 97
comp-object, 222, 254, 330
comparison functor, 235
compatible relation, 78
complete lattice, 108, 111
complex stacks, 58
complex values, 50–51, 119
eliminability of, 54
in CBV, 307
in Jump-With-Argument, 155
compositionality, 83, 102
computability, 66
computation, xxxiv
computation type, xxxvi
computation-unaffecting, 47, 49, 164
computational λ-calculus, *see* λ_c-calculus
computational effects , *see* effects
computations
categorical semantics of, 222
concatenation of stacks, 36, 58, 221
formal definition, 256
configuration, 152
of C-machine for JWA, 152
of CK-machine, 32, 35
of CK-machine with control, 98
conservative extension, 59
contents of cell, 120
context (list of typed identifiers), 28
context (term with hole), 17, 18
context extension, 263
context morphism, 216
continuation, 32, 36, 103
continuation semantics, *see* control effects, 101–106, 238
continuation-passing-style, *see* StkPS
continuous function, 13
control effects, 95–106, 171
combined with other effects, 105–106
control flow, xxxvi, 148, 195
coproduct-jumpwith, 190, 243, 246, 265
copycat, 188, 190, 193
copycat, 190
counit of adjunction, 235

Cpo, 13
cpo, 13
pointed, 22
$\mathbf{Cpo}^{\perp}$ (cppos and strict continuous functions), 70, 71
cpo semantics,
seeScott semantics13
cppo, *see* cpo, pointed
cppo enriched, *see* enriched
staggered Freyd category, 331
CPS transforms, 149–151
current stack, 95
currying, 61
cusl, 84

data of a jump, 176
decomposition
$+_{\mathbf{CBN}}$ into CBPV, 44
$\to_{\mathbf{CBN}}$ into CBPV, 44, 104
$\to_{\mathbf{CBN}}$ into linear logic, 294
$\to_{\mathbf{CBV}}$ into CBPV, 42
$\to_{\mathbf{CBV}}$ into monadic metalanguage, 294
definability
pointer games, 194
dependent types, 296–297
descended evenly, 185, 193
descended oddly, 184, 193
deterministic strategy, 185, 187
dinatural, 130
$\mathbf{Direct}_{\tau}$, 253
direct model
for $\times$-calculus, 253–254
direct model/categorical model equivalence, 249–260
dismantling a stack, 36, 37, 58, 222
formal definition, 256
distributive coproduct, 16, 218
in CBPV judgement model, 232
in JWA judgement model, 231
with left modules, 231, 273
diverge, 67
divergence, xxviii, 65
with cell generation, 133–135
domain, 83, 289
domain enriched, *see* enriched
staggered Freyd category, 331
domain equation, 69, 139
DomStr (domains and strict continuous functions), 83, 225
duality
between CBV and CBN, 288
between values and stacks, 98, 104, 239, 288

E-set, 113
edge in multigraph, 251
effect-free language, 216

effects, xxviii
Eilenberg-Moore, 213, 236, 283, 294
element style definition, 16, 265, 267
empty stack, 98
end, 295
enriched
    in $[\mathcal{C}^{op}, \mathbf{Set}]$, 211
    in **Cpo**, 67
    judgement model, 190
    judgement models, 223, 260
    staggered Freyd category, 331
environment, xxxiii
(e,p)-pair, 70
equality testing of cells, 118
equational theory
    CBPV + control, 162
    CBPV with stacks, 58
    CBPV without stacks, 51–52, 328
    for CBN, 311
    for CBV, 307
    for JWA, 155–156
erratic choice, 107–109, 239
    finite, 109, 294
errors, 111–114
evaluation context, 32
exception handling, 46, 111
exponential, 16, 218, 265
    in CBPV judgement model, 230
    in staggered Freyd category, 331
    Kleisli, 17, 237
    with right modules, 229, 271
exponential-product, 265
    in staggered Freyd category, 331
    with right modules, 271

F_S, 138
families
    of arenas, 175, 178
families construction, 243–248, 288–290
fglow, 110
fibre of locally indexed category, 211
finite approximant, 78, 79, 179
finitely generated lower set, 110
fold, 69
force as jump, xxxvi, 148, 195
forcing a thunk, xxxiv
forest, 176
Freyd category, 329
full abstraction
    of pointer game model, 171, 194
    of translation from CBN to CBPV, xxxii, 320
    of translation from CBV to CBPV, xxxii, 314
full completeness, 169
full reflection, 333
function types in jumbo λ-calculus, 301–302

functor
    contravariant, denoted by computation type, 129, 131
    covariant, denoted by value type, 124
    locally continuous, 73

global morphism, 211
global representing object, 267
global store, 91–95, 241, 296
Grothendieck construction, 264
ground context, 17, 18, 39
    in JWA, 155
ground returner, 13, 29
ground term, 22
ground type, 4, 29, 301
ground value, 29
grow, 264
growall, 264
$\mathsf{growhom}_{\times\mathcal{CN}}$, 272

head normal form, 20
hom, 209, 214
$\mathsf{hom}_{\times\mathcal{C}}$, 268
$\mathsf{hom}_{\times\mathcal{CN}}$, 272
$\mathsf{hom}^{\mathcal{OD}}$, 269, 270
homomorphism
    between $\mathcal{A}$-sets, 41
    between algebras, 41
    of E-sets, 114
homset functor, 209, 214
hyper-arity, 300
hypercongruence, 252, 257

$\mathcal{I}$ (categorical semantics), 253
ideal, 110, 111
Idealized Algol, 136
idempotents split, 237, 284
identifier, xxxiii
ignored, 46
·**ignored**, 161, 167
included in a strategy, 185
indexed category, 211
indifference theorem, 151
infinite play, 185, 187
infinitely deep syntax, 77–82, 140
infinitely wide syntax, xxxvi, 65, 132
initial object
    distributive, 17
initialization, 118, 121, 192
innocence, 174
innocence condition, 171
input, 90, 294
interaction sequence, 191, 193
interpretation of a theory, 335
invariant
    for functor, 73

invisible constructs, 40, 45
  thunk and **force**, 42
isomorphism of arena families, 178
isomorphism of arenas, 177
  Q/A-respecting, 196

j (isomorphism), 277
join-semilattice, 111
joint naturality, 262
judgement model
  CBPV, 222
  JWA, 220
Jumbo λ-calculus, 300–302
jump, xxxvi, 143
jump-point, 103, 143
Jump-With-Argument, 141–168, 295
  as type theory for classical logic, 157–160
  C-machine, 151–154
    with printing, 153
  dependent types, 296
  embedding into CBPV+Ans, 144
  equational theory, 155
  graphical syntax, 144
  jumping machine, 145–148, 154
  polymorphic types, 297
jumpabout, 145
  code, 145
  trace, 145
jumping implementation
  of CBPV, 148
jumping machine, 145–148, 154
jumpwith, 226, 265
justified sequence, 184
JWA, *see* Jump-With-Argument
JWA judgement model, 220
JWA module model, 234
JWA pre-families model, 246

${}^{k}Y$, 269, 270
kernel, 335
Kleisli, 236
  strong adjunction, 285
Kleisli adjunction, 284
Kleisli exponential, 17, 237
Kleisli part, 284
  of a strong adjunction, 286, 287
Kleisli product, 237

lazy, 20, 312–314
left adjunctive, 227, 267
left coproduct-adjunctive, 244, 247, 267
left module, 190, 208, 220
letstk, 95
LICat, 216
lift, 23
lift-reflecting, 67

lluf subcategory, 72
locally continuous, 73, 75, 138
locally indexed
  functor, 212
  natural transformation, 216
locally indexed category, 211
locally indexed functor, 212
logic
  classical, 157–160
  intuitionistic, 26, 158
  linear, xxxi, 110, 173, 295
    drawbacks of decomposition, 295
lower powerdomain, 110

may-testing equivalent, 109
metalanguage, CBPV as a, 45, 94, 105
minimal invariant, 138
  for functor, 73
mkpseudo, 172
model morphism, 335
model of a theory, 335
module
  left, *see* left module
  right, *see* right module
monad, 235
  definition of, 15
  drawbacks of decomposition, 294
  lifting, 17
  printing, 17
  semantics for CBV, 17
  strong, 15, 24, 114
  $\mathcal{T}$ on MGraph_{τ}, 252, 255
monadic metalanguage, xxxi, 83, 294, 307
move, 174
move with pointer, 183
move without pointer, 183
multicusl, 84
multigraph, 251

$\mathcal{N}$ (set of cell types), 122
$\mathcal{N}_A$, 210
$\mathcal{N}_{\mathsf{c}}$, 232
$\mathcal{N}_{\mathsf{k}}$, 232
nat, 65
naturality
  joint vs. separate, 262
naturality style definition, 265, 267
negation
  classical, 157
new, 118
newrec, 193
newrec, 118, 193
nil, 33, 98
${}^{\mathtt{nil}}P$, 98
no thunk storage, 123
non-closed
  CBPV computation, evaluation of, 33

JWA command, evaluation of, 143
nonreturning command, 46, 180
nonreturning commands
categorical semantics of, 220
nonreturning-command game, 183
nonreturning-command play, 185
nonreturning-command strategy, 185
nStrat, 185

O-move, 182, 185, 187
obj, 210
object
cell represented as, 171
object structure
×, 250
CBPV, 254
JWA, 259
observational equivalence
for CBN and lazy, 18
for CBPV, 39
may-testing, 109
with cell generation, 122
with control, 100
with global store, 92
with printing, 39
for CBV, 17
for JWA, 155
OpBimod, 111
opbimodule, 110, 240
pointwise ideal, 110, 240
operation symbol, 251
opGr, 213
opGrothendieck construction, 213
Opponent-awaiting play, 185, 187
Opponent-move, *see* O-move

P-bracketed play, 201
P-move, 182, 185, 187
parametricity, 137
parametrized invariant, 74
parametrized minimal invariant, 74
parametrized representability, 242, 273–277, 280
partial term, 78
partial-on-minimals, 84
Partmin, 84
pattern-match, 4, 45
pCpo, 72
pending Q-move, 200
plain map, 24
play
awaiting Opponent, 185, 187
awaiting Player, 185, 187
infinite, 185, 187
nonreturning-command (default), 185
value, 186
Player-awaiting play, 185, 187
Player-move, *see* P-move
pm (pattern-match), 4
PO-bracketed play, 201
point
code, 145
trace, 145
pointed cpo, *see* cpo, pointed
pointer game semantics, 169–203, 246–248, 296
problems with, 169
pointwise ideal opbimodule, 240
polarity, xxxi, 174
polymorphism, 297
pop, xxxiii, 33
possible worlds, 117–140, 241, 277
powerdomain
lower (Hoare), 110
pre-families model
of CBPV, 247
of JWA, 190, 246
predecessor of a token, 177
Predom, 84
predomain, 84, 111, 289, 294
prefix-closed, 185, 187
premonoidal category, 329
print, 8
printing semantics
for CBN, 22
for CBPV, 39
for CBV, 11
product, 4, 218
categorical, 14, 265
in CBPV judgement model, 228
in staggered Freyd category, 331
Kleisli, 237
pattern-match, 5, 30, 300, 301
projection, 5, 30, 300, 301
with right modules, 227, 271
projection, 4
provably equivalent sequents, 158
pseudo-computation-type in CBV, 316
pseudo-value-type in CBN, 321
pseudoref, 172, 173
$\mathsf{pt}_{i\in I}$ (arena operation), 177
push, xxxiii, 33

Q/A-labelled arena, 195
Q/A-respecting isomorphism, 196
question-move, 195

read, 192
read_i, 192
reading, 91, 118
$\mathsf{read}_{i,r}$, 192
realizability, 66
recursion, 66–76
type, 68–76, 295
$\mathsf{ref}_{i\in I}$, 192

ref, 118
ref A (type of A-storing cells), 118
$\mathbf{ref}_C$ (type of C-cells), 123
reflection, full, 333
reindexed morphism, 211
reindexing functor, 211
relation model, 107–109, 239
representable object, 261
representation, 261
representing object for functor, 265
Res, 235
resolution of monad, 235
result type of operation symbol, 251
return as jump, xxxvi, 148, 195
returner, xxxiv, 29
  in CBV, 11
reversible derivation, 7, 60, 218
  converted into functor, 266, 268
right adjunctive, 225, 266
  in staggered Freyd category, 331
right exponential-adjunctive, 245, 266
  in staggered Freyd category, 332
right exponential-product-adjunctive, 245, 266
  in staggered Freyd category, 332
right module, 208
  locally indexed, 215
right product-adjunctive, 245, 247, 266
  in staggered Freyd category, 332
root of an arena, 177
root represents data, 176
root, 177
rt , 176

safe returner, 310
Scott semantics, 40, 65–86
  for CBN, 23
  for CBV, 13
Scott-Ershov, *see* domain
Scott-Ershov/Abramsky-McCusker, *see* predomain
SECD machine, 32
selector of a jump, 176
self, 212
self $\mathcal{CN}$, 271
semantics of types
  ×-calculus, 250
separate naturality, 262
sequenced computation, xxxiv
sequent, 251, 255, 259, 332
SFL, xxxi
SFPL, xxxi
Sig, 251, 255
signature, 251, 255, 259, 332
simple fibration, 212
single-threaded, 186
soundness w.r.t. operational semantics
  for CBN
    with printing, 22
  for CBPV
    pointer games, 195
    with cell generation, 131
    with cell generation + divergence, 134
    with cell generation + printing, 133
    with control, 101
    with control + printing, 106
    with erratic choice, 108
    with errors, 114
    with global store, 92
    with global store + printing, 95
    with printing, 40
    with recursion, 67
  for CBV
    with printing, 13
  for JWA
    pointer games, 194
    with printing, 154
source arena, 186
source move, 186
source object, 330
source-to-target function, 330
stack, xxxiii, 32, 33, 96
  empty, 33
stack-passing-style, *see* StkPS
stackfree-in-stack arena, 199, 248
stackfree-in-value arena, 199, 248
stacks
  categorical semantics, 221
  continuation semantics, 104
  equational theory, 58
  erratic choice semantics, 109
  error semantics for, 114
  global store semantics, 93
  possible world semantics, 129
  printing semantics, 41
staggered category, 330
staggered Freyd category, 330
stk, 96, 98
StkPS transform, 141, 148, 194, 238
  is an equivalence, 160–168
StkPS transforms, other, *see* CPS
store, xxxiii, 117
  general, *see* cell generation
  global, *see* global store
strategy
  nonreturning command (default), 185
  value, 187
strength of monad, 15
strict continuous function, 41
strong adjunction, 234
strong monad, *see* monad, strong, 17, 216
structure

of $\mathcal{A}$-set, 21, 129
of algebra, 24
sub-arena, 177, 196
sub-arena-family, 179
sub-bilimit-compact, 84
subject reduction
for C-machine, 152
for CK-machine, 35
substitution into stack, 36
substitution lemma, 217, 220
subsumptive translation, xxx, xxxii
successor of a token, 177
sum type, 4, 25
Sw (stores in world w), 125

T type in CBV, 305
$\mathcal{T}$, monad on MGraph_{τ}, 252, 255
tag, xxxvi
target arena, 186
target move, 186
target object, 330
teacher, 145
homomorphism, 148
terminal term, 8
in CBN, 18
termination
for CBN, 18
for CBPV, 30
for CBPV + control, 100
for CBV, 10
for JWA, 153
without thunk storage, 123
terms built from signature, 252
theory, 335
thread-pointer, 191, 193
thunk, xxxiv, 103
thunk storage, 137–140
thunk storage disallowed, 123
tok, 176
token, 174
total term, 78
trace jumpabout, 145
translation
from CBN to CBPV, 45, 319–325
from CBPV+control to JWA, *see* StkPS transform
from CBV to CBPV, 42
from CG-CBV to FG-CBV, 308–310
from FG-CBV to CBPV, 314–319
from lazy to FG-CBV, 313–314
trivial model, 63, 236
trivialization transform, 236
tuple types in jumbo λ-calculus, 300–301
type canonical form
for CBPV, 61, 199
for JWA, 156, 157, 179
type-represent, 177, 178

U_S, 138
under, 177
unfold, 69
unit of adjunction, 235
universal arena, 174
universal element, 261
universal model, 169
untyped λ-calculus, 20

${}^{\not\mathsf{v}}p$, 271
${}^{\mathsf{v}}Y$, 271
val-object, 222, 254, 330
value, xxxiv, 90
in CBV, 10, 11
in JWA, 143
value game, 186
value play, 186
value strategy, 187
value type, xxxvi
variable, xxxii
vertex, 14
of cocone, 70
of representing object, 265, 267
visibility condition, 171
vStrat, 187

$\mathcal{W}$, 117
w-computation, 120
w-store, 120
w-value, 120
weakening, xxxiii, 98
where (stack formation), 58
where (stack formation), 161
world, 117, 119
world-store, 117, 120
write_i, 192
$\mathsf{written}_i$, 192
$\mathsf{written}_{i,r}$, 192

Kolibri 360 GmbH
Ferdinand-Jühlke-Straße 7
99095 Erfurt, Deutschland
produktsicherheit@kolibri360.de